Protective Groups in
Organic Synthesis

Protective Groups in Organic Synthesis

Theodora W. Greene

Harvard University

A Wiley-Interscience Publication

JOHN WILEY & SONS

New York • Chichester • Brisbane • Toronto • Singapore

Library of Congress Cataloging in Publication Data:

Greene, Theodora W   1931–
  Protective groups in organic synthesis.

  "A Wiley-Interscience publication."
  Includes index.
    1. Chemistry, Organic—Synthesis. 2. Protective
groups (Chemistry) I.   Title. [DNLM: 1.   Chemistry,
Organic. QD 262 G812p]

QD262.G665   547′.2   80-25348
ISBN 0-471-05764-9

Printed in the United States of America

10 9 8 7 6 5 4

To Fritz

Preface

The selection of a protective group is an important step in synthetic methodology, and reports of new protective groups appear regularly. This book presents information on the synthetically useful protective groups (~500) for five major functional groups: —OH, —NH, —SH, —COOH, and >C=O. References through 1979, the best method(s) of formation and cleavage, and some information on the scope and limitations of each protective group are given. The protective groups that are used most frequently and that should be considered first are listed in Reactivity Charts, which give an indication of the reactivity of a protected functionality to 108 prototype reagents.

The first chapter discusses some aspects of protective group chemistry: the properties of a protective group, the development of new protective groups, how to select a protective group from those described in this book, and an illustrative example of the use of protective groups in a synthesis of brefeldin. The book is organized by functional group to be protected. At the beginning of each chapter are listed the possible protective groups. Within each chapter protective groups are arranged in order of increasing complexity of structure (e.g., methyl, ethyl, *t*-butyl, . . . benzyl). The most efficient methods of formation or cleavage are described first. Emphasis has been placed on providing recent references, since the original method may have been improved. Consequently, the original reference may not be cited; my apologies to those whose contributions are not acknowledged. Chapter 8 explains the relationship between reactivities, reagents, and the Reactivity Charts that have been prepared for each class of protective groups.

This work has been carried out in association with Professor Elias J. Corey, who suggested the study of protective groups for use in computer-assisted synthetic analysis. I appreciate his continued help and encouragement. I am grateful to Dr. J. F. W. McOmie (Ed., *Protective Groups in Organic Chemistry,* Plenum Press, New York and London, 1973) for his interest in the project and for several exchanges of correspondence, and to Mrs. Mary Fieser, Professor Frederick D. Greene, and Professor James A. Moore for reading the manuscript. Special thanks are also due to Halina and Piotr Starewicz for drawing the structures, and to Kim Chen, Ruth Emery, Janice Smith, and Ann Wicker for typing the manuscript.

THEODORA W. GREENE

Harvard University
September 1980

Contents

Abbreviations[a]

A	adenine
AIBN	2,2′-azobisisobutyronitrile
Ac	acetyl
Ar	aryl
B	nucleoside base (e.g. adenine, cytosine, guanine, thymine, or uracil)
BOC	t-butoxycarbonyl (t-C_4H_9OCO—)
Bu	butyl
c-	cyclo
CBZ	benzyloxycarbonyl ($C_6H_5CH_2OCO$—)
CySH	cysteine
DABCO	1,4-diazabicyclo[2.2.2]octane
DBN	1,5-diazabicyclo[4.3.0]nonene-5
DBU	1,8-diazabicyclo[5.4.0]undecene-7
DCC	dicyclohexylcarbodiimide ($C_6H_{11}N{=}C{=}NC_6H_{11}$)
DMAP	4-N,N-dimethylaminopyridine
DME	dimethoxyethane
DMF	N,N-dimethylformamide
DMSO	dimethyl sulfoxide
Et	ethyl
FG	functional group
G	guanine
h	hour
HMPA=HMPT	hexamethylphosphoramide
LDA	lithium diisopropylamide, $LiN(i{-}C_3H_7)_2$
M	metal
Me	methyl
NBS	N-bromosuccinimide

[a]For a list of abbreviations, not used in this book, of less widely applied protective groups for amino acids, see, *Spec. Period. Rep.: Amino-Acids, Peptides, and Proteins,* **10,** xvii–xix (1979).

Nu	nucleophile
(P)—	polymer substituent
PG	protective group. This is the name that is indexed by Chemical Abstracts, not protect*ing* group or blocking group.
Ph	phenyl
Pr	propyl
Py	pyridine
T	thymine
TBDMS	*t*-butyldimethylsilyl
THF	tetrahydrofuran
THP	tetrahydropyran
TMEDA	*N*,*N*,*N'*,*N'*-tetramethylethylenediamine ($Me_2NCH_2CH_2NMe_2$)
Ts	*p*-toluenesulfonyl (p-$CH_3C_6H_4SO_2^-$)
TsOH	*p*-toluenesulfonic acid
Tyr-OH	tyrosine
Tyr-OMe	tyrosine methyl ester
U	uracil

Conventions

In each chapter of this book, every protective group has been assigned a number that corresponds with the number in the table of contents of that chapter. The protective groups in each chapter are numbered independently (i.e., the first protective group in each chapter is **1**). The same number is used to refer to a protective group included in a reactivity chart. A few protective groups appear without a number; these are cases from recent literature, included after the completion of the writing of a chapter. In some cases a letter designation has been used to refer to such a compound (e.g., **S**).

A few generalized structures have been used in this book. They are defined below.

Protection for Glycols

Substrate   **Convention**

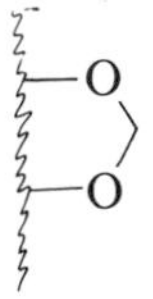

Indicates protection of a 1,2- or 1,3-diol.

Protection for the Carbonyl Group

Substrate   **Convention**

RR′CO   Indicates an aldehyde or ketone unless specified otherwise.

Protection for the Amino Group

Substrate   **Convention**

RNH_2 Only primary amines can be protected.

R^1R^2NH Primary or secondary amines can be protected.

RCHCO—   A peptide is being protected.
|
NH_2

Protective Groups in
Organic Synthesis

1

The Role of Protective Groups in Organic Synthesis

Properties of a Protective Group

When a chemical reaction is to be carried out selectively at one reactive site in a multifunctional compound, other reactive sites must be temporarily blocked. Many protective groups have been, and are being, developed for this purpose. A protective group must fulfill a number of requirements. It must react selectively in good yield to give a protected substrate that is stable to the projected reactions. The protective group must be selectively removed in good yield by readily available, preferably nontoxic reagents that do not attack the regenerated functional group. The protective group should form a crystalline derivative (without the generation of new chiral centers) that can be easily separated from side products associated with its formation or cleavage. The protective group should have a minimum of additional functionality to avoid further sites of reaction.

Historical Development

Since a few protective groups cannot satisfy all these criteria for elaborate substrates, a large number of mutually complementary protective groups are needed and, indeed, are becoming available. In early syntheses the chemist chose a standard derivative known to be stable to the subsequent reactions. In a synthesis of callistephin chloride the phenolic —OH group in **1** was selectively protected as an acetate.[1] In the presence of silver ion the aliphatic hydroxyl group in **2** dis-

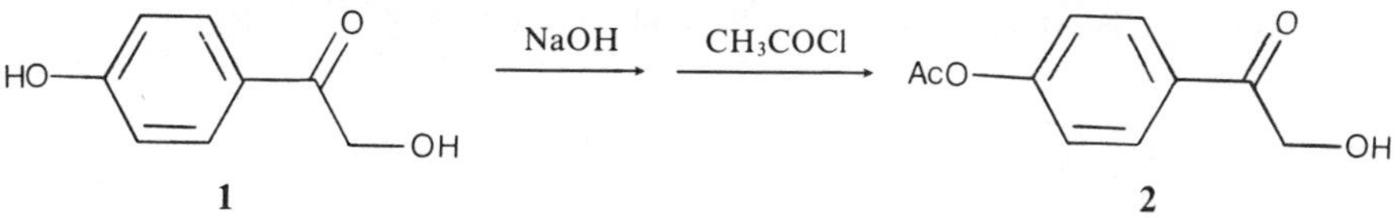

$$\text{HO}\!-\!\!\bigcirc\!\!-\!\text{C(O)}\!-\!\text{CH}_2\text{OH} \xrightarrow[\text{CH}_3\text{COCl}]{\text{NaOH}} \text{AcO}\!-\!\!\bigcirc\!\!-\!\text{C(O)}\!-\!\text{CH}_2\text{OH}$$

1   **2**

placed the bromide ion in a bromoglucoside. In a final step the acetate group was removed by basic hydrolysis. Other classical methods of cleavage include acidic hydrolysis (eq. 1), reduction (eq. 2), and oxidation (eq. 3):

(1) $ArO-R \rightarrow ArOH$

(2) $RO-CH_2Ph \rightarrow ROH$

(3) $RNH-CHO \rightarrow [RNHCOOH] \rightarrow RNH_2$

Some of the original work in the carbohydrate area in particular reveals extensive protection of carbonyl and hydroxyl groups. For example, a cyclic diacetonide of glucose was selectively cleaved to the monoacetonide.[2a] A more recent summary[2b] describes the selective protection of primary and secondary hydroxyl groups in a synthesis of gentiobiose, carried out in the 1870s, as triphenylmethyl ethers.

Development of New Protective Groups

As chemists proceeded to synthesize more complicated structures, they developed more satisfactory protective groups and more effective methods for the formation and cleavage of protected compounds. At first a tetrahydropyranyl acetal was prepared,[3a] by an acid-catalyzed reaction with dihydropyran, to protect a hydroxyl group. The acetal is readily cleaved by mild acid hydrolysis, but formation of this acetal introduces a new chiral center. Formation of the 4-methoxytetrahydropyranyl acetal[3b] eliminates this problem.

Catalytic hydrogenolysis of an O-benzyl protective group is a mild, selective method introduced by Bergmann and Zervas[4] to cleave a benzyl carbamate ($>NCO-OCH_2C_6H_5 \rightarrow >NH$) prepared to protect an amino group during peptide syntheses. The method also has been used to cleave alkyl benzyl ethers, stable compounds prepared to protect alkyl alcohols; benzyl esters are cleaved by catalytic hydrogenolysis under neutral conditions.

Three selective methods to remove protective groups are receiving much attention: "assisted," electrolytic, and photolytic removal. Four examples illustrate "assisted removal" of a protective group. A stable allyl group can be converted to a labile vinyl ether group (eq. 4)[5]; a β-haloethoxy (eq. 5)[6] or a β-silylethoxy (eq. 6)[7] derivative is cleaved by attack at the β-substituent; and a stable o-nitrophenyl derivative can be reduced to the o-amino compound, which undergoes cleavage by nucleophilic displacement (eq. 7)[8]:

(4) $ROCH_2CH=CH_2 \xrightarrow{\text{}^-O\text{-}t\text{-Bu}} [ROCH=CHCH_3] \xrightarrow{H_3O^+} ROH$

(5) $RO-CH_2-CCl_3 + Zn \rightarrow RO^- + CH_2=CCl_2$

(6) $RO-CH_2-CH_2-SiMe_3$ $\xrightarrow{F^-}$ $RO^- + CH_2{=}CH_2 + FSiMe_3$

R = alkyl, aryl, R'CO—, or R'NHCO—

(7)

The design of new protective groups that are cleaved by "assisted removal" is a challenging and rewarding undertaking.

Removal of a protective group by electrolytic oxidation or reduction can be very satisfactory. The equipment required ranges from a minimum of two electrodes, a potentiostat, and a source of DC current to quite sophisticated systems. A suitable electrolyte/solvent system is needed, and the deprotected product must not undergo further electrochemistry under the experimental conditions. The use and subsequent removal of chemical oxidants or reductants (e.g., Cr or Pb salts; Pt— or Pd—C) are eliminated. Reductive cleavages have been carried out in high yield at -1 to -3 V (vs. SCE) depending on the group; oxidative cleavages in good yield have been realized at 1.5–2 V (vs. SCE). For systems possessing two or more electrochemically labile protective groups, selective cleavage is possible when the half-wave potentials, $E_{1/2}$, are sufficiently different; excellent selectivity can be obtained with potential differences on the order of 0.25 V. Protective groups that have been removed by electrolytic oxidation or reduction are described at the appropriate places in this book; a review article by Mairanovsky[9] discusses electrochemical removal of protective groups.[10]

Photolytic cleavage reactions (e.g., of o-nitrobenzyl, phenacyl, and nitrophenylsulfenyl derivatives) take place in high yield on irradiation of the protected compound for a few hours at 254–350 nm. For example, the o-nitrobenzyl group, used to protect alcohols,[11] amines,[12] and carboxylic acids,[13] has been removed by irradiation. Protective groups that have been removed by photolysis are described at the appropriate places in this book; in addition, the reader may wish to consult three review articles.[14–16]

One widely used method of formation of protected compounds involves polymer-supported reagents,[17–20] with the advantage of simple workup by filtration and automated syntheses, especially of polypeptides. Polymer-supported reagents are used to protect a terminal —COOH group as a polymer-bound ester (RCOOR'—Ⓟ) during peptide syntheses[17]; to protect primary alcohols as Ⓟ-trityl ethers[21]; and to protect 1,2- and 1,3-diols as Ⓟ-phenyl boronates.[22] Monoprotection of symmetrical dialdehydes and diacid chlorides has been reported[19]; some diprotection occurs with diols and diamines.[19]

Internal protection, used by van Tamelen in a synthesis of colchicine, may be appropriate[23]:

Selection of a Protective Group from This Book

To select a specific protective group, the chemist must consider in detail all the reactants, reaction conditions, and functionalities involved in the proposed synthetic scheme. First he or she must evaluate all functional groups in the reactant to determine those that will be unstable to the desired reaction conditions and require protection. The chemist should then examine reactivities of possible protective groups, listed in the Reactivity Charts, to determine compatibility of protective group and reaction conditions. The protective groups listed in the Reactivity Charts (see Chapter 8) have been used most widely; consequently, considerable experimental information is available for them. He or she should consult the complete list of protective groups in the relevant chapter and consider their properties. It will frequently be advisable to examine the use of one protective group for several functional groups (i.e., a 2,2,2-trichloroethyl group to protect a hydroxyl group as an ether, a carboxylic acid as an ester, and an amino group as a carbamate). When several protective groups are to be removed simultaneously, it may be advantageous to use the same protective group to protect different functional groups (e.g., a benzyl group, removed by hydrogenolysis, to protect an alcohol and a carboxylic acid). When selective removal is required, different classes of protection must be used (e.g., a benzyl ether, cleaved by hydrogenolysis but stable to basic hydrolysis, to protect an alcohol, and an alkyl ester, cleaved by basic hydrolysis but stable to hydrogenolysis, to protect a carboxylic acid).

If a satisfactory protective group has not been located, the chemist has a number of alternatives: rearrange the order of some of the steps in the synthetic scheme so that a functional group no longer requires protection or a protective group that was reactive in the original scheme is now stable; redesign the synthesis, possibly making use of latent functionality[24] (i.e., a functional group in a precursor form; e.g., anisole as a precursor of cyclohexanone). Or, it may be necessary to include the synthesis of a new protective group in the overall plan.

A number of standard synthetic reference books are available.[25,26] A review article by Kössell and Seliger[27] discusses protective groups used in oligonucleotide syntheses, including protection for the phosphate group, which is not included in this book; and a series of articles[28] describe various aspects of protective group chemistry.

A Synthesis of Brefeldin: An Example of the Selection and Use of Protective Groups

Total syntheses of complex natural products require a variety of protective groups; nine were used in a synthesis of 5'-O-triphosphoryladenylyl(2' → 5')-adenylyl-(2' → 5')-adenosine.[29] A description of the synthesis[30,31] of (±)-brefeldin A, **A**, illustrates the use of a variety of hydroxyl protective groups and their selective removal.

Total Synthesis of (±)-Brefeldin A, A

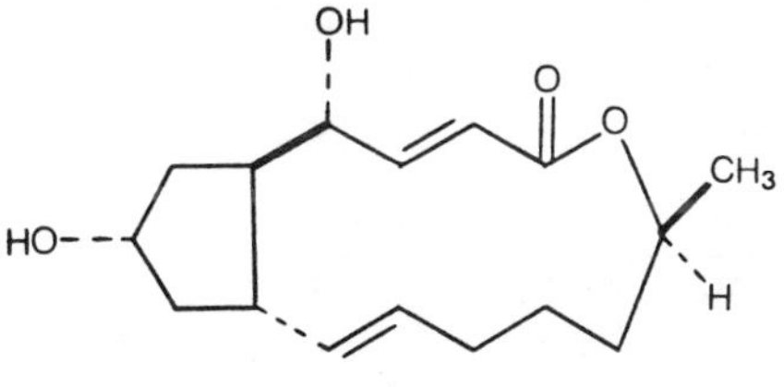

A

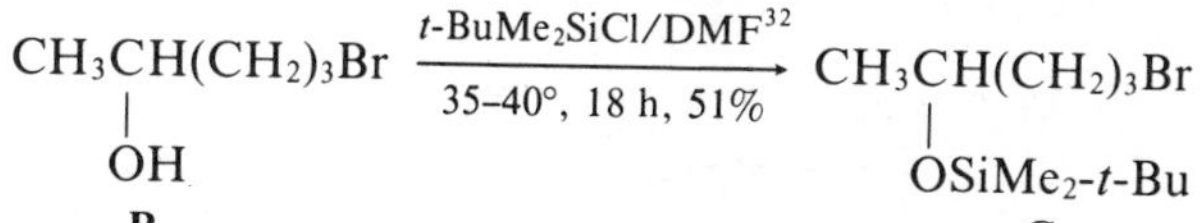

$$CH_3CH(CH_2)_3Br \xrightarrow[\text{35–40°, 18 h, 51\%}]{t\text{-BuMe}_2SiCl/DMF^{32}} CH_3CH(CH_2)_3Br$$

with OH below the first structure (**B**) and OSiMe$_2$-t-Bu below the second structure (**C**).

$$C \rightarrow \rightarrow Li^+C_3H_7C\equiv C\bar{C}CH=CH(CH_2)_3CHCH_3$$

OSiMe$_2$-t-Bu

D

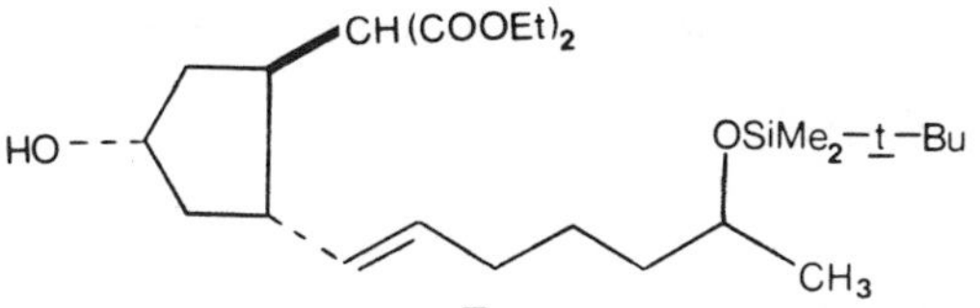

E

$$E \xrightarrow[\text{25°, 3 h, quant}]{CH_3OCH_2CH_2OCH_2Cl/Py, \ CH_2Cl_2^{33}}$$

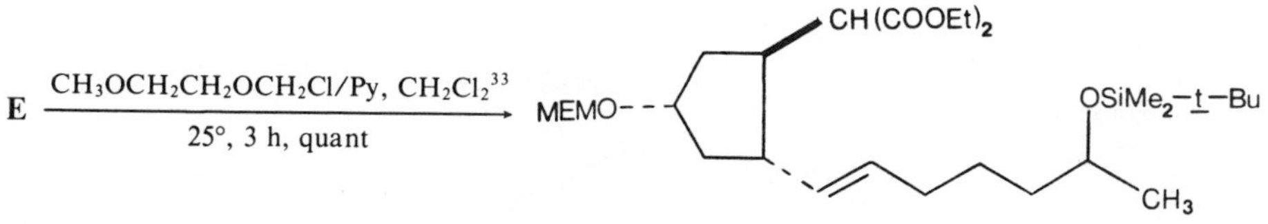

F

F $\xrightarrow[25°]{\text{2 N NaOH, MeOH}^{34}}$

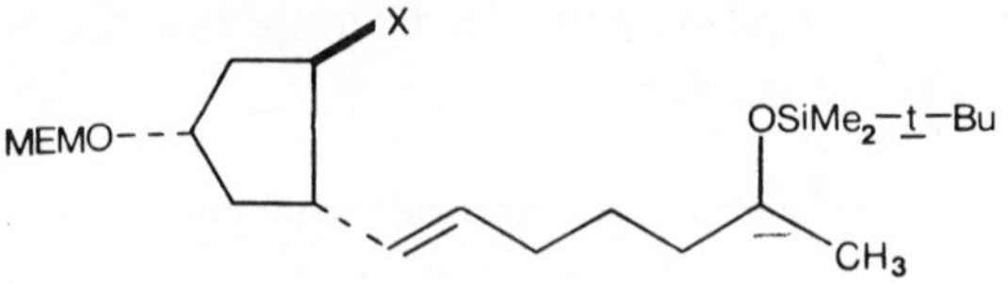

G, X = CH(COOH)₂

$\xrightarrow[\text{1 h, 100\%}]{\text{Py, reflux}}$ G $\xrightarrow[-78° \to 25°]{\text{LiN-}i\text{-Pr}_2}$ $\xrightarrow[25°, 1 \text{ h}]{(\text{MeO})_3\text{P/O}_2, \text{THF}}$

X = CH₂COOH

[G] $\xrightarrow[25°, 10 \text{ min}]{\text{Pb(OAc)}_4, \text{C}_6\text{H}_6}$ G

X = CH(OH)COOH   X = CHO

HC≡CCH₂OH $\xrightarrow[\text{DME}]{\text{NaH}}$ $\xrightarrow[0°, 1 \text{ h} \to 20°, 12 \text{ h}, 96\%]{\text{ClCH}_2\text{SCH}_3{}^{35}}$ HC≡CCH₂OCH₂SCH₃ → →

H   I

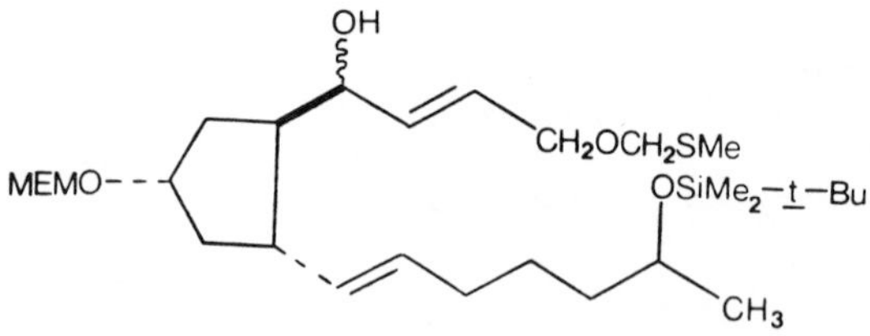

J

$\xrightarrow[0 \to 25°, 4 \text{ h}, 100\%]{\text{dihydropyran/CH}_2\text{Cl}_2, \text{TsOH}^{36}}$

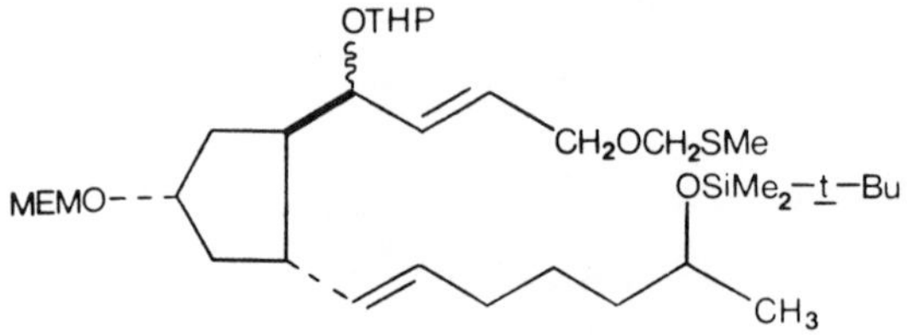

K

$\xrightarrow[25°, 4 \text{ h}, 95\%]{\text{HgCl}_2/\text{CH}_3\text{CN-H}_2\text{O (4:1), CaCO}_3{}^{35}}$

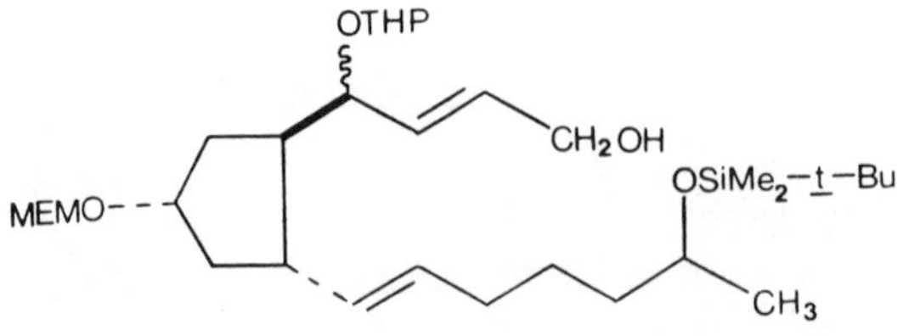

L

Collins oxidn   Ag₂O

$\xrightarrow[25°, 1 \text{ h}, 100\%]{n\text{-Bu}_4\text{NF/THF}^{32}}$

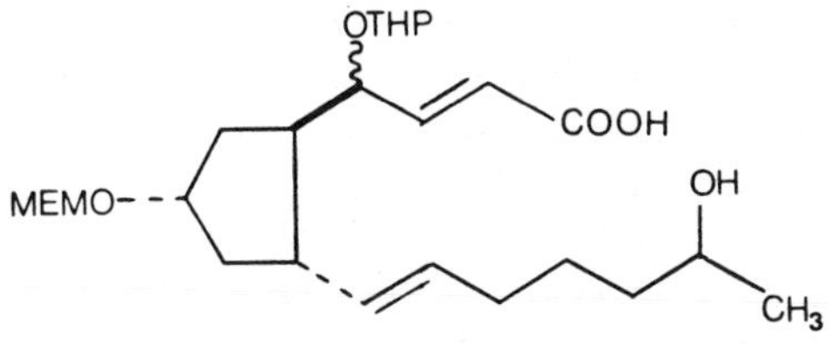

M

2,2'-dipyridyl[37]
disulfide

xylene[37]
reflux, 10 h

N

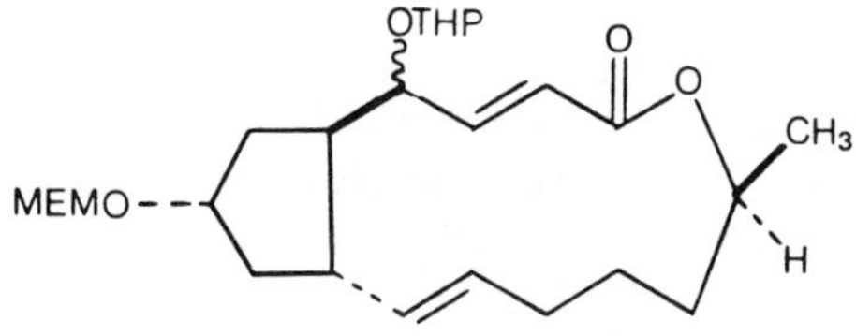

HOAc, H$_2$O, THF (3:3:1)[31]
50°, 4 h, 95%

O

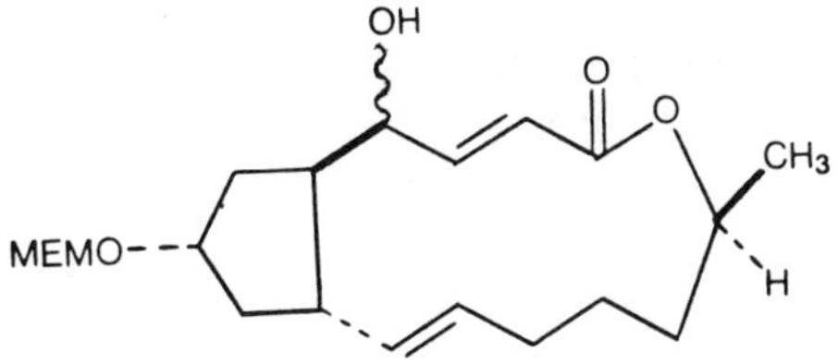

Collins oxidn
0°

NaBH$_4$[38]
−45°

P

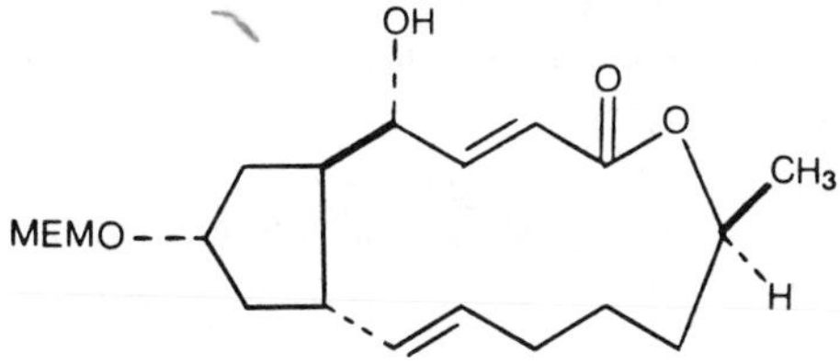

TiCl$_4$/CH$_2$Cl$_2$[33]
0°, 15 min

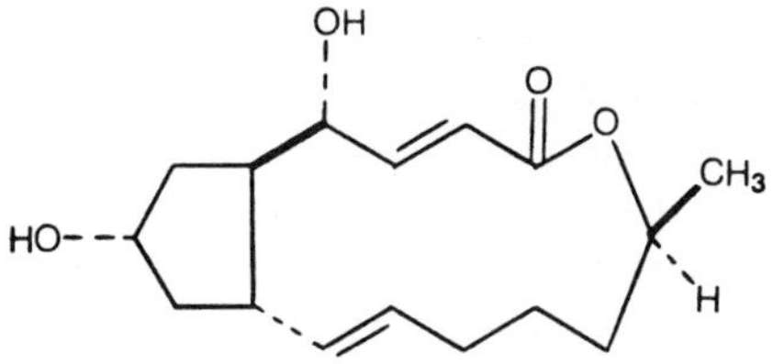

A

An early step in the synthesis of (±)-brefeldin A, **A**, was to protect[32] the hydroxyl group in compound **B** as a *t*-butyldimethylsilyl ether to give compound **C**. Compound **C** was converted in several steps to a vinyl Gilman reagent, **D**. Addition of **D** to diethyl 4-oxo-2-cyclopentenylmalonate followed by reduction of the carbonyl group gave compound **E**. The hydroxyl group in this compound was protected[33] as a methoxyethoxymethyl (MEM) ether to give compound **F**. Basic hydrolysis[34] of the ester groups in compound **F** gave a dicarboxylic acid, **G** [X = $CH(COOH)_2$]. Decarboxylation to give **G** (X = CH_2COOH) followed by oxidation gave an α-hydroxy acid, **G** [X = CH(OH)COOH], that was not isolated but oxidized directly to an aldehyde, **G** (X = CHO). To prepare compound **J**, the hydroxyl group in propargyl alcohol, **H**, was protected[35] as a methylthiomethyl (MTM) ether to give compound **I**. Conversion of compound **I** to a vinyllithium reagent followed by addition to compound **G** gave compound **J**. The hydroxyl group in compound **J** was protected[36] as a tetrahydropyranyl (THP) ether to give compound **K**, a compound that has four hydroxyl groups protected as four different ethers. The methylthiomethyl group was now removed[35] to give compound **L**. The free hydroxyl group was oxidized to a carboxyl group followed by removal[32] of the *t*-butyldimethylsilyl group by fluoride ion to give compound **M**. To effect lactonization the carboxyl group in **M** was converted[37] to an activated ester by reaction with 2,2'-dipyridyl disulfide followed by heating to give compound **N**. The tetrahydropyranyl group was removed[31] by acidic hydrolysis to give compound **O**. Collins oxidation of the hydroxyl group in compound **O** followed by reduction[38] of the resulting carbonyl group gave compound **P** with the desired stereochemistry of the hydroxyl group (i.e., cis to the lower side chain). Finally the methoxyethoxymethyl group was removed[33] to give ($\pm$)-brefeldin A, **A**.

[1] A. Robertson and R. Robinson, *J. Chem. Soc.,* 1460 (1928).

[2a] E. Fischer, *Ber.,* **28**, 1145–1167 (1895), see p. 1165.

[2b] B. Helferich, *Angew. Chem.,* **41**, 871 (1928).

[3a] W. E. Parham and E. L. Anderson, *J. Am. Chem. Soc.,* **70**, 4187 (1948).

[3b] C. B. Reese, R. Saffhill, and J. E. Sulston, *J. Am. Chem. Soc.,* **89**, 3366 (1967).

[4] M. Bergmann and L. Zervas, *Chem. Ber.,* **65**, 1192 (1932).

[5] J. Cunningham, R. Gigg, and C. D. Warren, *Tetrahedron Lett.,* 1191 (1964).

[6] R. B. Woodward, K. Heusler, J. Gosteli, P. Naegeli, W. Oppolzer, R. Ramage, S. Ranganathan, and H. Vorbruggen, *J. Am. Chem. Soc.,* **88**, 852 (1966).

[7] P. Sieber, *Helv. Chim. Acta,* **60**, 2711 (1977).

[8] I. D. Entwistle, *Tetrahedron Lett.,* 555 (1979).

[9] V. G. Mairanovsky, *Angew. Chem., Inter. Ed. Engl.,* **15**, 281 (1976).

[10] See also: M. F. Semmelhack and G. E. Heinsohn, *J. Am. Chem. Soc.,* **94**, 5139 (1972).

[11] S. Uesugi, S. Tanaka, E. Ohtsuka, and M. Ikehara, *Chem. Pharm. Bull.,* **26**, 2396 (1978).

[12] S. M. Kalbag and R. W. Roeske, *J. Am. Chem. Soc.,* **97**, 440 (1975).

[13] L. D. Cama and B. G. Christensen, *J. Am. Chem. Soc.,* **100**, 8006 (1978).

[14] V. N. R. Pillai, *Synthesis,* 1–26 (1980).

[15] P. G. Sammes, *Quart. Rev., Chem. Soc.,* **24**, 37–68 (1970), see pp. 66–68.

[16] B. Amit, U. Zehavi, and A. Patchornik, *Isr. J. Chem.,* **12**, 103–113 (1974).

[17] R. B. Merrifield, *J. Am. Chem. Soc.,* **85**, 2149 (1963).

[18] P. Hodge, *Chem. Ind. (London),* 624 (1979).

[19] C. C. Leznoff, *Acc. Chem. Res.,* **11**, 327 (1978).

[20] *Solid-Phase Synthesis,* E. C. Blossey and D. C. Neckers, Eds., Halsted, New York, 1975; *Polymer-Supported Reaction in Organic Synthesis,* P. Hodge and D. C. Sherrington, Eds., Wiley-Inter-

science, 1980. A comprehensive review of polymeric protective groups by J. M. J. Fréchet is included in this book.

[21]   J. M. J. Fréchet and K. E. Haque, *Tetrahedron Lett.,* 3055 (1975).

[22]   J. M. J. Fréchet, L. J. Nuyens, and E. Seymour, *J. Am. Chem. Soc.,* **101,** 432 (1979).

[23]   E. E. van Tamelen, T. A. Spencer, Jr., D. S. Allen, Jr., and R. L. Orvis, *Tetrahedron,* **14,** 8 (1961).

[24]   D. Lednicer, *Adv. Org. Chem.,* **8,** 179–293 (1972).

[25]   *Protective Groups in Organic Chemistry,* J. F. W. McOmie, Ed., Plenum, New York and London, 1973.

[26]   *Organic Syntheses,* Wiley-Interscience, New York, *Collect. Vols. I–V,* 1941–1973, **50–59,** 1970–1980; *Synthetic Methods of Organic Chemistry,* W. Theilheimer, Ed., S. Karger, Basel, Vols. 1–33, 1946–1979; *Methoden der Organischen Chemie* (Houben-Weyl), E. Müller, Ed., G. Thieme Verlag, Stuttgart, Vols. 1–15, 1958–1974; *Spec. Period. Rep.: General and Synthetic Methods,* **1** (1978), **2** (1979), **3** (1980); *The Chemistry of Functional Groups,* S. Patai, Ed., Wiley-Interscience, Vols. 1–20, 1964–1980.

[27]   H. Kössel and H. Seliger, *Prog. Chem. Org. Nat. Prod.,* **32,** 297–366 (1975).

[28]   J. F. W. McOmie, *Chem. Ind. (London),* 603–609 (1979); E. Haslam, *Chem. Ind. (London),* 610–617 (1979); P. M. Hardy, *Chem. Ind. (London),* 617–624 (1979); P. Hodge, *Chem. Ind. (London),* 624–627 (1979).

[29]   S. S. Jones and C. B. Reese, *J. Am. Chem. Soc.,* **101,** 7399 (1979).

[30]   R. H. Wollenberg, Ph.D. Thesis, Harvard University, 1976.

[31]   E. J. Corey, R. H. Wollenberg, and D. R. Williams, *Tetrahedron Lett.,* 2243 (1977).

[32]   E. J. Corey and A. Venkateswarlu, *J. Am. Chem. Soc.,* **94,** 6190 (1972).

[33]   E. J. Corey, J.-L. Gras, and P. Ulrich, *Tetrahedron Lett.,* 809 (1976).

[34]   E. J. Corey and R. H. Wollenberg, *Tetrahedron Lett.,* 4705 (1976).

[35]   E. J. Corey and M. G. Bock, *Tetrahedron Lett.,* 3269 (1975).

[36]   E. J. Corey, K. C. Nicolaou, and L. S. Melvin, Jr., *J. Am. Chem. Soc.,* **97,** 654 (1975).

[37]   E. J. Corey and K. C. Nicolaou, *J. Am. Chem. Soc.,* **96,** 5614 (1974).

[38]   E. J. Corey and R. H. Wollenberg, *Tetrahedron Lett.,* 4701 (1976).

2

Protection for the Hydroxyl Group Including 1,2- and 1,3-Diols

---

*Included in Reactivity Chart 1.

**Included in Reactivity Chart 2.

**Included in Reactivity Chart 2.

Hydroxyl groups are present in a number of compounds of biological and synthetic interest including nucleosides, carbohydrates, steroids, and the side chain of some amino acids. During oxidation, acylation, halogenation with phosphorus or hydrogen halides, or dehydration reactions of these compounds, a hydroxyl group must be protected. Ethers, acetals and ketals (cleaved by mild acidic hydrolysis), and esters (cleaved by basic hydrolysis) can be prepared to protect isolated hydroxyl groups; 1,2- and 1,3-diols can be protected as cyclic ethers (e.g., acetonides), cleaved by acidic hydrolysis, and as cyclic esters (e.g., carbonates and boronates), cleaved by basic hydrolysis. Simple *n*-alkyl ethers are stable compounds that are resistant to mild cleavage conditions. Benzyl and benzyl-

---

***Included in Reactivity Chart 3.

type ethers, however, can be cleaved by hydrogenolysis, silyl ethers by reaction with fluoride ion or by mild acidic hydrolysis, and various 2-substituted ethyl ethers by "assisted removal." Attempts to monoprotect a symmetrical diol as a polymer-supported ether have not been entirely satisfactory.[a]

ETHERS

In general an alcohol is converted to an ether by reaction with an alkylating agent in the presence of base; for example, see the preparation of methyl ethers, following compound 1. Equation 1 illustrates a modification of this method. Iodo-trimethylsilane (eq. 2) can be used to cleave a variety of ethers under mild conditions. (It also cleaves esters.[c,d])

(1) $ROH + TlOEt \xrightarrow[20°, \text{quant}]{C_6H_6} ROTl \xrightarrow[20-60°]{R'X, CH_3CN} ROR'$ [b]

high yields

R'X = primary alkyl halides

(2) $ROR' \xrightarrow[1-40 \text{ h}, 25-80°]{Me_3SiI^{c,d}} ROH + HOR'$, high yields

Ethers that have been used extensively to protect alcohols are included in Reactivity Chart 1.[e]

---

[a]   C. C. Leznoff, Acc. Chem. Res., **11**, 327 (1978).

[b]   H.-O. Kalinowski, D. Seebach, and G. Crass, Angew. Chem., Inter. Ed. Engl., **14**, 762 (1975).

[c]   M. E. Jung and M. A. Lyster, J. Org. Chem., **42**, 3761 (1977).

[d]   G. A. Olah, S. C. Narang, B. G. B. Gupta, and R. Malhotra, J. Org. Chem., **44**, 1247 (1979).

[e]   See also: C. B. Reese, "Protection of Alcoholic Hydroxyl Groups and Glycol Systems," in *Protective Groups in Organic Chemistry*, J. F. W. McOmie, Ed., Plenum, New York and London, 1973, pp. 95–143; H. M. Flowers, "Protection of the Hydroxyl Group," in *The Chemistry of the Hydroxyl Group*, S. Patai, Ed., Wiley-Interscience, New York, 1971, Vol. 10/2, pp. 1001–1044; C. B. Reese, *Tetrahedron*, **34**, 3143–3179 (1978), see pp. 3145–3150; V. Amarnath and A. D. Broom, *Chem. Rev.*, **77**, 183–217 (1977), see pp. 184–194.

1.   Methyl Ether: $ROCH_3$, 1

Formation

A methyl ether can be prepared by the following conditions:

   i. $Me_2SO_4/NaOH$, $n\text{-}Bu_4N^+I^-$, org solvent, 60–90% yield[a]
   ii. CH_2N_2/silica gel, 0–10°, 100% yield[b]
   iii. CH_2N_2/HBF_4, CH_2Cl_2, Et_3N, 25°, 1 h, 95% yield[c]
   iv.   MeI/solid KOH, DMSO, 20°, 5–30 min, 85–90% yield[d]

v.   (MeO)$_2$POH/cat. TsOH, 90–100°, 12 h, 60% yield. Under these conditions, the C$_5$-double bond in a 3-hydroxy steroid does not shift to the C$_4$-
position.[e]

vi.   Me$_3$O$^+$BF$_4^-$, 3 days, 55% yield[f]

vii.   carbohydrate + CF$_3$SO$_2$OMe, CH$_2$Cl$_2$, Py, 80°, 2.5 h, 85–90%[g]

Cleavage

(1)   ROMe $\xrightarrow[\text{25°, 6 h}]{\text{Me}_3\text{SiI, CHCl}_3{}^{h}}$ [ROSiMe$_3$, 95%, + MeOSiMe$_3$ +

RI + MeI] $\xrightarrow{\text{H}_2\text{O}}$ ROH + MeOH, high yields

Selective cleavage, due to rate differences, is possible for triphenylmethyl, benzyl,
or t-butyl ethers in the presence of methyl, ethyl, isopropyl, or cyclohexyl ethers.
Dialkyl ethers are selectively cleaved (90–95% yield) in the presence of aryl alkyl
ethers. A methyl ether is selectively cleaved in the presence of a methyl ester.
Acetylenic, olefinic, carbonyl, and aryl halide compounds are stable.[h]

(2)

$$\xrightarrow[\text{70–80\%}]{\text{BF}_3 \cdot \text{Et}_2\text{O, R'SH, 25°, few days}^{i}}$$

ROMe   ROH

$$\xrightarrow[\text{95–98\%}]{\text{AlX}_3, \text{EtSH, 25°, 0.5–3 h}^{j}}$$

[i] or [j]

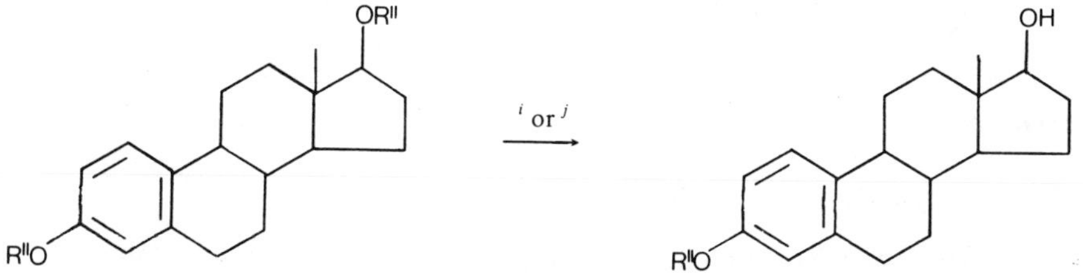

R″ = Me

R = HS(CH$_2$)$_2$, R = Me

93% yield[i]

R″ = H[j]

(3)

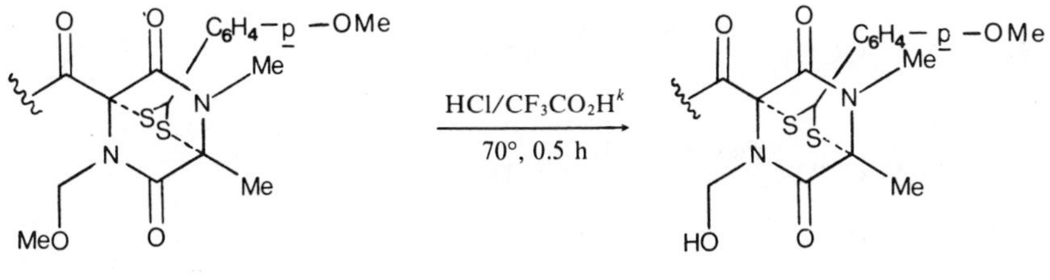

$$\xrightarrow[\text{70°, 0.5 h}]{\text{HCl/CF}_3\text{CO}_2\text{H}^{k}}$$

[a] A. Merz, *Angew. Chem., Inter. Ed. Engl.*, **12**, 846 (1973).

[b] K. Ohno, H. Nishiyama, and H. Nagase, *Tetrahedron Lett.*, 4405 (1979).

[c] M. Neeman and W. S. Johnson, *Org. Synth., Collect. Vol. V*, 245 (1973).

[d] R. A. W. Johnstone and M. E. Rose, *Tetrahedron*, **35**, 2169 (1979).

[e] Y. Kashman, *J. Org. Chem.*, **37**, 912 (1972).

[f] H. Meerwein, G. Hinz, P. Hofmann, E. Kroning, and E. Pfeil, *J. Prakt. Chem.*, **147**, 257 (1937).

[g] J. Arnarp and J. Lönngren, *Acta Chem. Scand., Ser. B*, **32**, 465 (1978).

[h] M. E. Jung and M. A. Lyster, *J. Org. Chem.*, **42**, 3761 (1977).

[i] M. Node, H. Hori, and E. Fujita, *J. Chem. Soc., Perkin Trans. 1*, 2237 (1976).

[j] M. Node, K. Nishide, M. Sai, K. Ichikawa, K. Fuji, and E. Fujita, *Chem. Lett.*, 97 (1979).

[k] Y. Kishi, S. Nakatsuka, T. Fukuyama, and M. Havel, *J. Am. Chem. Soc.*, **95**, 6493 (1973).

Substituted Methyl Ethers

2.   Methoxymethyl Ether (MOM Ether): $ROCH_2OCH_3$, 2

Formation

(1) $ROH + ClCH_2OMe \xrightarrow[\text{i-Pr$_2$NEt, 0°, 1 h $\rightarrow$ 25°, 8 h, 86\%^{b}}]{\text{NaH, THF, 80\%^{a} or}} 2$

ROH = allylic alcohol

(2) $ROH + CH_2(OMe)_2 \xrightarrow[\text{25°, 30 min, 95\%}]{\text{cat. P$_2$O$_5$, CHCl$_3$^{c}}} 2$

Cleavage

(1) $ROCH_2OCH_3 \xrightarrow{\text{HCl}} ROH$

Trace concd HCl/MeOH, 62°, 15 min^d

6 *M* HCl/aq THF, 50°, 6–8 h, 95% yield^e

(2) $ArOCH_2OCH_3 \xrightarrow[\text{reflux, 10–15 min, 80\%}]{\text{50\% HOAc–cat. concd H$_2$SO$_4$^{f}}} ArOH$

(3) $ROCH_2OCH_3 \xrightarrow[\text{98\%}]{\text{PhSH, BF$_3$·Et$_2$O^g}} ROH$

(4) $ROCH_2OCH_3 \xrightarrow[\text{25°}]{\text{Ph$_3$C$^+$BF$_4^-$, CH$_2$Cl$_2$^{h}}} ROH$

[a] A. F. Kluge, K. G. Untch, and J. H. Fried, *J. Am. Chem. Soc.,* **94,** 7827 (1972).

[b] G. Stork and T. Takahashi, *J. Am. Chem. Soc.,* **99,** 1275 (1977).

[c] K. Fuji, S. Nakano, and E. Fujita, *Synthesis,* 276 (1975).

[d] J. Auerbach and S. M. Weinreb, *J. Chem. Soc., Chem. Commun.,* 298 (1974).

[e] A. I. Meyers, J. L. Durandetta, and R. Munavu, *J. Org. Chem.,* **40,** 2025 (1975).

[f] F. B. Laforge, *J. Am. Chem. Soc.,* **55,** 3040 (1933).

[g] G. R. Kieczykowski and R. H. Schlessinger, *J. Am. Chem. Soc.,* **100,** 1938 (1978).

[h] T. Nakata, G. Schmid, B. Vranesic, M. Okigawa, T. Smith-Palmer, and Y. Kishi, *J. Am. Chem. Soc.,* **100,** 2933 (1978).

3.   Methylthiomethyl Ether (MTM Ether): $ROCH_2SCH_3$, 3

Formation

(1) $ROH \xrightarrow[\text{DME}]{\text{NaH}} \xrightarrow[0°, 1 \text{ h} \rightarrow 25°, 1.5 \text{ h}, > 86\%]{\text{ClCH}_2\text{SCH}_3/\text{NaI}^a} ROCH_2SCH_3$

ROH = primary alcohol

(2) $ROH \xrightarrow[20°, 12 \text{ h}, 80-90\%]{\text{MeSCH}_2\text{I, DMSO, Ac}_2\text{O}^b} \text{ or } \xrightarrow[20°, 1-2 \text{ days}, 80\%]{\text{DMSO, Ac}_2\text{O, HOAc}^c} 3$

ROH = primary, secondary, or tertiary alcohol

(3) $ROH \xrightarrow[22-80°, 4-24 \text{ h}, 60-80\%]{\text{ClCH}_2\text{SMe, AgNO}_3, \text{Et}_3\text{N, C}_6\text{H}_6{}^d} ROCH_2SCH_3$

Cleavage

(1) $3 \xrightarrow[25°, 1-2 \text{ h}, 88-95\%]{\text{HgCl}_2, \text{CH}_3\text{CN}-\text{H}_2\text{O}^a} \text{ or } \xrightarrow[25°, 45 \text{ min}, 88-95\%]{\text{AgNO}_3, \text{THF}-\text{H}_2\text{O}, 2,6\text{-lutidine}^a} ROH$

Methylthiomethyl ethers are stable to the mildly acidic conditions used to remove O,O'-acetonides or O-tetrahydropyranyl ethers. Silyl and O-tetrahydropyranyl ethers, and 1,3-dithianes are stable to the neutral cleavage conditions used here.[a]

(2) $3 \xrightarrow[\text{heat, few h}, 80-95\%]{\text{MeI/acetone}-\text{H}_2\text{O, NaHCO}_3{}^c} ROH$

ROH = primary, secondary, or tertiary alcohol

A variety of alcohols, R′OH, have been protected as hemithioacetals by reaction with a sulfonium salt, a^c:

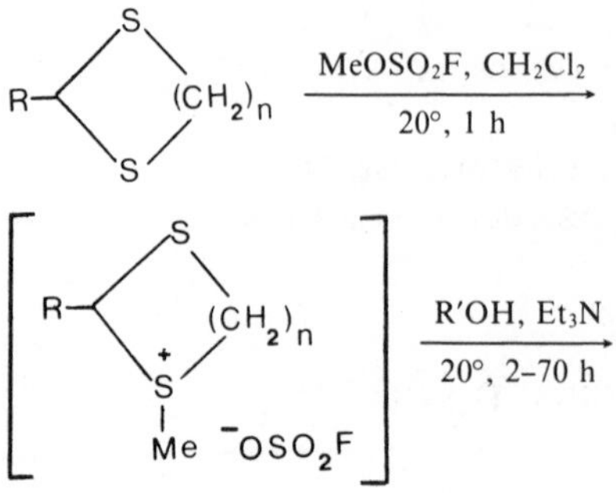

$$R'OCH(R)S(CH_2)_nSMe, \ 30\text{--}75\%$$

$R' = n\text{-Pr}, \ i\text{-Pr}, \ Ph, \ Me_3SiCH_2, \ \ldots$

[a] E. J. Corey and M. G. Bock, *Tetrahedron Lett.,* 3269 (1975).
[b] K. Yamada, K. Kato, H. Nagase, and Y. Hirata, *Tetrahedron Lett.,* 65 (1976).
[c] P. M. Pojer and S. J. Angyal, *Aust. J. Chem.,* **31,** 1031 (1978).
[d] K. Suzuki, J. Inanaga, and M. Yamaguchi, *Chem. Lett.,* 1277 (1979).
[e] T. A. Hase and R. Kivikari, *Synth. Commun.,* **9,** 107 (1979).

4.   Benzyloxymethyl Ether: $ROCH_2OCH_2C_6H_5$, **4**

Formation (→) / Cleavage (←)^a

$$\xrightarrow[\text{10--20°, 12 h, 95\%}]{\text{PhCH}_2\text{OCH}_2\text{Cl, EtN-}i\text{-Pr}_2}$$

ROH   **4**

$$\xleftarrow[\qquad\qquad]{\text{Na/NH}_3, \text{EtOH}}$$

ROH = allylic alcohol

[a] G. Stork and M. Isobe, *J. Am. Chem. Soc.,* **97,** 6260 (1975).

5.   *t*-Butoxymethyl Ether: $ROCH_2OC(CH_3)_3$, **5**

Formation (→) / Cleavage (←)^a

$$\xrightarrow[-20 \to 20°, \ 3 \ h, \ 54\text{--}80\%]{\text{ClCH}_2\text{O-}t\text{-Bu, Et}_3\text{N}}$$

ROH   **5**

$$\xleftarrow[20°, \ 48 \ h, \ 85\text{--}90\%]{\text{aq CF}_3\text{COOH}}$$

ROH = alkyl, benzylic, allylic alcohol

Compound **5** is stable to some acidic conditions (e.g., hot glacial acetic acid; aq HOAc, 20°; anhyd CF_3COOH, 20°).[a]

---

[a] H. W. Pinnick and N. H. Lajis, *J. Org. Chem.*, **43**, 3964 (1978).

6.   2–Methoxyethoxymethyl Ether (MEM Ether): $ROCH_2OCH_2CH_2OCH_3$, 6

Formation[a]

(1) $RO^- + CH_3OCH_2CH_2OCH_2Cl \xrightarrow[\text{0°, 10–60 min, } > 95\%]{\text{THF or DME}} 6$

(2) $ROH + MEMN^+Et_3Cl^- \xrightarrow[\text{reflux, 30 min, } > 90\%]{CH_3CN} 6$

(3) $ROH + MEMCl \xrightarrow[\text{25°, 3 h, quant}]{\text{EtN-}i\text{-Pr}_2, CH_2Cl_2} 6$

Cleavage[a]

$6 \xrightarrow[\text{25°, 2–10 h, 90\%}]{ZnBr_2, CH_2Cl_2}$ or $\xrightarrow[\text{0°, 20 min, 95\%}]{TiCl_4, CH_2Cl_2} ROH$

Compound **6** was designed to protect primary, secondary, or tertiary alcohols with formation and cleavage under aprotic conditions. MEM ethers are stable to mild acidic hydrolysis (e.g., $HOAc–H_2O$, 35°, 4 h; cat. TsOH, MeOH, 23°, 3 h), conditions that cleave tetrahydropyranyl and silyl (including *t*-butyldimethylsilyl) ethers.[a]

Compound **6** can be cleaved by fluoroboric acid (HBF_4, CH_2Cl_2, 0°, 3 h, 50–60% yield).[b]

---

[a] E. J. Corey, J.-L. Gras, and P. Ulrich, *Tetrahedron Lett.*, 809 (1976).
[b] N. Ikota and B. Ganem, *J. Chem. Soc., Chem. Commun.*, 869 (1978).

7.   2,2,2-Trichloroethoxymethyl Ether: $ROCH_2OCH_2CCl_3$, 7

Formation[a]

(1) $RO^- + Cl_3CCH_2OCH_2Cl \xrightarrow[\text{5 h, 70–90\%}]{LiI/THF} 7$

(2) $ROH + Cl_3CCH_2OCH_2Cl \xrightarrow[\text{30–60\%}]{\text{EtN-}i\text{-Pr}_2,\ CH_2Cl_2}$ **7**

Cleavage[a]

(1)   **7** $\xrightarrow[\text{reflux, 97\%}]{\text{Zn–Cu or Zn–Ag/MeOH}}$ ROH

(2)   **7** $\xrightarrow[\text{reflux, 4 h, 90–100\%}]{\text{Zn/MeOH, HOAc, Et}_3\text{N}}$ ROH

(3)   **7** $\xrightarrow{\text{Li/NH}_3}$ ROH

Methoxymethyl ethers are stable to these cleavage conditions.[a]

[a] R. M. Jacobson and J. W. Clader, *Synth. Commun.*, **9**, 57 (1979).

8.   Bis(2-chloroethoxy)methyl Ether: ROCH(OCH₂CH₂Cl)₂, 8

Formation (→) / Cleavage (←)[a]

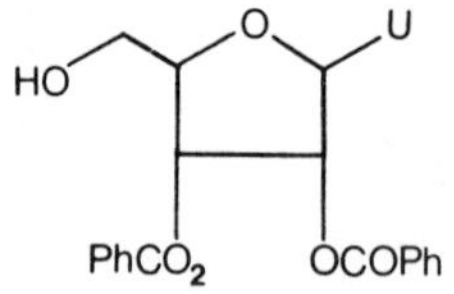

$$\xrightarrow[\text{100°, 10 min to 2 h, 76\%}]{\text{HC(OCH}_2\text{CH}_2\text{Cl)}_3}$$

8a

$$\xleftarrow[\text{20°, 1 h}]{\text{80\% HOAc}}$$

In this synthesis an acetal or unsubstituted ortho ester was too labile to acidic hydrolysis.[a]

[a] T. Hata and J. Azizian, *Tetrahedron Lett.*, 4443 (1969).

2-(Trimethylsilyl)ethoxymethyl Ether (SEM Ether): ROCH₂OCH₂CH₂Si(CH₃)₃, A

Formation[a]

$$ROH + Me_3SiCH_2CH_2OCH_2Cl \xrightarrow[\text{35–40°, 1–5 h, 86–100\%}]{\text{EtN-}i\text{-Pr}_2,\ CH_2Cl_2} RO\text{—SEM}$$

Cleavage[a]

$$RO\text{—SEM} \xrightarrow[\text{45°, 8–12 h, 85–96\%}]{n\text{-Bu}_4\text{N}^+\text{F}^-,\ \text{THF or HMPA}} ROH$$

ROH = primary, secondary, tertiary, aromatic alcohols

A SEM ether is stable to acetic acid (HOAc/H$_2$O/THF, 45°, 7 h) and to BuLi.[a]

[a] B. H. Lipshutz and J. J. Tegram, *Tetrahedron Lett.,* **21,** 3343 (1980).

9. Tetrahydropyranyl Ether (THP Ether):

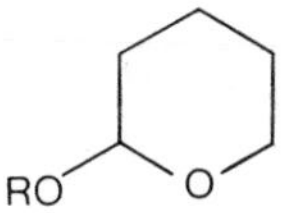

, 9

Since an asymmetric center is formed when a THP ether is prepared, the product may contain a mixture of diastereomers. No new asymmetric center is formed by reaction of an alcohol with 5,6-dihydro-4-methoxy-2*H*-pyran (e.g., see compound **12**).

Formation (→) / Cleavage (←)

(1)

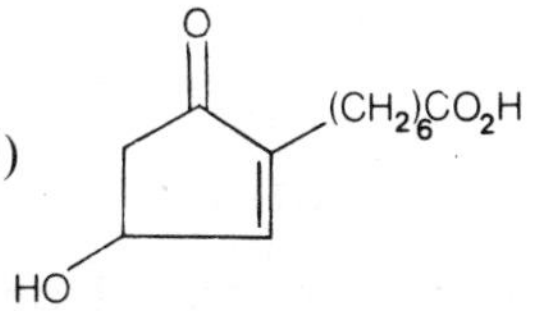

dihydropyran, TsOH, CH$_2$Cl$_2$[a]
20°, 1.5 h, 100%

HOAc, THF–H$_2$O (4:2:1)[a]
45°, 3.5 h

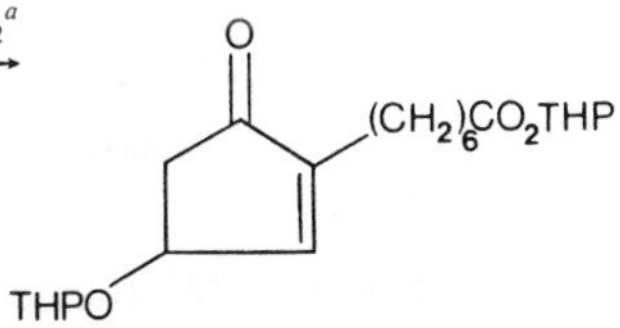

(2)

dihydropyran, PPTS, CH$_2$Cl$_2$[b]
20°, 4 h, 94–100%

ROH   ROTHP

PPTS, EtOH (pH 3.0)[b]
55°, 3 h, 95–100%

PPTS = pyridinium *p*-toluenesulfonate

An epoxide is stable to formation of compound **9** under these conditions.[b]

(3)

dihydropyran, Amberlyst H-15, hexane[c]
1–2 h, 95%

ROH   ROTHP

Amberlyst H-15, MeOH[c]
45°, 1 h, 95%

Amberlyst H-15 is an ion-exchange resin that contains —SO$_3$H substituents.

Compound **9** has been cleaved by treatment with methanol/Dowex-50W-X8 (25°, 1 h, 99% yield).[d]

Cleavage

(1)   Glyceride–OTHP $\xrightarrow[\text{90°, 2 h, 80–95\%}]{\text{boric acid, EtOCH}_2\text{CH}_2\text{OH}^e}$ ROH

(2)   ROTHP $\xrightarrow[\text{25°, 1 h, 94\%}]{\text{TsOH, MeOH}^f}$ ROH

ROH = a primary, allylic alcohol

Explosions have been reported on distillation of compounds containing a tetra-hydropyranyl ether after a reaction with $B_2H_6/H_2O_2\text{–OH}^-$, and with 40% CH_3CO_3H:

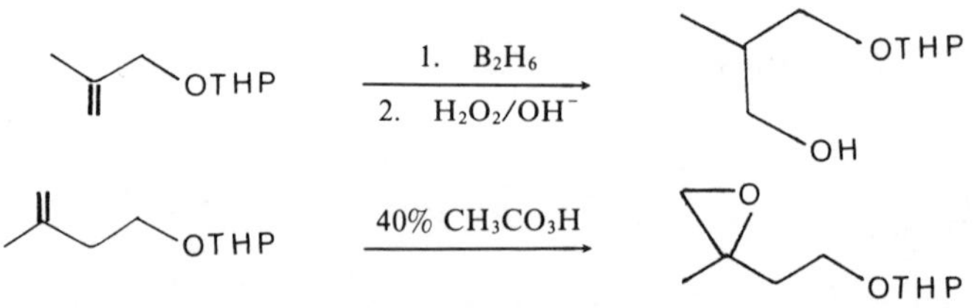

It was thought that the acetal might have reacted with peroxy reagents, forming explosive peroxides. It was suggested that this could also occur with compounds such as tetrahydrofuranyl acetals, 1,3-dioxolanes, and methoxymethyl ethers.[g]

[a] K. F. Bernady, M. B. Floyd, J. F. Poletto, and M. J. Weiss, *J. Org. Chem.*, **44**, 1438 (1979).

[b] M. Miyashita, A. Yoshikoshi, and P. A. Grieco, *J. Org. Chem.*, **42**, 3772 (1977).

[c] A. Bongini, G. Cardillo, M. Orena, and S. Sandri, *Synthesis*, 618 (1979).

[d] R. Beier and B. P. Mundy, *Synth. Commun.*, **9**, 271 (1979).

[e] J. Gigg and R. Gigg, *J. Chem. Soc. C*, 431 (1967).

[f] E. J. Corey, H. Niwa, and J. Knolle, *J. Am. Chem. Soc.*, **100**, 1942 (1978).

[g] A. I. Meyers, S. Schwartzman, G. L. Olson, and H.-C. Cheung, *Tetrahedron Lett.*, 2417 (1976).

10.   3-Bromotetrahydropyranyl Ether: RO-3-bromotetrahydropyranyl, 10

Compound **10** was prepared from a C_{17}-hydroxy steroid and 2,3-dibromopyran (Py/C_6H_6, 20°, 24 h); it was cleaved by zinc/ethanol.[a]

[a] A. D. Cross and I. T. Harrison, *Steroids*, **6**, 397 (1965).

11.   Tetrahydrothiopyranyl Ether: , 11

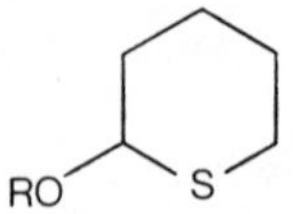

Compound **11** was prepared from a C_3-hydroxy steroid and dihydrothiopyran ($CF_3COOH/CHCl_3$, 35% yield); it can be cleaved under neutral conditions (AgNO₃, aq acetone, 85% yield).[a]

^a L. A. Cohen and J. A. Steele, *J. Org. Chem.*, **31**, 2333 (1966).

12. 4-Methoxytetrahydropyranyl Ether:
CH₃O   OR

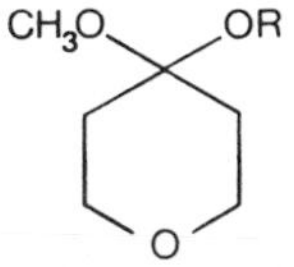

, **12**

Protection of an alcohol as compound **12** avoids the formation of an asymmetric center that occurs when a tetrahydropyranyl ether is formed (e.g., see compound **9**). This ether (**12**) is used extensively in oligonucleotide syntheses.

Formation (→) / Cleavage (←)[a]

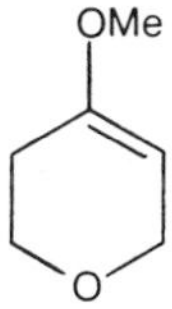

OMe

, TsOH, 20°, 4 h

ROH   **12**

0.01 *N* HCl, 1 h

ROH = uridine, $t_{1/2}$ = 24 min[a]

^a C. B. Reese, R. Saffhill, and J. E. Sulston, *J. Am. Chem. Soc.*, **89**, 3366 (1967).

13. 4-Methoxytetrahydrothiopyranyl Ether:
CH₃O   OR

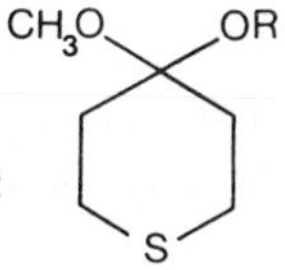

, **13**

14. 4-Methoxytetrahydrothiopyranyl Ether *S,S*-Dioxide:

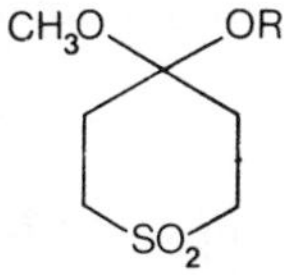

, **14**

Compound **13,** used in nucleotide syntheses, is hydrolyzed by acid 5 times faster than the oxygen analog (e.g., compound **12**). A sulfone, **14,** prepared from compound **13** by oxidation with *m*-ClC₆H₄CO₃H, is hydrolyzed 2000 times more slowly than the oxygen analog, **12.**[a]

^a J. H. van Boom, P. van Deursen, J. Meeuwse, and C. B. Reese, *J. Chem. Soc., Chem. Commun.*, 766 (1972).

15.   Tetrahydrofuranyl Ether: ⟨structure⟩ , **15**

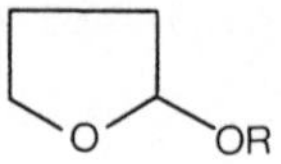

Formation

(1)   ROH + [tetrahydrofuran $\xrightarrow[\text{25}°, 0.5 \text{ h}, 85\%]{\text{SO}_2\text{Cl}_2/\text{THF}}$] $\xrightarrow[\text{30 min, 82–98\%}]{\text{Et}_3\text{N}}$ **15**[a]

(2)   ROH + Ph$_2$CHCOO-2-tetrahydrofuranyl $\xrightarrow[\text{20}°, 30 \text{ min}, 90–99\%]{\text{1\% TsOH, CCl}_4{}^{a,b}}$ **15**

Cleavage

(1)   **15** $\xrightarrow[\text{25}°, 30 \text{ min}, 90\%]{\text{HOAc-H}_2\text{O-THF (3:1:1)}^a}$ or $\xrightarrow[\text{25}°, 10 \text{ min}, 50\%]{\text{0.01 } N \text{ HCl/THF (1:1)}^a}$ ROH

(2)   **15** $\xrightarrow[\text{25}°, 3 \text{ h}, 90\%]{\text{pH 5}^a}$ ROH

The authors report[a] that formation of compound **15** by reaction with 2-chlorotetrahydrofuran (eq. 1, *Formation*) avoids a laborious procedure[c] that is required when dihydrofuran is used. The tetrahydrofuranyl ester (used in eq. 2, *Formation*) is reported[a] to be a readily available, stable solid. A tetrahydropyranyl ether is not cleaved at pH 5 (eq. 2, *Cleavage*).[a]

[a]   C. G. Kruse, F. L. Jonkers, V. Dert, and A. van der Gen, *Recl. Trav. Chim. Pays-Bas*, **98**, 371 (1979).
[b]   C. G. Kruse, E. K. Poels, F. L. Jonkers, and A. van der Gen, *J. Org. Chem.*, **43**, 3548 (1978).
[c]   E. L. Eliel, B. E. Nowak, R. A. Daignault, and V. G. Badding, *J. Org. Chem.*, **30**, 2441 (1965).

16.   Tetrahydrothiofuranyl Ether: ⟨structure⟩ , **16**

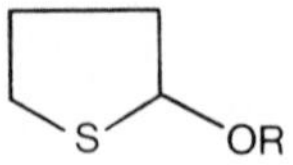

Formation (→) / Cleavage (←)

(1) $\xrightarrow[\text{reflux, 6 days, 75\%}]{\text{dihydrothiofuran/CHCl}_3, \text{ CF}_3\text{CO}_2\text{H}^a}$

ROH   **16**

$\xleftarrow[\text{boil, few min, 90\%}]{\text{AgNO}_3, \text{ aq acetone}^a}$

(2)

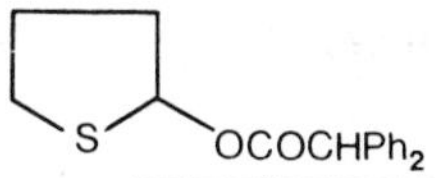

$$\text{ROH} \quad \xrightarrow[\text{20°, 5 h, 85-95\%}]{\text{(thiofuranyl)OCOCHPh}_2 \text{ , cat. TsOH, CHCl}_3{}^b} \quad \textbf{16}$$

$$\textbf{16} \quad \xleftarrow[\text{25°, 10 min, quant}]{\text{HgCl}_2, \text{CH}_3\text{CN-H}_2\text{O}^b} \quad \text{ROH}$$

Tetrahydrothiofuranyl ethers are used in ribonucleotide syntheses since they can be removed under neutral conditions, avoiding the alkaline conditions that cleave esters and acidic conditions that cleave tetrahydropyranyl ethers.[a]
Tetrahydrothiofuranyl ethers decompose above 100°.[b]

[a] L. A. Cohen and J. A. Steele, *J. Org. Chem.,* **31**, 2333 (1966).
[b] C. G. Kruse, E. K. Poels, F. L. Jonkers, and A. van der Gen, *J. Org. Chem.,* **43**, 3548 (1978).

Substituted Ethyl Ethers

17. 1-Ethoxyethyl Ether: $\text{ROCH(OC}_2\text{H}_5)\text{CH}_3$, 17

Formation

$$\text{ROH} + \text{CH}_2{=}\text{CHOEt} \quad \xrightarrow{\text{anhyd HCl}^a} \text{ or } \xrightarrow[\text{25°, 1 h}]{\text{TsOH·H}_2\text{O}^b} \quad \textbf{17}$$

ROH = nucleosides[a]; maytansinoid precursor[b]

Cleavage

$$\textbf{17} \quad \xrightarrow[\text{20°, 2 h, 100\%}]{\text{5\% HOAc}^a} \text{ or } \xrightarrow[\text{0°, 100\%}]{\text{0.5 } N \text{ HCl/THF}^b} \quad \text{ROH}$$

Compound **17** was used in preference to a THP ether (i.e. compound **9**) in nucleotide syntheses since **17** is more readily cleaved by acidic hydrolysis (e.g., **17**: 100% yield, no isomerization; **9**: 37% yield, some isomerization).[a]

[a] S. Chládek and J. Smrt, *Chem. Ind. (London),* 1719 (1964).
[b] A. I. Meyers, D. L. Comins, D. M. Roland, R. Henning, and K. Shimizu, *J. Am. Chem. Soc.,* **101**, 7104 (1979).

18.   1-Methyl-1-methoxyethyl Ether: $ROC(OCH_3)(CH_3)_2$, 18

Formation $(\rightarrow)$ / Cleavage $(\leftarrow)$[a]

$$\begin{array}{ccc}
 & \underrightarrow{\quad CH_2{=}C(Me)OMe, \text{ cat. } POCl_3 \quad} & \\
 & 20°, 30 \text{ min, } 100\% & \\
ROH & & \textbf{18} \\
 & \overleftarrow{\quad 20\% \text{ HOAc} \quad} & \\
 & 20°, 10 \text{ min} &
\end{array}$$

ROH = an allylic alcohol used in prostaglandin model studies

[a] A. F. Klug, K. G. Untch, and J. H. Fried, *J. Am. Chem. Soc.,* **94,** 7827 (1972).

19.   1-(Isopropoxy)ethyl Ether: $ROCH(CH_3)OCH(CH_3)_2$, 19

Compound **19** was prepared from a cyanohydrin and isopropyl vinyl ether (cold, trace HCl); it was cleaved by acidic hydrolysis (1 N HCl, warm, 30 min).[a]

[a] B. Tchoubar, *C. R. Hebd. Seances Acad. Sci., Ser. C,* **237,** 1006 (1953).

20.   2,2,2-Trichloroethyl Ether: $ROCH_2CCl_3$, 20

Compound **20,** prepared to protect a hydroxyl group in a carbohydrate, is cleaved[a] by Zn/HOAc, NaOAc (3 h, 92% yield).

[a] R. U. Lemieux and H. Driguez, *J. Am. Chem. Soc.,* **97,** 4069 (1975).

21.   2-(Phenylselenyl)ethyl Ether: $ROCH_2CH_2SeC_6H_5$, 21

Compound **21** has been prepared from an alcohol and the ethyl bromide (AgNO₃, CH₃CN, 20°, 10–15 min, 80–90% yield); it is cleaved by oxidation (H₂O₂, 1 h; O₃; or IO₄⁻), followed by acidic hydrolysis (dil HCl, 65-70% yield).[a]

[a] T.-L. Ho and T. W. Hall, *Synth. Commun.,* **5,** 367 (1975).

22.   *t*-Butyl Ether: $ROC(CH_3)_3$, 22

A *t*-butyl group is used for selective protection of primary hydroxyl groups.

Formation

$$ROH + CH_2{=}CMe_2 \xrightarrow{\text{cat.}} \textbf{22, high yields}$$

cat. = concd H_2SO_4[a]; BF_3/H_3PO_4[b]

Cleavage

Compound **22** is cleaved by the following conditions:

i. Anhyd CF_3COOH, 0–20°, 1–16 h, 80–90%[a,b]

ii. HBr/HOAc, 20°, 30 min[c]

iii. 4 N HCl/dioxane, reflux, 3 h; it was stable to 10 N HCl/MeOH, 0–5°, 30 h[d]

iv. Me_3SiI/CCl_4 or $CHCl_3$, 25°, < 0.1 h, 100% yield[e]

[a] H. C. Beyerman and J. S. Bontekoe, *Proc. Chem. Soc.*, 249 (1961).

[b] H. C. Beyerman and G. J. Heiszwolf, *J. Chem. Soc.*, 755 (1963).

[c] F. M. Callahan, G. W. Anderson, R. Paul, and J. E. Zimmerman, *J. Am. Chem. Soc.*, **85**, 201 (1963).

[d] U. Eder, G. Haffer, G. Neef, G. Sauer, A. Seeger, and R. Wiechert, *Chem. Ber.*, **110**, 3161 (1977).

[e] M. E. Jung and M. A. Lyster, *J. Org. Chem.*, **42**, 3761 (1977).

23.   Allyl Ether: $ROCH_2CH{=}CH_2$, 23

Allyl ethers are stable to moderately acidic conditions (1 N HCl, reflux, 10 h).[a]

Formation

$$ROH + CH_2{=}CHCH_2Br \xrightarrow[\text{NaH, }C_6H_6\text{, 90–100\%}^c]{\text{NaOH, }C_6H_6\text{, reflux, 1.5 h}^b\text{ or}} \textbf{23}$$

Cleavage

$$(1)\quad \textbf{23} \xrightarrow[\text{100°, 15 min}]{\text{KO-}t\text{-Bu/DMSO}^d} ROCH{=}CHCH_3 \xrightarrow{\text{i, ii, iii, or iv}} ROH$$

i. 0.1 N HCl/acetone–H_2O, reflux, 30 min[d]

ii. $KMnO_4$/NaOH-H_2O, 10°, 100% yield. These basic conditions avoid acid-catalyzed acetonide cleavage.[a]

iii. $HgCl_2$/HgO, acetone–H_2O, 5 min, 100% yield[e]

iv. O_3[d]

$$(2)\quad \textbf{23} \xrightarrow[\text{reflux, 3 h}]{(Ph_3P)_3RhCl,\ DABCO,\ EtOH} \xrightarrow{\text{Hg(II), pH 2}} ROH,\ > 90\%^c$$

Allyl ethers are isomerized by $(Ph_3P)_3RhCl$, and KO-t-Bu/DMSO in the following order[f]:

$(Ph_3P)_3RhCl$: allyl > 2-methylallyl > but-2-enyl

KO-t-Bu: but-2-enyl > allyl > 2-methylallyl

$$(3) \quad \textbf{23} \xrightarrow[\text{60--80}°,\ 24\ \text{h},\ 80\text{--}95\%]{\text{Pd/C--H}_2\text{O, MeOH, cat. TsOH or HClO}_4} \text{ROH}^g$$

Benzyl ethers, nitriles, epoxides, esters, and α,β-unsaturated carbonyl compounds are stable to these cleavage conditions.[g]

$$(4) \quad \textbf{23} \xrightarrow[\text{reflux, 1 h, 50\%}]{\text{SeO}_2/\text{HOAc--dioxane}^h} [\text{ROCH(OH)CH}\!=\!\text{CH}_2] \rightarrow \text{ROH}$$

[a] J. Cunningham, R. Gigg, and C. D. Warren, *Tetrahedron Lett.*, 1191 (1964).
[b] R. Gigg and C. D. Warren, *J. Chem. Soc. C*, 2367 (1969).
[c] E. J. Corey and W. J. Suggs, *J. Org. Chem.*, **38**, 3224 (1973).
[d] J. Gigg and R. Gigg, *J. Chem. Soc. C*, 82 (1966).
[e] R. Gigg and C. D. Warren, *J. Chem. Soc. C*, 1903 (1968).
[f] P. A. Gent and R. Gigg, *J. Chem. Soc., Chem. Commun.*, 277 (1974).
[g] R. Boss and R. Scheffold, *Angew. Chem., Inter. Ed. Engl.*, **15**, 558 (1976).
[h] K. Kariyone and H. Yazawa, *Tetrahedron Lett.*, 2885 (1970).

24. Cinnamyl Ether: $\text{ROCH}_2\text{CH}\!=\!\text{CHC}_6\text{H}_5$, 24

Formation ($\rightarrow$) / Cleavage ($\leftarrow$)[a]

$$\text{ROH} \xrightarrow[\text{50}°,\ 5\ \text{h}]{\text{PhCH}\!=\!\text{CHCH}_2\text{Br, HgCN, C}_6\text{H}_6} \xrightarrow[\text{H}_2\text{O}]{\text{H}_2\text{S, NaHCO}_3} \textbf{24}$$

$$\xleftarrow[\text{80--90\%}]{-2.58\ \text{V, Et}_4\text{N}^+\text{I}^-}$$

A cinnamyl ether is stable to acidic hydrolysis (10 N HCl, 20°, 20 h). Electrolytic treatment of an N-cinnamyl group would probably lead to reduction of the double bond, rather than to cleavage of the protective group, although a cinnamyl group can probably be removed electrolytically from RCOO-cinnamyl.[a]

[a] A. Ya. Veinberg, V. G. Mairanovskii, and G. I. Samokhavalov, *J. Gen. Chem. USSR*, **38**, 643 (1968).

25. p-Chlorophenyl Ether: $\text{ROC}_6\text{H}_4\text{-}p\text{-Cl}$, 25

Formation ($\rightarrow$) / Cleavage ($\leftarrow$)[a]

$$\text{ROH} \xrightarrow{\text{MeSO}_2\text{Cl, Py}} \xrightarrow{p\text{-ClC}_6\text{H}_4\text{ONa}} \textbf{25}$$

$$\xleftarrow{\text{H}_3\text{O}^+} \xleftarrow{\text{Li/NH}_3}$$

A *p*-chlorophenyl ether was used in this synthesis to minimize ring sulfonation during cyclization of a diketo ester with conc $H_2SO_4/HOAc$.[a]

[a] J. A. Marshall and J. J. Partridge, *J. Am. Chem. Soc.,* **90,** 1090 (1968).

26. Benzyl Ether: $ROCH_2C_6H_5$, 26

Formation

Benzyl ethers were originally prepared under rather drastic conditions (e.g., ROH + $PhCH_2Cl$, powdered KOH, 130–140°, 86% yield).[a] Some more recent milder methods are as follows:

(1) ROH $\xrightarrow{\text{NaH/THF}}$ $\xrightarrow[\text{20°, 3 h}]{\text{PhCH}_2\text{Br, Bu}_4\text{N}^+\text{I}^{-b}}$ $ROCH_2Ph$, 100%

This method was used to protect hindered hydroxyl groups in glucosides.[b]

(2) ROH + $PhCH_2Cl$ $\xrightarrow[\text{20°, 1–3 h, 85–95\%}]{\text{NaH, NaNH}_2 \text{ or NaOH/DMSO}^c}$ $ROCH_2Ph$

(3) ROH $\xrightarrow[\text{25°}]{\text{PhCH}_2\text{X, Ag}_2\text{O/DMF}}$ $\xrightarrow{\text{H}_3\text{O}^+}$ **26,** good yields[d]

   X = Cl, Br

This method was used to avoid deacylation in the starting carbohydrate.[e]

(4) ROH $\xrightarrow[\text{reflux, 3 h, 80–90\%}]{\text{PhCH}_2\text{Cl, Ni(acac)}_2{}^f}$ $ROCH_2Ph + HCl \uparrow$

Cleavage

In general benzyl ethers are cleaved by catalytic (eqs. 1 and 2) or chemical (eq. 3) reduction. Some other mild methods are also described.

(1) $ROCH_2Ph$ $\xrightarrow[\text{95\%}]{\text{H}_2/\text{Pd–C, EtOH}^g}$ $ROH + PhCH_3$

Pd is used since Pt would hydrogenate the aromatic ring.[g]
    An isolated double bond has been selectively reduced without cleaving a benzyl ether (H_2/5% Pd–C, 97% yield).[h]

(2) $ROCH_2Ph + Pd/C \xrightarrow[\text{1-8 h, 80-90\%}]{\text{cyclohexene}^i}$ or $\xrightarrow[\text{25°, 2 h, good yields}]{\text{cyclohexadiene}^j}$ ROH

Transfer hydrogenation may be used to remove benzyl ethers from S-containing peptides; t-butoxycarbonyl protective groups are stable. However benzyl esters, N-benzyloxycarbonyl, N-benzyl, and nitro groups are unstable to this method.[i,j]

(3) $ROCH_2Ph \xrightarrow{\text{Na/NH}_3{}^k \text{ or EtOH}^l} ROH$

(4) Benzyl ethers can be cleaved by electrolytic reduction $(-3.1 \text{ V}, R_4N^+F^-, DMF)$.[m]

(5) $ROCH_2Ph \xrightarrow[\text{25°, 15 min, 100\%}]{\text{Me}_3\text{SiI, CH}_2\text{Cl}_2{}^n} ROH$

(6) Benzyl ethers have been cleaved by lithium aluminum hydride (THF, reflux, 10 h → 25°, 12 h, 82% yield).[o]

(7) $ROCH_2Ph \xrightarrow[\text{25°, 50\%}]{\text{CrO}_3/\text{HOAc}^p} ROCOPh \; [\rightarrow ROH + HOOCPh]$

This method has been used to remove benzyl ethers from carbohydrates that contain functional groups sensitive to catalytic hydrogenation or dissolving metals. Esters are stable, but glycosides or acetals are cleaved.[p]

(8) Benzyl ethers have been cleaved by some other oxidants (e.g., $Ph_3C^+BF_4^-$, 25°, 0.5 h, 65% yield[q]; UF_6/FCl_2CClF_2, 44-69% yield[r]).

(9) Benzyl ethers are cleaved oxidatively by a cation radical under homogeneous electron transfer conditions:

$ROCH_2Ph + [Ar_3N \xrightarrow[\text{20°, CH}_3\text{CN, LiClO}_4]{\text{1.4-1.7 V}} Ar_3N^{\ddot{+}}] \xrightarrow[\text{80-90\%}]{\text{Py or NaHCO}_3}$

ROH^s

Ar = mono- or dibromophenyl

Benzyl or p-methoxybenzyl ethers can be selectively cleaved by arylamine cation radicals with different substituents.[s]

(10) Benzyl[t] and p-methoxybenzyl[t,u] ethers are cleaved by electrolytic oxidation (1.9 V, CH_3CN, $LiClO_4$, 60-70% yield).

(11) $ROCH_2Ph \xrightarrow[0°]{BBr_3/CCl_4{}^{v}}$ or $\xrightarrow[20°, \text{ 2–24 h, 85–90%}]{BF_3 \cdot Et_2O, \text{ EtSH}^{w}}$ ROH

(12) $ROCH_2Ph \xrightarrow[0\text{–}25°]{Br_2/CCl_4, h\nu} \xrightarrow{aq \text{ } Na_2CO_3}$ ROH, good yields[x]

Free radical bromination of the benzyl hydrogen followed by hydrolysis was used to remove the benzyl ether protective group from a carbohydrate with a functional group unstable to catalytic hydrogenation. This procedure could not be used selectively in the presence of trityl ethers or benzylidene acetals.[x]

[a] H. G. Fletcher, *Methods Carbohydr. Chem.*, **II**, 166 (1963).

[b] S. Czernecki, C. Georgoulis, and C. Provelenghiou, *Tetrahedron Lett.*, 3535 (1976).

[c] T. Iwashige and H. Saeki, *Chem. Pharm. Bull.*, **15**, 1803 (1967).

[d] R. Kuhn, I. Löw, and H. Trischmann, *Chem. Ber.*, **90**, 203 (1957).

[e] I. Croon and B. Lindberg, *Acta Chem. Scand.*, **13**, 593 (1959).

[f] M. Yamashita and Y. Takegami, *Synthesis,* 803 (1977).

[g] C. H. Heathcock and R. Ratcliffe, *J. Am. Chem. Soc.*, **93**, 1746 (1971).

[h] J. S. Bindra and A. Grodski, *J. Org. Chem.*, **43**, 3240 (1978).

[i] G. M. Anantharamaiah and K. M. Sivanandaiah, *J. Chem. Soc., Perkin Trans. 1*, 490 (1977).

[j] A. M. Felix, E. P. Heimer, T. J. Lambros, C. Tzougraki, and J. Meienhofer, *J. Org. Chem.*, **43**, 4194 (1978).

[k] C. M. McCloskey, *Adv. Carbohydr. Chem.*, **12**, 137 (1957).

[l] E. J. Reist, V. J. Bartuska, and L. Goodman, *J. Org. Chem.*, **29**, 3725 (1964).

[m] V. G. Mairanovsky, *Angew. Chem., Inter. Ed. Engl.*, **15**, 281 (1976).

[n] M. E. Jung and M. A. Lyster, *J. Org. Chem.*, **42**, 3761 (1977).

[o] J. P. Kutney, N. Abdurahman, C. Gletsos, P. LeQuesne, E. Piers, and I. Vlattas, *J. Am. Chem. Soc.*, **92**, 1727 (1970).

[p] S. J. Angyal and K. James, *Carbohydr. Res.*, **12**, 147 (1970).

[q] D. H. R. Barton, P. D. Magnus, G. Smith, G. Streckert, and D. Zurr, *J. Chem. Soc., Perkin Trans. 1*, 542 (1972).

[r] G. A. Olah, J. Welch, and T.-L. Ho, *J. Am. Chem. Soc.*, **98**, 6717 (1976).

[s] W. Schmidt and E. Steckhan, *Angew. Chem., Inter. Ed. Engl.*, **18**, 801 (1979).

[t] E. A. Mayeda, L. L. Miller, and J. F. Wolf, *J. Am. Chem. Soc.*, **94**, 6812 (1972).

[u] S. M. Weinreb, G. A. Epling, R. Comi, and M. Reitano, *J. Org. Chem.*, **40**, 1356 (1975).

[v] J. P. Kutney, N. Abdurahman, P. LeQuesne, E. Piers, and I. Vlattas, *J. Am. Chem. Soc.*, **88**, 3656 (1966).

[w] K. Fuji, K. Ichikawa, M. Node, and E. Fujita, *J. Org. Chem.*, **44**, 1661 (1979).

[x] J. N. BeMiller, R. E. Wing, and C. Y. Meyers, *J. Org. Chem.*, **33**, 4292 (1968).

27. *p*-Methoxybenzyl Ether: $ROCH_2C_6H_4$-*p*-OCH_3, 27

The methods used to form and cleave benzyl ethers (compound **26**) should be consulted; the conditions described in **Cleavage** eqs. 9 and 10 have been used to cleave *p*-methoxybenzyl ethers.

28. o-Nitrobenzyl Ether: $ROCH_2C_6H_4$-o-NO_2, 28

29. p-Nitrobenzyl Ether: $ROCH_2C_6H_4$-p-NO_2, 29

Compounds 28 and 29 can be prepared and cleaved by many of the methods described for benzyl ethers (compound 26). In addition, compound 28 can be cleaved by irradiation (320 nm, 10 min, quant yield of carbohydrate[a]; 280 nm, 95% yield of nucleotide[b]). Compound 29 has cleaved by electrolytic reduction (−1.1 V, DMF, $R_4N^+X^-$, 60% yield).[c]

[a] U. Zehavi, B. Amit, and A. Patchornik, *J. Org. Chem.*, 37, 2281 (1972); U. Zehavi and A. Patchornik, *J. Org. Chem.*, 37, 2285 (1972).

[b] E. Ohtsuka, S. Tanaka, and M. Ikehara, *J. Am. Chem. Soc.*, 100, 8210 (1978).

[c] V. G. Mairanovsky, *Angew. Chem., Inter. Ed. Engl.*, 15, 281 (1976).

30. p-Halobenzyl Ether: $ROCH_2C_6H_4$-p-X, X = Br, Cl, 30

p-Halobenzyl ethers have been prepared to protect side chain hydroxyl groups in amino acids. They are stable to the conditions of acidic hydrolysis (50% CF_3CO_2H) used to remove amine protective groups; they are cleaved by HF (0°, 10 min).[a]

[a] D. Yamashiro, *J. Org. Chem.*, 42, 523 (1977).

31. p-Cyanobenzyl Ether: $ROCH_2C_6H_4$-p-CN, 31

Compound 31, prepared from an alcohol and the benzyl bromide in the presence of sodium hydride (74% yield), can be cleaved by electrolytic reduction (−2.1 V, 71% yield). It is stable to electrolytic removal (−1.4 V) of a tritylone ether [i.e., 9-(9-phenyl-10-oxo)anthryl ether].[a]

[a] C. van der Stouwe and H. J. Schäfer, *Tetrahedron Lett.*, 2643 (1979).

32. 3-Methyl-2-picolyl N-Oxido Ether: , 32

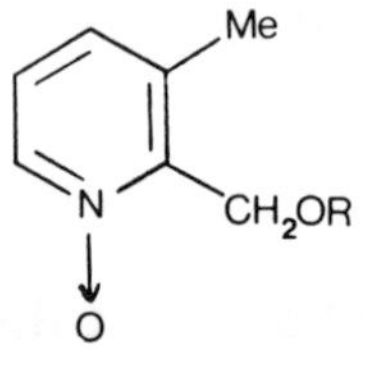

The authors prepared a number of substituted 2-diazomethylene derivatives of picolyl oxide to use for monoprotection of the cis glycol system in nucleosides. The 3-methyl derivative proved most satisfactory.[a]

Formation (→) / Cleavage (←)[a]

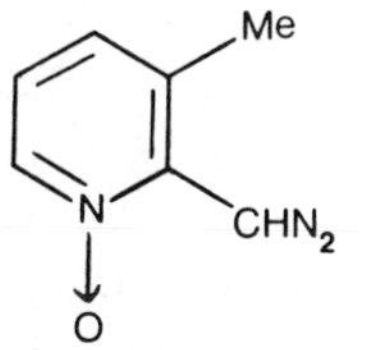

$$\overset{\text{, SnCl}_2}{\underset{63-91\%}{\xrightarrow{\hspace{4cm}}}}$$

ROH   **32**

$$\overset{\text{aq HOAc, 70°, 3 h}}{\underset{\text{quant}}{\xleftarrow{\hspace{3cm}}}}$$

ROH = ribonucleosides

[a] Y. Mizuno, T. Endo, and K. Ikeda, *J. Org. Chem.,* **40,** 1385 (1975); Y. Mizuno, T. Endo, and T. Nakamura, *J. Org. Chem.,* **40,** 1391 (1975).

33. Diphenylmethyl Ether: ROCH(C$_6$H$_5$)$_2$, 33

See also the methods used to prepare and cleave benzyl ethers (compound **26**).

Formation

(1) ROH $\xrightarrow[\text{reflux, 4–9 h, 65–92\%}]{\text{(Ph}_2\text{CHO)}_3\text{PO, cat. CF}_3\text{COOH, CH}_2\text{Cl}_2{}^a}$ ROCHPh$_2$

R = *n*-Pr, *i*-Pr, *n*-Bu, PhCH$_2$, Ph$_2$CH

(2) ROH $\xrightarrow[\text{12 h, 70\%}]{\text{Ph}_2\text{CHOH, concd H}_2\text{SO}_4{}^b}$ ROCHPh$_2$

R = Me

Cleavage

(1) ROCHPh$_2$ $\xrightarrow[\text{reflux, 24 h, 91\%}]{\text{Pd–C/AlCl}_3, \text{ cyclohexene}^c}$ ROH

(2) ROCHPh$_2$ $\xrightarrow{\text{−3.0 V, DMF, R}_4\text{N}^+\text{X}^{-b}}$ ROH

[a] L. Lapatsanis, *Tetrahedron Lett.,* 3943 (1978).
[b] V. G. Mairanovsky, *Angew. Chem., Inter. Ed. Engl.,* **15,** 281 (1976).
[c] G. A. Olah, G. K. S. Prakash, and S. C. Narang, *Synthesis,* 825 (1978).

34. 5-Dibenzosuberyl Ether:

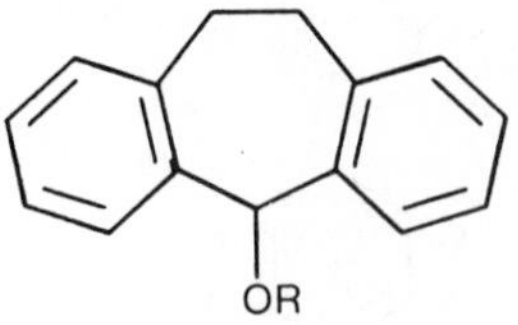

, 34

Compound **34** is prepared from an alcohol and the suberyl chloride (which has also been used to protect amines, thiols, and carboxylic acids) in the presence of triethylamine (CH_2Cl_2, 20°, 3 h, 75% yield). It is cleaved by acidic hydrolysis (1 N HCl/dioxane, 20°, 6 h, 80% yield).[a]

[a] J. Pless, *Helv. Chim. Acta*, **59**, 499 (1976).

35. Triphenylmethyl Ether: ROC(C$_6$H$_5$)$_3$, 35

The bulky triphenylmethyl group has been used to protect, selectively, the primary hydroxyl groups in carbohydrates and nucleosides.

Formation

(1)

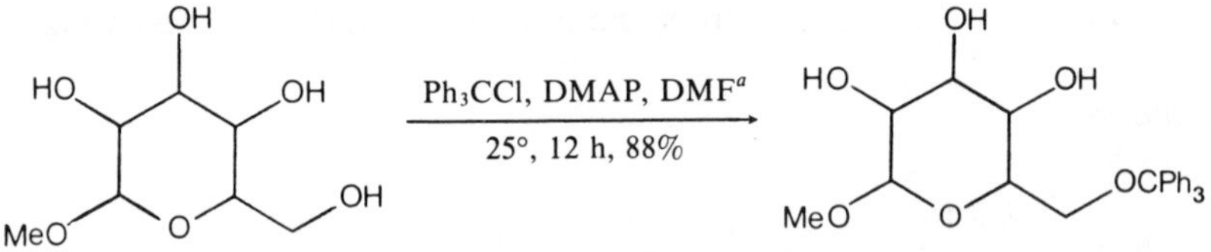

Ph$_3$CCl, DMAP, DMF[a]
25°, 12 h, 88%

DMAP = 4-N,N-dimethylaminopyridine

A secondary alcohol reacts more slowly (40–45°, 18–24 h, 68–70% yield).[a]

(2)   ROH + C$_5$H$_5$N$^+$CPh$_3$BF$_4^-$ $\xrightarrow[\text{60–70°, 75–90\%}]{\text{CH}_3\text{CN, Py}^b}$ ROCPh$_3$

Triphenylmethyl ethers can be prepared more readily with triphenylmethylpyridinium fluoroborate than with triphenylmethyl chloride/pyridine.[b]

(3)   ROH + ClCPh$_2$C$_6$H$_4$-p-Ⓟ $\xrightarrow[\text{5 days, 90\%}]{\text{Py, 25}^{oc}}$ ROCPh$_2$C$_6$H$_4$-p-Ⓟ

Ⓟ = styrene-divinylbenzene polymer

Triarylmethyl ethers of primary hydroxyl groups in gluocopyranosides have been prepared using a polymeric form of triphenylmethyl chloride. Although the yields are not improved, workup is simplified.[c]

Cleavage

Triphenylmethyl ethers have been cleaved by a number of methods, including aqueous or anhydrous acidic hydrolysis, and catalytic, chemical, or electrolytic reduction.

(1) $ROCPh_3 \xrightarrow[96\%]{\text{HOAc, 56°, 7.5 h}^d} ROH$

ROH = C_3-Δ^6-hydroxy steroid

(2) $ROCPh_3 \xrightarrow[20°, 2\text{-}30 \text{ min}]{90\% \text{ CF}_3\text{COOH, }t\text{-BuOH}} \xrightarrow[\text{resin}]{\text{Bio-Rad 1} \times 2 \text{ (OH}^-)} ROH^e, 67\text{-}100\%$

ROH = 5'-OH of a nucleoside. Bio-Rad resin neutralizes the acidic hydrolysis reaction and minimizes cleavage of glycosyl bonds.[e]

(3) $ROCPh_3 \xrightarrow[25°, 16 \text{ h, } 81\%]{\text{SiO}_2, \text{ C}_6\text{H}_6{}^f} ROH$

ROH = carbohydrate

This cleavage reaction is carried out on a column.[f]

(4) $ROCPh_3 \xrightarrow[0°, 1 \text{ h, } 91\%]{\text{HCl(g), CHCl}_3{}^g} ROH$

ROH = carbohydrate

(5)

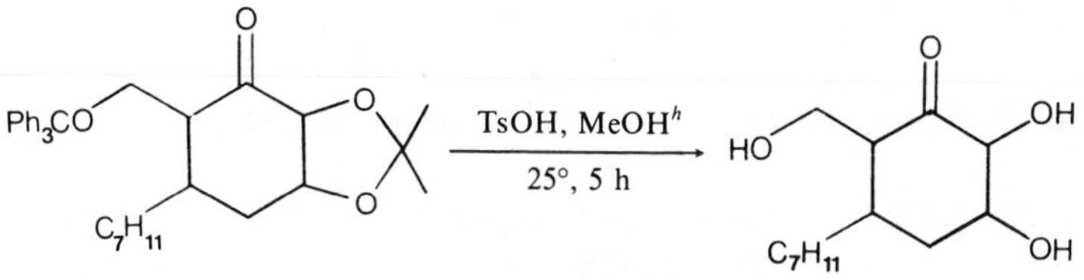

(6)

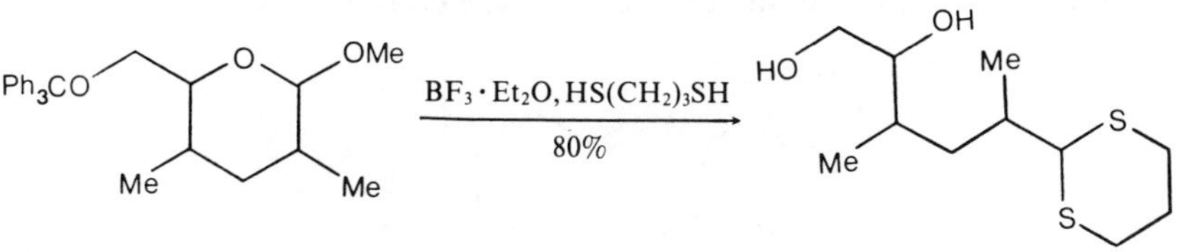

(7) $ROCPh_3 \xrightarrow[20°, 14 \text{ h, } 80\%]{\text{H}_2/\text{Pd, EtOH}^{j,k}} ROH$

(8)

$$\xrightarrow{\text{Na/NH}_3{}^l}$$

(9) $\text{ROCPh}_3 \xrightarrow{-2.9\ \text{V},\ \text{R}_4\text{N}^+\text{X}^-,\ \text{DMF}^m} \text{ROH}$

(10) $\text{CH}_3\text{CH(OCPh}_3)(\text{CH}_2)_4\text{CH}_2\text{OCPh}_3 \xrightarrow[20°,\ 15\ \text{min},\ 91\%]{\text{Ph}_3\text{C}^+\text{BF}_4{}^-,\text{CH}_2\text{Cl}_2{}^n} \text{CH}_3\text{CO(CH}_2)_4\text{CH}_2\text{OH}$

Since a secondary alcohol is oxidized in preference to a primary alcohol by triphenylmethyl tetrafluoroborate, this reaction results in selective protection of a primary alcohol.[n]

[a] S. K. Chaudhary and O. Hernandez, *Tetrahedron Lett.*, 95 (1979).

[b] S. Hanessian and A. P. A. Staub, *Tetrahedron Lett.*, 3555 (1973).

[c] J. M. J. Fréchet and K. E. Haque, *Tetrahedron Lett.*, 3055 (1975).

[d] R. T. Blickenstaff, *J. Am. Chem. Soc.*, **82**, 3673 (1960).

[e] M. MacCoss and D. J. Cameron, *Carbohydr. Res.*, **60**, 206 (1978).

[f] J. Lehrfeld, *J. Org. Chem.*, **32**, 2544 (1967).

[g] Y. M. Choy and A. M. Unrau, *Carbohydr. Res.*, **17**, 439 (1971).

[h] A. Ichihara, M. Ubukata, and S. Sakamura, *Tetrahedron Lett.*, 3473 (1977).

[i] P.-E. Sum and L. Weiler, *Can. J. Chem.*, **56**, 2700 (1978).

[j] R. N. Mirrington and K. J. Schmalzl, *J. Org. Chem.*, **37**, 2877 (1972).

[k] S. Hanessian and G. Rancourt, *Pure Appl. Chem.*, **49**, 1201 (1977).

[l] P. Kováč and Š. Bauer, *Tetrahedron Lett.*, 2349 (1972).

[m] V. G. Mairanovsky, *Angew. Chem., Inter. Ed. Engl.*, **15**, 281 (1976).

[n] M. E. Jung and L. M. Speltz, *J. Am. Chem. Soc.*, **98**, 7882 (1976).

36. α-Naphthyldiphenylmethyl Ether: ROC(C₆H₅)₂-α-C₁₀H₇, 36

Compound **36** has been prepared to protect, selectively, the 5′-OH group in nucleosides. It is prepared from α-naphthyldiphenylmethyl chloride in pyridine (65% yield), and cleaved selectively in the presence of a *p*-methoxyphenyldiphenylmethyl ether with sodium anthracenide, **a**, (THF, 97% yield).[a]

a

[a] R. L. Letsinger and J. L. Finnan, *J. Am. Chem. Soc.*, **97**, 7197 (1975).

37. p-Methoxyphenyldiphenylmethyl Ether: $ROC(C_6H_5)_2C_6H_4$-p-OCH_3, 37

In his work with nucleosides and nucleotides, Khorana[a,b] required a protective group that would be selective for primary hydroxyl groups. However, it had to be more easily cleaved by acid hydrolysis than the triphenylmethyl ether, since acid cleavage of the latter group also cleaved glycosyl bonds. Introduction of p-methoxy groups increased the rate of hydrolysis by about one order of magnitude for each p-methoxy substituent. For 5'-protected uridine derivatives in 80% HOAc, 20°, the time for hydrolysis was as follows:

$ROC(C_6H_5)_m(C_6H_4$-p-$OMe)_n$

n = 0, m = 3, 48 h

n = 1, m = 2, 2 h (therefore the most useful compound)

n = 2, m = 1, 15 min

n = 3, m = 0, 1 min

Cleavage

Compound **37** is cleaved by sodium naphthalenide in HMPA (90% yield).[c] It is not cleaved by sodium anthracenide, used to cleave α-naphthyldiphenylmethyl ethers.[d]

[a] H. G. Khorana, *Pure Appl. Chem.*, **17**, 349 (1968).

[b] M. Smith, D. H. Rammler, I. H. Goldberg, and H. G. Khorana, *J. Am. Chem. Soc.*, **84**, 430 (1962).

[c] G. L. Greene and R. L. Letsinger, *Tetrahedron Lett.*, 2081 (1975).

[d] R. L. Letsinger and J. L. Finnan, *J. Am. Chem. Soc.*, **97**, 7197 (1975).

38. p-(p'-Bromophenacyloxy)phenyldiphenylmethyl Ether: $ROC(C_6H_5)_2C_6H_4$-p-$OCH_2COC_6H_4$-p-Br, 38

Several substituted triphenylmethyl ethers were developed to provide selective protection for the 5'-OH group in nucleosides. Compound **38** proved the most satisfactory. It is prepared from the corresponding triarylmethyl chloride, and cleaved by reductive cleavage (Zn/HOAc) of the phenacyl ether to the p-hydroxyphenyldiphenylmethyl ether followed by acidic hydrolysis with HCOOH.[a]

[a] A. T.-Rigby, Y.-H. Kim, C. J. Crosscup, and N. A. Starkovsky, *J. Org. Chem.*, **37**, 956 (1972).

39. 9-Anthryl Ether: RO-9-anthryl, 39

Compound **39,** formed by reaction of the anion of 9-hydroxyanthracene and an O-tosylate, can be cleaved by oxidation with singlet oxygen, a new way to remove a protective group.

Cleavage[a]

39 $\xrightarrow[-30°]{(PhO)_3P, O_3}$ $\xrightarrow{H_2/Raney\ Ni}$

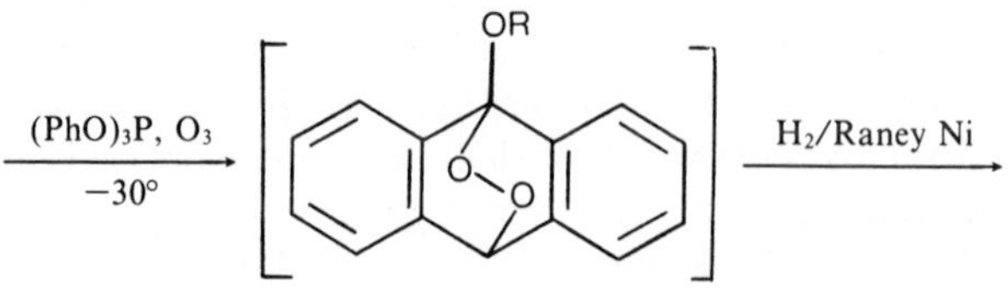

ROH + 9-hydroxyanthrone

quant

[a] W. E. Barnett and L. L. Needham, *J. Chem. Soc., Chem. Commun.*, 1383 (1970); *J. Org. Chem.*, **36**, 4134 (1971).

40. 9-(9-Phenyl)xanthenyl Ether: 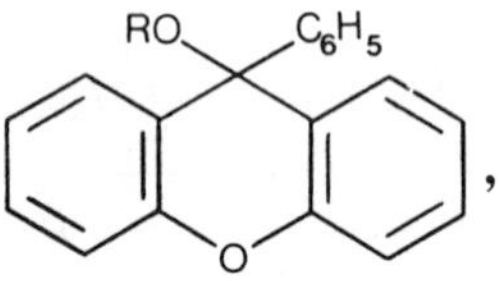   , **40**

Compound **40** has been prepared, from the xanthenyl chloride, 68–87% yield, to protect 5'-OH groups in nucleosides; it is readily cleaved by acidic hydrolysis (80% HOAc, 20°, 8–15 min, 100% yield). Compound **40** forms better crystalline derivatives than the corresponding di(*p*-methoxyphenyl)phenylmethyl ether, which is cleaved in 16 minutes under similar conditions.[a]

[a] J. B. Chattopadhyaya and C. B. Reese, *J. Chem. Soc., Chem. Commun.*, 639 (1978).

41. 9-(9-Phenyl-10-oxo)anthryl Ether (Tritylone Ether): 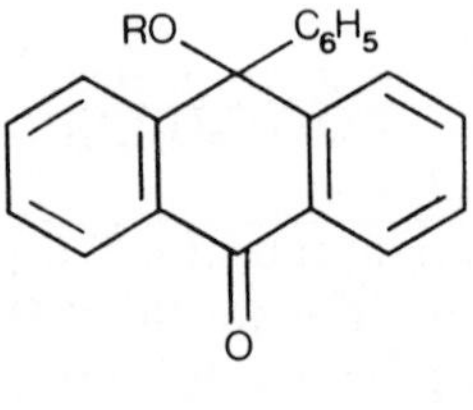   , **41**

Compound **41** has been prepared to protect primary hydroxyl groups in the presence of secondary hydroxyl groups, by reaction of an alcohol with 9-phenyl-9-hydroxyanthrone (cat. TsOH, C_6H_6, reflux, 55–95% yield).[a,b] It can be cleaved under the harsh conditions of Wolff-Kishner reduction (H_2NNH_2, NaOH, 200°, 88% yield),[a] and by electrolytic reduction (−1.4 V, LiBr, MeOH, 80–85% yield).[b] It is stable to acidic hydrolysis (10% HCl, 54 h).[a]

[a] W. E. Barnett, L. L. Needham, and R. W. Powell, *Tetrahedron*, **28**, 419 (1972).
[b] C. van der Stouwe and H. J. Schäfer, *Tetrahedron Lett.*, 2643 (1979).

42.   **Benzisothiazolyl** *S,S*-**Dioxido Ether:** , **42**

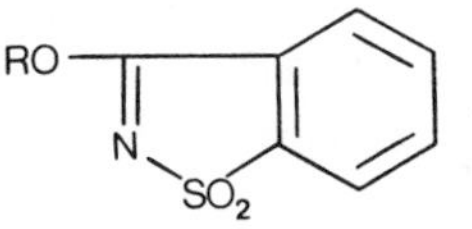

Formation (→) */ Cleavage* (←)[a]

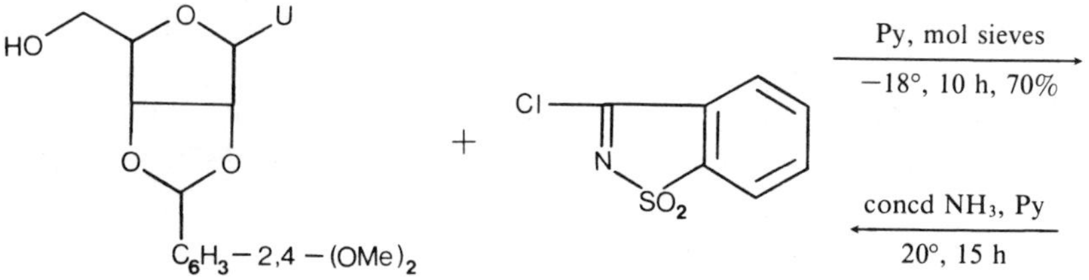

Note that compound **42** is an imino ether that can be cleaved under basic conditions.

[a] H. Sommer and F. Cramer, *Chem. Ber.,* **107**, 24 (1974).

Silyl Ethers

An active hydrogen can be protected as a silyl derivative; the order of reactivity is reported[a] as ROH > ArOH > COOH > NH > CONH > SH. Reaction of a perfluorinated resin of trimethylsilyl trifluoromethanesulfonate with a variety of functional groups to form trimethylsilyl derivatives indicates a different order of reactivity[b]: EtSH, 29°, Et_3N, 2 h, 91% yield > CH_3COOH, 25°, Et_3N, 3 h, 54% yield > Et_2NH, 28°, 6 h, 94% yield > EtOH, 29°, 12 h, 100% yield > PhOH, 23°, Et_3N, 18 h, 86% yield.

A series of silyl groups (e.g., methyldiisopropyl-, tetramethyleneisopropyl-, *t*-butyldimethyl-, triisopropyl-, and tetramethylene-*t*-butyl-) was prepared to study selective protection of 2′-, 3′-, and 5′-hydroxyl groups in ribonucleosides.[c] The *t*-butyldimethylsilyl ether proves to be one of the most useful silyl derivatives for a wide range of alcohols.[d]

Trimethylsilyl ethers of primary hydroxyl groups are readily hydrolyzed (K_2CO_3, or HOAc/MeOH, 0°); the rate constant for hydrolysis of a trimethylsilyl ether of a secondary hydroxyl group is smaller by a factor of 25.[e] The order of hydrolysis of some other trimethylsilyl (TMS) derivatives is as follows[a]: > NTMS > -COOTMS > ArOTMS > ROTMS; -STMS > -OTMS.

[a] B. E. Cooper, *Chem. Ind.* (*London*), 794 (1978).
[b] S. Murata and R. Noyori, *Tetrahedron Lett.,* **21**, 767 (1980).
[c] K. K. Ogilvie, E. A. Thompson, M. A. Quilliam, and J. B. Westmore, *Tetrahedron Lett.,* 2865 (1974).

[d] For a discussion of the use of this ether in nucleoside syntheses, see K. K. Ogilvie, S. L. Beaucage, A. L. Schifman, N. Y. Theriault, and K. L. Sadana, *Can. J. Chem.*, **56**, 2768 (1978).

[e] A. G. McInnes, *Can. J. Chem.*, **43**, 1998 (1965).

43.   Trimethylsilyl Ether (TMS Ether): ROSi(CH₃)₃, 43

Many reagents (e.g., trimethylchlorosilane, hexamethyldisilazane, N,O-bistrimethylsilylacetamide, bistrimethylsilylurea, N-trimethylsilyl-N,N'-diphenylurea, trimethylsilylimidazole, trimethylsilyldiethylamine, and monotrimethylsilylacetamide) form trimethylsilyl derivatives of compounds with an active hydrogen.[a,b] Some conditions that have been used to prepare trimethylsilyl ethers are shown below.

Formation

(1) $\text{ROH} + \text{Me}_3\text{SiCl} \xrightarrow[\text{25°, 8 h, 90\%}]{\text{Et}_3\text{N, THF}^c} \text{ROSiMe}_3$

(2) $\text{ROH} + \text{Me}_3\text{SiCl} \xrightarrow[\text{25°, 12 h, 75–95\%}]{\text{Li}_2\text{S, CH}_3\text{CN}^d} \text{ROSiMe}_3$

Silylation occurs under neutral conditions with this combination of reagents.[d]

(3) $\text{ROH} + (\text{Me}_3\text{Si})_2\text{NH} \xrightarrow[\text{20°, 5 min, } \sim 100\%]{\text{Me}_3\text{SiCl, Py}^e} \text{ROSiMe}_3$

ROH = carbohydrate

(4) $\text{ROH} + (\text{Me}_3\text{Si})_2\text{O} \xrightarrow[\text{reflux, 4 days, 80–90\%}]{\text{C}_5\text{H}_5\overset{+}{\text{N}}\text{H OTs}^-\text{, C}_6\text{H}_6\text{, mol sieves}^f} \text{ROSiMe}_3$

These are only mildly acidic conditions, suitable for acid-sensitive alcohols.[f]

(5) $\text{ROH} + \text{Me}_3\text{SiNEt}_2{}^g \rightarrow \text{ROSiMe}_3$

Trimethylsilyldiethylamine selectively silylates equatorial hydroxyl groups in quantitative yield (4–10 h, 25°). The report indicated no reaction at axial hydroxyl groups.[g]

In the prostaglandin series the order of reactivity of trimethylsilyldiethylamine is $C_{11} > C_{15} \gg C_9$ (no reaction). These trimethylsilyl ethers of secondary hydroxyl groups were hydrolyzed with aqueous methanol containing a trace of acetic acid.[h]

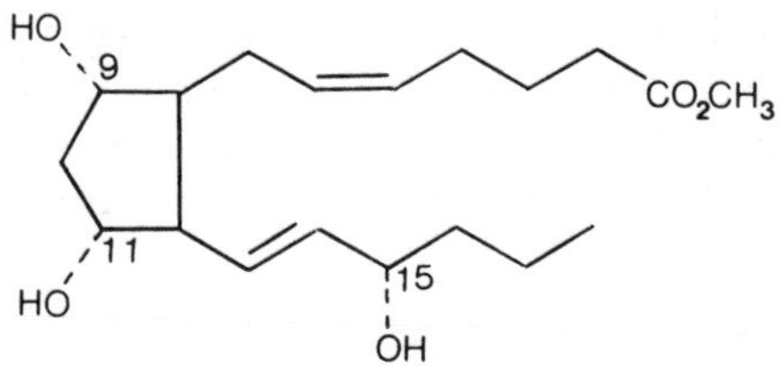

(6) $ROH + CH_3C(OSiMe_3)=NSiMe_3 \xrightarrow[78°.]{DMF^i} ROSiMe_3$

$ROH = C_{14}$-hydroxy steroid

N,O-Bis(trimethylsilyl)acetamide was used to protect a sterically hindered, tertiary hydroxyl group. The resulting trimethylsilyl ether, stable to a Grignard reaction, was slowly cleaved with 0.1 N HCl/10% aq THF, 25°.[i]

(7) $ROH + Me_3SiCH_2CO_2Et \xrightarrow[25°, \ 1-3 \ h, \ 90\%]{cat. \ n\text{-}Bu_4N^+F^{-j}} ROSiMe_3$

Use of ethyl trimethylsilylacetate/tetra-n-butylammonium fluoride allows isolation of pure products under nonaqueous conditions. This reagent also converts aldehydes and ketones to trimethylsilyl enol ethers.[j]

(8) $ROH + Me_3SiNHCO_2SiMe_3 \xrightarrow[rapid, \ 80-95\%]{THF^k} ROSiMe_3$

This reagent also silylates phenols and carboxyl groups.[k]

(9) $ROH + Me_3SiNHSO_2OSiMe_3 \xrightarrow[30°, \ 0.5 \ h, \ 92-98\%]{CH_2Cl_2{}^l} ROSiMe_3$

Higher yields of trimethylsilyl derivatives are realized by reaction of aliphatic, aromatic, and carboxylic hydroxyl groups with N,O-bis(trimethylsilyl) sulfamate than by reaction with N,O-bis(trimethylsilyl)acetamide.[l]

(10) $ROH + MeCH=C(OMe)OSiMe_3 \xrightarrow[50°, \ 30-50 \ min, \ 83-99\%]{CH_3CN \ or \ CH_2Cl_2{}^m} ROSiMe_3$

This reagent also silylates phenols, thiols, amides, and carboxyl groups.[m]

(11) $ROH + Me_3SiCH_2CH=CH_2 \xrightarrow[70-80°, \ 1-2 \ h]{TsOH/CH_3CN^n} ROSiMe_3, \ 90-95\%$

Formation of silyl derivatives may be effected by reaction with an allylsilane under acid catalysis. This silylating reagent is stable to moisture [Me₃SiCl and

$(Me_3Si)_2NH$ are readily hydrolyzed]. Allylsilanes can be used to protect alcohols, phenols, and carboxylic acids; there is no reaction with thiophenol. The method is also applicable to the formation of *t*-butyldimethylsilyl derivatives; the silyl ether of cyclohexanol was prepared in 95% yield from allyl-*t*-butyldimethyl-silane.[n]

Cleavage

(1) $ROSiMe_3 \xrightarrow[\text{aprotic conditions}]{n\text{-}Bu_4N^+F^-,\ THF^c} ROH$

(2) $ROSiMe_3 \xrightarrow[0°,\ 45\ min,\ 100\%]{K_2CO_3,\ anhyd\ MeOH^o} ROH$

ROH = a carbohydrate

(3) $ROSiMe_3 \xrightarrow[20°,\ 10\ min,\ 100\%]{citric\ acid/MeOH^p} ROH$

ROH = 9-hydroxy prostaglandin derivative

Like other ethers with an α-hydrogen, trimethylsilyl ethers are oxidized by NBS/$h\nu$ or Br_2^q:

$$RCH_2OSiMe_3 + NBS/h\nu \xrightarrow{0°,\ 5\ h} RCO_2R,\ 80\%$$

$$C_6H_5CH_2OSiMe_3 + NBS/h\nu \xrightarrow{-20°,\ 2.5\ h} C_6H_5CHO,\ 48\%$$

$$RR'CHOSiMe_3 + NBS/h\nu \xrightarrow[Py]{20°,\ 3.5\ h} RR'C{=}O,\ 55\text{--}75\%$$

[a] B. E. Cooper, *Chem. Ind.* (*London*), 794 (1978).

[b] A. E. Pierce, *Silylation of Organic Compounds,* Pierce Chem. Co., Rockford, Illinois, 1968.

[c] E. J. Corey and B. B. Snider, *J. Am. Chem. Soc.,* **94**, 2549 (1972).

[d] G. A. Olah, B. G. B. Gupta, S. C. Narang, and R. Malhotra, *J. Org. Chem.,* **44**, 4272 (1979).

[e] C. C. Sweeley, R. Bentley, M. Makita, and W. W. Wells, *J. Am. Chem. Soc.,* **85**, 2497 (1963).

[f] H. W. Pinnick, B. S. Bal, and N. H. Lajis, *Tetrahedron Lett.,* 4261 (1978).

[g] I. Weisz, K. Felföldi, and K. Kovács, *Acta Chim. Acad. Sci. Hung.,* **58**, 189 (1968).

[h] E. W. Yankee, U. Axen, and G. L. Bundy, *J. Am. Chem. Soc.,* **96**, 5865 (1974); E. L. Cooper and E. W. Yankee, *J. Am. Chem. Soc.,* **96**, 5876 (1974).

[i] M. N. Galbraith, D. H. S. Horn, E. J. Middleton, and R. J. Hackney, *J. Chem. Soc., Chem. Commun.,* 466 (1968).

[j] E. Nakamura, T. Murofushi, M. Shimizu, and I. Kuwajima, *J. Am. Chem. Soc.,* **98**, 2346 (1976).

[k] L. Birkofer and P. Sommer, *J. Organomet. Chem.,* **99**, C1 (1975).

[l] B. E. Cooper and S. Westall, *J. Organomet. Chem.*, **118**, 135 (1976).
[m] Y. Kita, J. Haruta, J. Segawa, and Y. Tamura, *Tetrahedron Lett.*, 4311 (1979).
[n] T. Morita, Y. Okamoto, and H. Sakurai, *Tetrahedron Lett.*, **21**, 835 (1980).
[o] D. T. Hurst and A. G. McInnes, *Can. J. Chem.*, **43**, 2004 (1965).
[p] G. L. Bundy and D. C. Peterson, *Tetrahedron Lett.*, 41 (1978).
[q] H. W. Pinnick and N. H. Lajis, *J. Org. Chem.*, **43**, 371 (1978).

44. Triethylsilyl Ether: $ROSi(C_2H_5)_3$, 44

Formation[a]

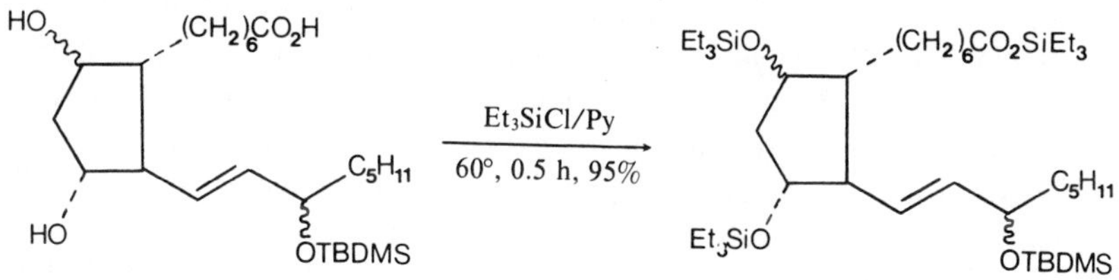

44a

Cleavage[a]

44a

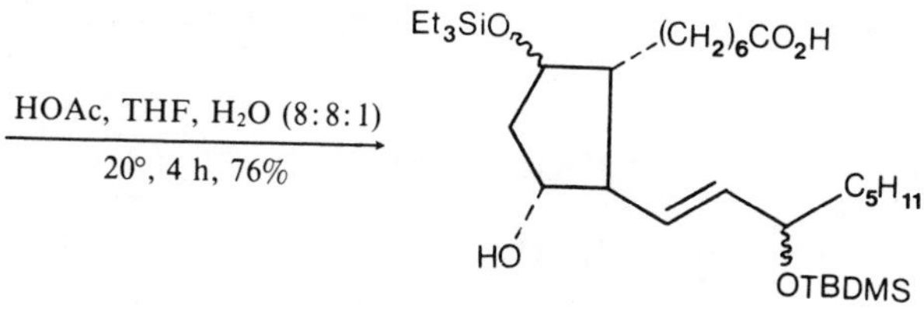

TBDMS = *t*-butyldimethylsilyl (compound **46**).

More acidic conditions [HOAc, THF, H_2O (6:1:3), 45°, 3 h] cleave all of the protective groups (76% yield).[a]

[a] T. W. Hart, D. A. Metcalfe, and F. Scheinmann, *J. Chem. Soc., Chem. Commun.*, 156 (1979).

45. Isopropyldimethylsilyl Ether: $ROSi(CH_3)_2$-*i*-C_3H_7, 45

Formation (→) / ***Cleavage*** (←)[a]

$$\mathrm{ROH} \underset{\text{HOAc/H}_2\text{O (3:1)},\ 35°,\ 10\ \text{min},\ 100\%\ (\leftarrow)}{\overset{\text{HN(SiMe}_2\text{-}i\text{-Pr)}_2/\text{ClSiMe}_2\text{-}i\text{-Pr},\ 25°,\ 48\ \text{h},\ 98\%\ (\rightarrow)}{\rightleftharpoons}} \mathbf{45}$$

ROH = PGE_2

An isopropyldimethylsilyl ether is more easily cleaved than a tetrahydropyranyl ether. It is *not* stable to Grignard or Wittig reactions, or to Jones oxidation.[a]

[a] E. J. Corey and R. K. Varma, *J. Am. Chem. Soc.*, **93**, 7319 (1971).

46.   *t*-Butyldimethylsilyl Ether (TBDMS Ether): $ROSi(CH_3)_2$-*t*-C_4H_9, 46

The *t*-butyldimethylsilyl group is one of the most useful silyl protective groups; a TBDMS ether is more stable to hydrolysis than a trimethylsilyl or dimethyliso-propylsilyl ether, but is still readily cleaved by a variety of selective conditions. TBDMS and tetrahydropyranyl (THP) ethers are cleaved by mild acidic hydroly-sis [$HOAc$-H_2O-THF (3:1:1), 25°] under similar conditions: THP: 4 h, 90% yield; TBDMS: 6 h, 96% yield. TBDMS esters are cleaved 20 times faster than TBDMS ethers by these conditions. In the course of prostaglandin synthetic stud-ies, the TBDMS group was developed as a hydroxyl protective group that could be selectively removed in the presence of an acetate ester, a benzyl ether, a 2,2,2-trichloroethyl ether, or a tetrahydropyranyl ether. It is stable to the Wittig reac-tion, to $Zn/MeOH$, H_2/Pd–C, Na/NH_3, $(Me_2CHCH_2)_2AlH$, CrO_3/Py, H_2O_2/OH^-, and MeI/Ag_2O.[a]

During fluoride cleavage of a *t*-butyldimethylsilyl ether (protecting one hy-droxyl group in a glycerol) an acyl group (protecting a second hydroxyl group) underwent migration[b]:

$$\begin{array}{c}CH_2OCOC_{17}H_{35}\text{-}n\\ |\\ CHOTBDMS\\ |\\ CH_2OCOC_{17}H_{35}\text{-}n\end{array} \xrightarrow{\ n\text{-}Bu_4N^+F^-\ } \begin{array}{c}CH_2OCOC_{17}H_{35}\text{-}n\\ |\\ CHOH\\ |\\ CH_2OCOC_{17}H_{35}\text{-}n\end{array} + \begin{array}{c}CH_2OH\\ |\\ CHOCOC_{17}H_{35}\text{-}n\\ |\\ CH_2OCOC_{17}H_{35}\text{-}n\end{array}$$

Examples of migration have been reported in other substrates (e.g., carbohy-drates,[c] prostaglandins,[d] and nucleosides[e,f]). Consequently the *t*-butyldimethyl-silyl group should not be used for hydroxyl protection when acyl migration is possible.

Formation

(1) $ROH + ClSiMe_2\text{-}t\text{-}Bu \xrightarrow[\text{25°, 10 h, high yields}]{\text{imidazole, DMF}^a} ROTBDMS$

(2) $ROH + t\text{-}Bu\text{-}Me_2SiOClO_3 \xrightarrow[\text{20 min, 100\%}]{\text{CH}_3\text{CN, Py}^g} ROTBDMS$

ROH = tertiary alcohols

In this example the yields with *t*-butyldimethylchlorosilane were unsatisfactory (10% yield, 72 h).[g]

(3)   ROH + ClSiMe$_2$-*t*-Bu $\xrightarrow[\text{25°, 5-8 h, 75-95\%}]{\text{Li}_2\text{S, CH}_3\text{CN}^h}$ ROTBDMS

This reaction occurs under nearly neutral conditions.[h]

(4)   PhCH(OH)CH$_2$OH $\xrightarrow[\text{25°, 12 h}]{\text{ClSiMe}_2\text{-}t\text{-Bu, DMAP, Et}_3\text{N, DMF}^i}$

PhCH(OH)CH$_2$OTBDMS + PhCH(OTBDMS)CH$_2$OH +
                                        PhCH(OTBDMS)CH$_2$OTBDMS

      60-90%   minor products

(5)   *c*-C$_6$H$_{11}$OH + CH$_2$=CHCH$_2$SiMe$_2$-*t*-Bu $\xrightarrow[\text{70-80°, 2.5 h}]{\text{TsOH/CH}_3\text{CN}^j}$ ROTBDMS, 95%

See p. 41–42 for a discussion of allylsilanes.[j]

Cleavage

(1)   ROTBDMS $\xrightarrow[\text{25°, 30 min, 100\%}]{n\text{-Bu}_4\text{N}^+\text{F}^-, \text{THF}^a}$ ROH

(2)   ROTBDMS $\xrightarrow[\text{25°, 4 h, 95\%}]{n\text{-Bu}_4\text{N}^+\text{Cl}^-/\text{KF}\cdot2\text{H}_2\text{O, CH}_3\text{CN}^k}$ ROH

This method readily generates fluoride ion *in situ,* and is reported to be suitable for reactions that normally require anhydrous conditions.[k]

(3)   ROTBDMS $\xrightarrow[\text{20°, 1-3 h, 90-100\%}]{\text{aq HF, CH}_3\text{CN}^l}$ ROH

A solution of CH$_3$CN: 40% aq HF (95:5) efficiently hydrolyzes *t*-butyldimethylsilyl ethers under mild conditions.[l]

(4)   ROTBDMS $\xrightarrow[\text{25-80°, 15 min to 5 h}]{\text{HOAc-H}_2\text{O}^a}$ ROH, high yield

(5)   ROTBDMS $\xrightarrow[\text{20°, MeOH}]{\text{Dowex 50W-X8}^m}$ ROH

Dowex 50W-X8 is a carboxylic acid resin, H$^+$ form.

(6)   ROTBDMS $\xrightarrow[\text{0-25°, 15 min to 3 h, 70-90\%}]{\text{BF}_3\cdot\text{Et}_2\text{O, CHCl}_3{}^n}$ ROH

(7)   ROTBDMS $\xrightarrow[\text{0°, 15 min, 92\%}]{\text{FeCl}_3,\ \text{Ac}_2\text{O}^o}$ ROAc

ROH = 2-octanol. Methyl and benzyl ethers are also cleaved.^{o}

(8)   ROTBDMS $\xrightarrow[\text{25°, 60 h, 40–80\%}]{\text{Ph}_3\text{C}^+\text{BF}_4^-\ \text{or LiBF}_4,\ \text{CH}_2\text{Cl}_2{}^p}$ ROH

ROH = primary or secondary alcohol

(9)

46a

$\xrightarrow{\hspace{3cm}}$ thymidine

Compound **46a** was subjected to the following conditions^{e,q}:

  i.   0.5 N NaOH/EtOH–H$_2$O, 22°, 24 h, 80% yield of thymidine.
 ii.   15% NH$_4$OH/EtOH, 22°, 1.5 h (used to cleave -OAc), -OTBDMS stable.
iii.   9 M NH$_4$OH, 60°, 1 h (used to cleave -OAc), 6% yield of thymidine.
 iv.   H$_2$NNH$_2$/HOAc–Py, 22°, 24 h (used to cleave -NAc), -OTBDMS stable.
  v.   80% HOAc, 90°, 15 min, 100% yield of thymidine.
 vi. n-Bu$_4$N$^+$F$^-$/THF, 22°, 30 min, 100% yield of thymidine.

^{a} E. J. Corey and A. Venkateswarlu, *J. Am. Chem. Soc.*, **94**, 6190 (1972).

^{b} G. H. Dodd, B. T. Golding, and P. V. Ioannou, *J. Chem. Soc., Chem. Commun.*, 249 (1975).

^{c} F. Franke and R. D. Guthrie, *Aust. J. Chem.*, **31**, 1285 (1978).

^{d} Y. Torisawa, M. Shibasaki, and S. Ikegami, *Tetrahedron Lett.*, 1865 (1979).

^{e} K. K. Ogilvie, S. L. Beaucage, A. L. Schifman, N. Y. Theriault, and K. L. Sadana, *Can. J. Chem.*, **56**, 2768 (1978).

^{f} S. S. Jones and C. B. Reese, *J. Chem. Soc., Perkin Trans. 1*, 2762 (1979).

^{g} T. J. Barton and C. R. Tully, *J. Org. Chem.*, **43**, 3649 (1978).

^{h} G. A. Olah, B. G. B. Gupta, S. C. Narang, and R. Malhotra, *J. Org. Chem.*, **44**, 4272 (1979).

^{i} S. K. Chaudhary and O. Hernandez, *Tetrahedron Lett.*, 99 (1979).

^{j} T. Morita, Y. Okamoto, and H. Sakurai, *Tetrahedron Lett.*, **21**, 835 (1980).

^{k} L. A. Carpino and A. C. Sau, *J. Chem. Soc., Chem. Commun.*, 514 (1979).

^{l} R. F. Newton, D. P. Reynolds, M. A. W. Finch, D. R. Kelly, and S. M. Roberts, *Tetrahedron Lett.*, 3981 (1979).

^{m} E. J. Corey, J. W. Ponder, and P. Ulrich, *Tetrahedron Lett.*, **21**, 137 (1980).

^{n} D. R. Kelly, S. M. Roberts, and R. F. Newton, *Synth. Commun.*, **9**, 295 (1979).

^{o} B. Ganem and V. R. Small, Jr., *J. Org. Chem.*, **39**, 3728 (1974).

^{p} B. W. Metcalf, J. P. Burkhart, and K. Jund, *Tetrahedron Lett.*, **21**, 35 (1980).

^{q} K. K. Ogilvie and D. J. Iwacha, *Tetrahedron Lett.*, 317 (1973).

47.   (Triphenylmethyl)dimethylsilyl Ether: $ROSi(CH_3)_2C(C_6H_5)_3$, 47

Formation[a]

$$ROH + BrSi(Me)_2CPh_3 \xrightarrow[20°, 80-90\%]{AgNO_3/DMF} \text{ or } \xrightarrow[100°, 12\,h, 50-75\%]{DMF} 47$$

Cleavage[a]

(1) $47 \xrightarrow{K_2CO_3/MeOH} ROH$

(2) $47 \xrightarrow{n\text{-}Bu_4N^+F^-/dioxane} ROH$

(3) $47 \xrightarrow{acidic\ hydrolysis} ROH$

Acidic hydrolysis = $0.01\,M\,H_2SO_4$ or $0.01\,M\,HCl/THF-H_2O$ (3:2), 20°, $t_{1/2} = 4$ days; $0.2\,M\,H_2SO_4/MeOH-THF$ (2:3), reflux, $t_{1/2} = 3$ days; $TsOH \cdot H_2O/C_6H_6$, reflux, 15 h, 100% yield; CF_3CO_2H, 20°, 3 h, 100% yield.

Compound 47 is more stable to hydrolysis than is a *t*-butyldimethylsilyl ether. For example, it is stable to the following conditions: $HOAc/Et_2O$ (1:1), 20°, 96 h; $HOAc-H_2O-THF$ (6:4:1), reflux, 24 h; $TsOH/C_6H_6$, reflux, 20 h; $0.2\,M\,H_2SO_4/MeOH-Et_2O$ (2:1), 20°, 92 h; $NH_3/MeOH$, 20°, 24 h. It is stable in an ether solution that is washed with $0.1\,M$ NaOH.[a]

[a] D. J. Ager and I. Fleming, *J. Chem. Res., Synop.*, 6 (1977).

48.   *t*-Butyldiphenylsilyl Ether: $ROSi(C_6H_5)_2$-*t*-C_4H_9, 48

Compound 48 was prepared to protect a primary hydroxyl group during a synthesis of thromboxane B_2 from D-glucose.[a]

Formation[a]

$$ROH + ClSiPh_2\text{-}t\text{-}Bu \xrightarrow[25-60°, 75-87\%]{imidazole,\ DMF} 48$$

Cleavage

(1) $48 \xrightarrow[25°, 1-5\,h, >90\%]{n\text{-}Bu_4N^+F^-,\ THF^a} ROH$

(2) $48 \xrightarrow[25°, 3\,h, 71\%]{3\%\ methanolic\ hydrogen\ chloride^a} ROH$

(3) $48 \xrightarrow[\text{25°, 7 h, 93\%}]{\text{5 } N \text{ NaOH/EtOH}^{a}} \text{ROH}$

(4) $48 \xrightarrow{\text{C}_5\text{H}_5\overset{+}{\text{N}}\text{H F}^{-}, \text{THF}^{b}} \text{ROH}$

ROH = an allylic alcohol

t-Butyldiphenylsilyl ethers are stable to K_2CO_3/CH_3OH, to $9 M NH_4OH$, 60°, 2 h, and to $NaOCH_3$(cat.) $/CH_3OH$, 25°, 24 h. A t-butyldiphenylsilyl ether is stable to 80% acetic acid, used to cleave t-butyldimethylsilyl, triphenylmethyl, and tetrahydropyranyl ethers. It is also stable to HBr/HOAc, 12°, 2 min, and to 25–75% aq HCO_2H, 25°, 2–6 h. It is stable to concd HCl, 25°, 24 h, and to 50% aq CF_3CO_2H, 25°, 15 min (conditions used to form and cleave acetals). t-Butyldiphenylsilyl ethers are stable to oxidation [DMSO/1-ethyl-3-(3′-dimethylaminopropyl)carbodiimide·HCl, 25°, 12 h] and to reduction [20% $Pd(OH)_2$-C; $(i\text{-}C_4H_9)_2AlH$].^{a}

^{a} S. Hanessian and P. Lavallee, *Can. J. Chem.*, **53**, 2975 (1975); **55**, 562 (1977).
^{b} K. C. Nicolaou, S. P. Seitz, M. R. Pavia, and N. A. Petasis, *J. Org. Chem.*, **44**, 4011 (1979).

49.   Methyldiisopropylsilyl Ether: ROSiCH₃[CH(CH₃)₂]₂, 49

Compound **49** is one in a series of silyl ethers prepared to study selective protection of 2′-, 3′-, and 5′-OH groups in nucleosides.^{a}

Formation (→) / Cleavage (←)^{a}

$$\text{ROH} \quad \overset{\displaystyle \xrightarrow{\text{ClSiMe-}i\text{-Pr}_2/\text{imidazole, DMF}}}{\underset{\underset{\text{1.5–6 h, 100\%}}{\xleftarrow{\text{80\% HOAc, 20°, or 0.01 } N \text{ HCl, 90°}}}}{}} \quad \textbf{49}$$

^{a} K. K. Ogilvie, E. A. Thompson, M. A. Quilliam, and J. B. Westmore, *Tetrahedron Lett.*, 2865 (1974).

50.   Methyldi-t-butylsilyl Ether: ROSiCH₃(t-C₄H₉)₂, 50

Formation^{a}

$$\text{ROH} + t\text{-Bu}_2\text{MeSiOClO}_3 \xrightarrow[99\%]{\text{CH}_3\text{CN/Py}} \textbf{50}$$

Cleavage[a]

(1)   **50** $\xrightarrow{\text{BF}_3/\text{CH}_2\text{Cl}_2}$ [ROBF$_2$] $\xrightarrow[\text{0}°, \text{30 min}, 94\%]{\text{NaHCO}_3, \text{H}_2\text{O}}$ ROH

(2)   **50** + 1% HCl/95% EtOH $\xrightarrow[\text{0}\%]{\text{25}°, \text{3 days}}$ or $\xrightarrow[\text{50}\%]{\text{90}°, \text{24 h}}$ ROH

ROH = primary, secondary, or tertiary alcohol

Compound **50** is more stable to acidic hydrolysis than is a tetrahydropyranyl or *t*-butyldimethylsilyl ether prepared from a primary or secondary alcohol. It is stable to basic hydrolysis (e.g., 5% NaOH, EtOH, 80°, 3 days).[a]

[a] T. J. Barton and C. R. Tully, *J. Org. Chem.*, **43**, 3649 (1978).

51.   Tribenzylsilyl Ether: ROSi(CH$_2$C$_6$H$_5$)$_3$, 51

52.   Tri-*p*-xylylsilyl Ether: ROSi(CH$_2$C$_6$H$_4$-*p*-CH$_3$)$_3$, 52

To control the stereochemistry of epoxidation at the 10,11-double bond in intermediates in prostaglandin syntheses, a bulky protective group was used for the C$_{15}$-OH group. Epoxidation of compound **52** yielded 88% α-oxide; epoxidation of **51** was less selective.[a]

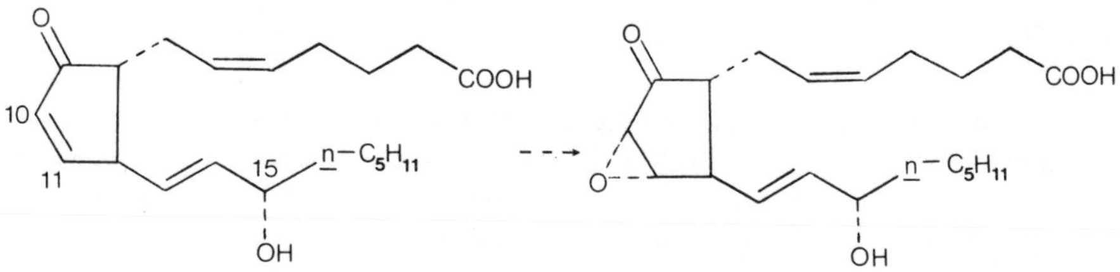

Formation ($\rightarrow$) / Cleavage ($\leftarrow$)[a]

ROH $\xrightarrow[\text{$-20$°, 24–36 h, 90–100\%}]{\text{ClSi(CH}_2\text{C}_6\text{H}_4\text{-}p\text{-Y})_3, \text{DMF, 2,6-lutidine}}$ **51 or 52**

$\xleftarrow[\text{26°, 6 h} \rightarrow \text{45°, 3 h, 85\%}]{\text{HOAc/THF/H}_2\text{O (3:1:1)}}$

ROH = PGA$_2$; Y = H, Me (**51** and **52**, respectively)

[a] E. J. Corey and H. E. Ensley, *J. Org. Chem.*, **38**, 3187 (1973)

53.   Triisopropylsilyl Ether: ROSi[CH(CH₃)₂]₃, 53

Compound **53**, like compound **49**, was designed to study selective protection of 2′-, 3′-, and 5′-hydroxy groups in nucleosides. Triisopropylchlorosilane reacts almost exclusively with the primary 5′-OH group to form compound **53** (imidazole, DMF, 82% yield). Compound **53** is cleaved by acidic hydrolysis (0.01 N HCl/EtOH, 90°, 15 min, 100% yield).[a]

[a] K. K. Ogilvie, E. A. Thompson, M. A. Quilliam, and J. B. Westmore, *Tetrahedron Lett.*, 2865 (1974).

54.   Triphenylsilyl Ether: ROSi(C₆H₅)₃, 54

Formation (→) / Cleavage (←)[a]

$$\text{ROH} \quad \underset{\underset{70°, 3\text{ h}, 70\%}{\overline{\text{HOAc–H}_2\text{O–THF (3:1:1)}}}}{\overset{\overset{\text{Ph}_3\text{SiBr, Py}}{\underline{-40°, 15\text{ min}}}}{\rightleftharpoons}} \quad \textbf{54}$$

ROH = prostaglandin derivative

[a] H. Nakai, N. Hamanaka, H. Miyake, and M. Hayashi, *Chem. Lett.*, 1499 (1979).

ESTERS

Esters and carbonates, in general prepared from an alcohol and an acid chloride or acid anhydride, or chloroformate, respectively, and cleaved by basic hydrolysis, complement ethers, cleaved by acidic hydrolysis, as protective group derivatives for alcohols. Some of the general methods used to prepare (see Chapter 5, Newer Methods of Formation, eqs. 2, 8, and 9) and cleave (Chapter 5, Newer Methods of Cleavage, eqs. 3 and 4) esters of carboxylic acids can be used to protect hydroxyl groups. Other general methods of formation (eqs. 1–3) and cleavage (eqs. 4–6) are as follows:

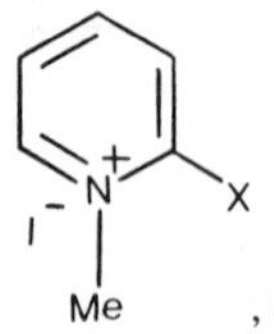

(1) $\text{ROH} + \text{R'COOH} \xrightarrow[\text{reflux, 3 h, 62–97\%}]{\text{(pyridinium), } n\text{-Bu}_3\text{N, CH}_2\text{Cl}_2{}^{a}} \text{R'COOR}$

R = Et, t-Bu, Ph, PhCH₂, PhCH(Me), PhCH=CHCH₂

R' = Me, *t*-Bu, Ph, PhCH$_2$, PhCH=CH

(2)

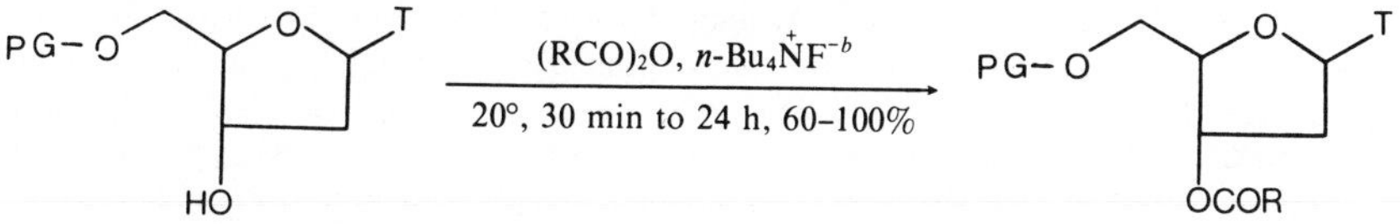

$$\text{(RCO)}_2\text{O, } n\text{-Bu}_4\overset{+}{\text{N}}\text{F}^{-b}$$
20°, 30 min to 24 h, 60–100%

R = Me, *t*-Bu, Ph

PG = protective group

(3) ROH + *o*-BrCH$_2$C$_6$H$_4$COOCOR' $\xrightarrow[85-100\%]{80-145°, \text{ 26 min to 10 h}^c}$ ROCOR'

R = Et, *t*-Bu, Ph, PhCH$_2$

R' = Et, Pr, *i*-Pr, *t*-Bu, Ph

(4) ROCOR' + H$_2$NNH$_2$ $\xrightarrow[\text{good yields}]{\text{HOAc/Py } (1:4)^d}$ ROH

R' = Me, Ph

ROH = nucleosides

(5) ROCOR' $\xrightarrow[\text{2 h}]{\text{Na or Li/NH}_3\text{, Et}_2\text{O}}$ $\xrightarrow{\text{NH}_4\text{Cl}}$ ROH, 70–90%^{e}

ROH = primary, secondary alcohol

R' = Ph, *t*-Bu

(6) ROCOR' $\xrightarrow{\text{LiAlH}_4{}^f}$ ROH, high yields

Cleavage of an ester by reduction with lithium aluminum hydride is a satisfactory method if no other functional group is present that is reactive to the reagent.

Esters that are most useful as alcohol protective groups are listed in Reactivity Chart 2.^{g}

[a] T. Mukaiyama, M. Usui, E. Shimada, and K. Saigo, *Chem. Lett.,* 1045 (1975).

[b] S. L. Beaucage and K. K. Ogilvie, *Tetrahedron Lett.,* 1691 (1977).

[c] H. Horimoto, S. Takimoto, T. Katsuki, and M. Yamaguchi, *Chem. Lett.,* 145 (1979).

[d] Y. Ishido, N. Nakazaki, and N. Sakairi, *J. Chem. Soc., Perkin Trans. 1,* 2088 (1979).

[e] H. W. Pinnick and E. Fernandez, *J. Org. Chem.,* **44,** 2810 (1979).

[f] For example, see S. P. Tanis and K. Nakanishi, *J. Am. Chem. Soc.,* **101,** 4398 (1979).

[g] See also: C. B. Reese, "Protection of Alcoholic Hydroxyl Groups and Glycol Systems," in *Protective Groups in Organic Chemistry,* J. F. W. McOmie, Ed., Plenum, New York and London, 1973, pp. 109–120; H. M. Flowers, "Protection of the Hydroxyl Group," in *The Chemistry of the Hydroxyl Group,* S. Patai, Ed., Wiley-Interscience, New York, 1971, Vol. 10/2, pp. 1012–1025; C. B. Reese, *Tetrahedron,* **34,** 3143–3179 (1978); V. Amarnath and A. D. Broom, *Chem. Rev.,* **77,** 183–217 (1977).

1. Formate Ester: ROCHO, 1

Formation

(1) $\text{ROH} + 85\% \text{ HCOOH} \xrightarrow[93\%]{60°, \text{ 1 h}^a} \text{ROCHO}$

ROH = steroid

(2) $\text{ROH} + 70\% \text{ HCOOH/cat. HClO}_4 \xrightarrow[\text{good yields}]{50\text{–}55^{ob}} \text{ROCHO}$

ROH = steroid

(3) $\text{ROH} + \text{MeCOOCHO} \xrightarrow[-20°, \text{ 80–100\%}]{\text{Py}^{c,d}} \text{ROCHO}$

ROH = steroid,[c] nucleoside[d]

Cleavage

(1) $\text{ROCHO} \xrightarrow[20°, \text{ 3 days}]{\text{KHCO}_3/\text{aq MeOH}^c} \text{ROH}$

(2) $\text{ROCHO} \xrightarrow[22°, \text{ 62\%}]{\text{dil NH}_3, \text{ pH } 11.2^e} \text{ROH}$

ROH = nucleoside

A formate ester can be cleaved selectively in the presence of an acetate [MeOH/ reflux[d] or dil NH₃ (formate 100 times faster than acetate)[e]] or benzoate ester (dil NH₃).[e]

[a] H. J. Ringold, B. Löken, G. Rosenkranz, and F. Sondheimer, *J. Am. Chem. Soc.,* **78,** 816 (1956).

[b] I. W. Hughes, F. Smith, and M. Webb, *J. Chem. Soc.,* 3437 (1949).

[c] F. Reber, A. Lardon, and T. Reichstein, *Helv. Chim. Acta,* **37,** 45 (1954).

[d] J. Žemlička, J. Beránek, and J. Smrt, *Collect. Czech. Chem. Commun.,* **27,** 2784 (1962).

[e] C. B. Reese and J. C. M. Stewart, *Tetrahedron Lett.,* 4273 (1968).

2. Benzoylformate Ester: $ROCOCOC_6H_5$, 2

Compound 2 can be prepared from the 3'-hydroxy group in a deoxyribonucleo-tide by reaction with benzoyl chloroformate (anhyd Py, 20°, 12 h, 86% yield); it is cleaved by aqueous pyridine (20°, 12 h, 31% yield), conditions that do not cleave an acetate ester.[a]

[a] R. L. Letsinger and P. S. Miller, *J. Am. Chem. Soc.*, **91**, 3356 (1969).

3. Acetate Ester: $ROCOCH_3$, 3

See also pp. 50–51.

Formation

(1) $ROH + Ac_2O \xrightarrow[20°, 12 h, \sim 100\%]{Py^a} ROAc$

ROH = 3'-hydroxynucleotide

(2) $ROH + CH_3COCl \xrightarrow[67-79\%]{25°, 16 h^b} ROAc$

ROH = a sugar

(3) $ROH + (R'CO)_2O$ or $R'COCl \xrightarrow[1-40 h, 72-95\%]{DMAP, 24-80°^c} ROCOR'$

ROH = hindered secondary or tertiary alcohol

DMAP = 4-*N,N*-dimethylaminopyridine, 10^4 times as active an acylation catalyst as pyridine

R' = Me, Et, Ph

(4) $ROH + CH_3CO_2C_6F_5 \xrightarrow[80°, 12-60 h, 72-95\%]{Et_3N^d} ROAc$

ROH = aminoethanols

This reagent reacts with an amino group (25°, no Et_3N) to form an *N*-acetyl derivative in 80–90% yield.[d]

(5) $ROH + $ $\cdot PtCl_2(C_2H_4) \xrightarrow[51-87\%]{23°, 0.5-144 h^e} ROAc$

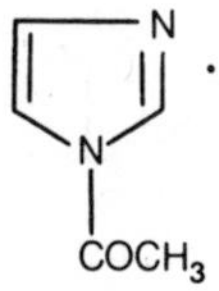

ROH = pyridinyl alcohol, $C_5H_4N(CH_2)_nCH_2OH$

Platinum(II) acts as a template to catalyze this acetylation.[e]

(6) $HOCH_2(CH_2)_nCH_2OH \xrightarrow[\text{30 h to 1 week, 60–95\%}]{HOAc/H_2SO_4,\ H_2O^{f}} AcOCH_2(CH_2)_nCH_2OH$

A monoacetate can be isolated by continuous extraction with organic solvents such as cyclohexane/CCl_4.[f]

Cleavage

(1) $ROAc \xrightarrow[\text{20°, 1 h, 100\%}]{K_2CO_3/\text{aq MeOH}^{g}} ROH$

ROH = secondary or allylic alcohol

(2)

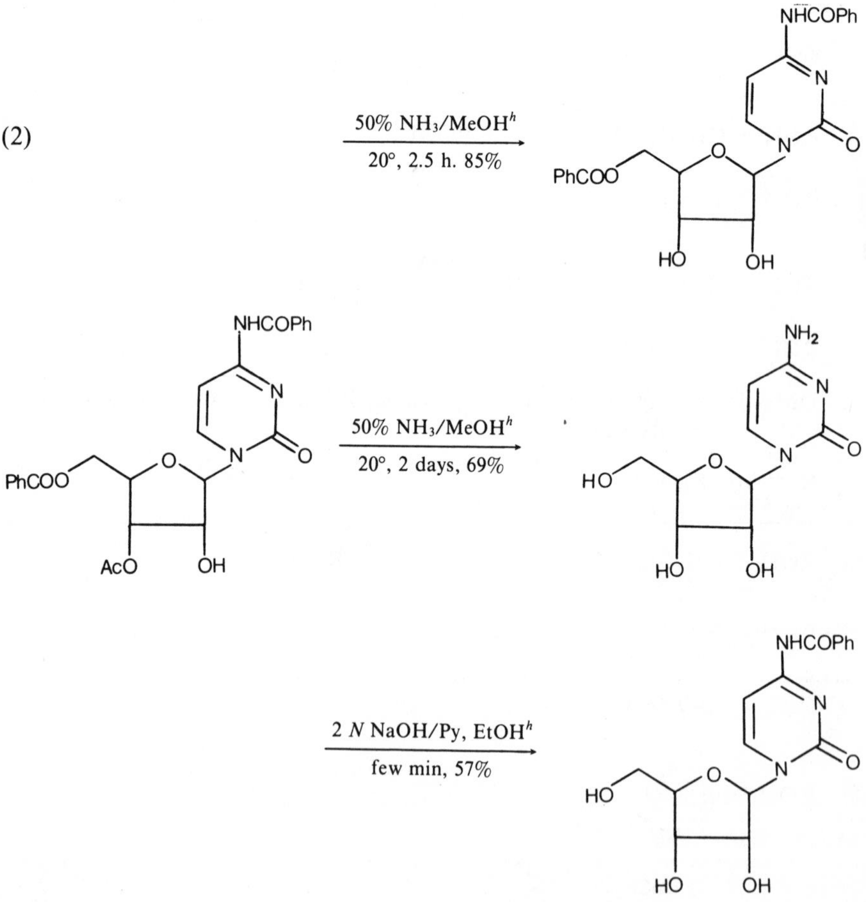

NHCOPh
N
N
O
O
PhCOO
HO
OH
50% NH3/MeOH[h]
20°, 2.5 h. 85%
NHCOPh
PhCOO
AcO
OH
50% NH3/MeOH[h]
20°, 2 days, 69%
NH2
HO
HO
OH
2 N NaOH/Py, EtOH[h]
few min, 57%
NHCOPh
HO
HO
OH

(3)   ROAc $\xrightarrow[\text{20° to reflux, 12 h}]{\text{KCN/95\% EtOH}^{i}}$ ROH, 93%[j]

Potassium cyanide is a mild transesterification catalyst, suitable for acid- or base-sensitive compounds.[i,j]

[a] H. Weber and H. G. Khorana, *J. Mol. Biol.,* **72,** 219 (1972); R. I. Zhdanov and S. M. Zhenodarova, *Synthesis,* 222 (1975).

[b] D. Horton, *Org. Synth., Collect. Vol. V,* 1 (1973).

[c] G. Höfle, W. Steglich, and H. Vorbrüggen, *Angew. Chem., Inter. Ed. Engl.,* **17,** 569 (1978).

[d] L. Kisfaludy, T. Mohacsi, M. Low, and F. Drexler, *J. Org. Chem.,* **44,** 654 (1979).

[e] J. C. Chottard, E. Mulliez, and D. Mansuy, *J. Am. Chem. Soc.,* **99,** 3531 (1977).

[f] J. H. Babler and M. J. Coghlan, *Tetrahedron Lett.,* 1971 (1979).

[g] J. J. Plattner, R. D. Gless, and H. Rapoport, *J. Am. Chem. Soc.,* **94,** 8613 (1972).

[h] T. Neilson and E. S. Werstiuk, *Can. J. Chem.,* **49,** 493 (1971).

[i] K. Mori, M. Tominaga, T. Takigawa, and M. Matsui, *Synthesis,* 790 (1973).

[j] K. Mori and M. Sasaki, *Tetrahedron Lett.,* 1329 (1979).

4.   Chloroacetate Ester: ROCOCH₂Cl, 4

Compound **4** can be prepared from an alcohol and the acid anhydride (nucleoside/Py, 0°, 2 h, 70–90% yield)[a] or acid chloride (prostaglandin/Py–Et₂O, 87% yield).[b] It is cleaved selectively in the presence of an acetate and benzoate ester by reaction with thiourea (NaHCO₃/EtOH, 70°, 5 h, 70% yield).[b] It can also be cleaved by reaction with HSCH₂CH₂NH₂, H₂NCH₂CH₂NH₂, or *o*-phenylenediamine,[a] and by basic hydrolysis (aq Py, pH 6.7, 20 h, 100% yield).[c]

[a] A. F. Cook and D. T. Maichuk, *J. Org. Chem.,* **35,** 1940 (1970).

[b] M. Naruto, K. Ohno, N. Naruse, and H. Takeuchi, *Tetrahedron Lett.,* 251 (1979).

[c] F. Johnson, N. A. Starkovsky, A. C. Paton, and A. A. Carlson, *J. Am. Chem. Soc.,* **86,** 118 (1964).

5.   Dichloroacetate Ester: ROCOCHCl₂, 5

Compound **5** was prepared, by reaction with the acid chloride, to protect the hydroxyl group in 7-mandelamido-3-cephem-4-carboxylic acids. It is stable to mild acid and is cleaved at pH 9–9.5 (20°, 30 min).[a]

[a] J. R. E. Hoover, G. L. Dunn, D. R. Jakas, L. L. Lam, J. J. Taggart, J. R. Guarini, and L. Phillips, *J. Med. Chem.,* **17,** 34 (1974).

6.   Trichloroacetate Ester: ROCOCCl₃, 6

Compound **6** was prepared to protect a C₁₇-hydroxy steroid (Cl₃CCOCl/Py, DMF, 20°, 2 days, 60–90% yield). A trichloroacetate ester can be cleaved in the presence of an acetate ester (NH_3/EtOH, $CHCl_3$, 20°, 6 h, 81% yield) or a formate ester (KOH/MeOH, 72% yield).[a]

[a] V. Schwarz, *Collect. Czech. Chem. Commun.*, **27**, 2567 (1962).

7. Trifluoroacetate Ester: ROCOCF₃, 7

Trifluoroacetates, prepared[a] by reaction of an alcohol with the anhydride, are very readily hydrolyzed. In a series of nucleoside esters a trifluoroacetate is hydrolyzed immediately in 100% yield at 20°, pH 7.[b] A hindered alcohol was protected as a trifluoroacetate by reaction with pertrifluoroacetic acid (CF_3CO_3H, 20°, 4 h, 83% yield); trifluoroacetic acid did not form a trifluoroacetate.[c]

[a] A. Lardon and T. Reichstein, *Helv. Chim. Acta*, **37**, 443 (1954).
[b] F. Cramer, H. P. Bär, H. J. Rhaese, W. Sänger, K. H. Scheit, G. Schneider, and J. Tennigkeit, *Tetrahedron Lett.*, 1039 (1963).
[c] G. W. Holbert and B. Ganem, *J. Chem. Soc., Chem. Commun.*, 248 (1978).

8. Methoxyacetate Ester: ROCOCH₂OCH₃, 8

Compound **8,** prepared by reaction with the acetyl chloride to protect nucleosides, is selectively cleaved (aq NH_3 or NH_3/MeOH, 78% yield) in the presence of acetate or benzoate esters; a methoxyacetate is cleaved 20 times faster than an acetate.[a]

[a] C. B. Reese and J. C. M. Stewart, *Tetrahedron Lett.*, 4273 (1968).

9. Triphenylmethoxyacetate Ester: ROCOCH₂OC(C₆H₅)₃, 9

Compound **9** was prepared in 53% yield from a nucleoside and the sodium acetate ($Ph_3COCH_2CO_2Na$/i-Pr_3-$C_6H_2SO_2Cl$, Py) as a derivative that could be easily detected (i.e., it has a distinct orange-yellow color after it is sprayed with ceric sulfate). It is readily cleaved by NH_3/MeOH (100% yield).[a]

[a] E. S. Werstiuk and T. Neilson, *Can. J. Chem.*, **50**, 1283 (1972).

10. Phenoxyacetate Ester: ROCOCH₂OC₆H₅, 10

Compound **10,** prepared to protect nucleosides by reaction with the anhydride, is selectively cleaved by aqueous or methanolic ammonia in the presence of acetate or benzoate esters; it is 50 times as labile to aqueous ammonia as an acetate ester.[a]

[a] C. B. Reese and J. C. M. Stewart, *Tetrahedron Lett.*, 4273 (1968).

11.   *p*-Chlorophenoxyacetate Ester: $ROCOCH_2OC_6H_4$-*p*-Cl, 11

Compound **11**, prepared to protect a nucleoside by reaction with the acetyl chloride, is cleaved by 0.2 *M* NaOH/dioxane–H_2O, 0°, 30 s.[a]

[a] S. S. Jones and C. B. Reese, *J. Am. Chem. Soc.,* **101**, 7399 (1979).

12.   2,6-Dichloro-4-methylphenoxyacetate Ester: $ROCOCH_2OC_6H_2$-2,6-Cl_2-4-CH_3, 12

13.   2,6-Dichloro-4-(1,1,3,3-tetramethylbutyl)phenoxyacetate Ester: $ROCOCH_2OC_6H_2$-2,6-Cl_2-4-$C(CH_3)_2CH_2C(CH_3)_3$, 13

14.   2,4-Bis(1,1-dimethylpropyl)phenoxyacetate Ester: $ROCOCH_2OC_6H_3$-2,4-$[C(CH_3)_2CH_2CH_3]_2$, 14

Compounds **12, 13,** and **14** were developed to protect nucleosides. They are prepared by reaction with the acetyl chloride and cleaved by dilute ammonia.[a]

[a] C. B. Reese, *Tetrahedron,* **34**, 3143 (1978).

15.   Chlorodiphenylacetate Ester: $ROCOCCl(C_6H_5)_2$, 15

Compound **15** was prepared in 87% yield from a nucleoside and the acetyl chloride. It is cleaved by methanolic ammonia (66 h, 68% yield).[a]

[a] A. F. Cook and D. T. Maichuk, *J. Org. Chem.,* **35**, 1940 (1970).

16.   *p*-Ⓟ-Phenylacetate Ester: $ROCOCH_2C_6H_4$-*p*-Ⓟ, 16

Monoprotection of a symmetrical diol can be effected by reaction with a polymer-supported phenylacetyl chloride. The free hydroxyl group is then converted to an ether and the phenylacetate cleaved by aqueous ammonia–dioxane, 48 h.[a]

$$HO(CH_2)_nOH + p\text{-}Ⓟ\text{-}C_6H_4CH_2COCl \xrightarrow{Py} HO(CH_2)_nOCOCH_2C_6H_4\text{-}p\text{-}Ⓟ$$

n = 2, 4, 6, 8, 10

Ⓟ = polystyrene-divinylbenzene copolymer

[a] J. Y. Wong and C. C. Leznoff, *Can. J. Chem.,* **51**, 2452 (1973).

17. 3-Phenylpropionate Ester: $ROCOCH_2CH_2C_6H_5$, 17

Compound **17** has been used in nucleoside syntheses.[a] It is cleaved by α-chymotrypsin[a] (37°, 8–16 h, 70–90% yield).[b]

[a] H. S. Sachdev and N. A. Starkovsky, *Tetrahedron Lett.*, 733 (1969).
[b] A. T.-Rigby, *J. Org. Chem.*, **38**, 977 (1973).

18. 3-Benzoylpropionate Ester: $ROCOCH_2CH_2COC_6H_5$, 18

Formation ($\rightarrow$) / Cleavage ($\leftarrow$)[a]

$$\text{ROH} \quad \frac{\overset{\text{PhCO(CH}_2)_2\text{CO}_2\text{H/DCC, Py}}{\longrightarrow}}{20°, 3 \text{ h}, 50–65\%} \quad \frac{\text{H}_2\text{NNH}_2, \text{HOAc, Py}}{\underset{20°, 12 \text{ h}, \sim100\%}{\longleftarrow}} \quad \textbf{18}$$

ROH = nucleoside

An acetate ester is stable to these cleavage conditions.[a]

[a] R. L. Letsinger and P. S. Miller, *J. Am. Chem. Soc.*, **91**, 3356 (1969).

19. Isobutyrate Ester: $ROCOCH(CH_3)_2$, 19

Isobutyric anhydride reacts with deoxyguanosine ($Et_4N^+OH^-$, 20°, 48 h) to form the $N,3',5'$-triisobutyryl derivative. The isobutyrate esters are cleaved by basic hydrolysis (2 M NaOH/EtOH, 0°, 15 min, 100% yield), conditions that do not cleave the N-isobutanamide. Isobutyrate esters are stable to potassium bicarbonate.[a]

[a] H. Büchi and H. G. Khorana, *J. Mol. Biol.*, **72**, 251 (1972).

20. Monosuccinoate Ester: $ROCOCH_2CH_2COOH$, 20

A $C_{3,12}$-dihydroxy steroid was treated with succinic anhydride/Py, 3 h, to form the C_3-monosuccinoate ester in 55% yield. The C_{12}-hydroxy group was oxidized (CrO_3/HOAc) and the succinate cleaved by basic hydrolysis (KOH/MeOH, reflux, 2 h, 78% yield).[a]

[a] P. L. Julian, C. C. Cochrane, A. Magnani, and W. J. Karpel, *J. Am. Chem. Soc.*, **78**, 3153 (1956).

21. 4-Oxopentanoate (Levulinate) Ester: $ROCOCH_2CH_2COCH_3$, 21

Formation

(1) $ROH + (CH_3COCH_2CH_2CO)_2O \xrightarrow[\text{24 h, 70–85\%}]{\text{Py, 25}^{\circ a}} \textbf{21}$

ROH = steroids, nucleosides, . . .

(2) $ROH + CH_3COCH_2CH_2CO_2H \xrightarrow[\text{96\%}]{\text{DCC/DMAP}^b} \textbf{21}$

ROH = nucleoside

Cleavage

(1) $\textbf{21} \xrightarrow[\text{20}^{\circ}\text{, 20 min, 80–95\%}]{\text{NaBH}_4/\text{H}_2\text{O, pH 5–8}^a} ROH +$

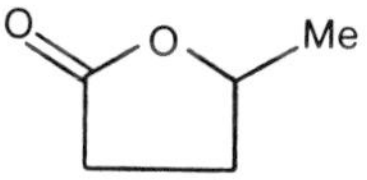

, water soluble

(2) $\textbf{21} \xrightarrow[\text{2 min, 100\%}]{\text{0.5 }M\text{ H}_2\text{NNH}_2/\text{H}_2\text{O, Py, HOAc}^b} ROH$

[a] A. Hassner, G. Strand, M. Rubinstein, and A. Patchornik, *J. Am. Chem. Soc.*, **97**, 1614 (1975).
[b] J. H. van Boom and P. M. J. Burgers, *Tetrahedron Lett.*, 4875 (1976).

22. Pivaloate Ester: $ROCOC(CH_3)_3$, 22

A pivaloate ester is formed selectively from a primary hydroxyl group (i.e., from the 5'-OH in a nucleoside) by reaction with pivaloyl chloride/Py (0–75°, 2.5 days, 99% yield).[a]

Cleavage

(1) $\textbf{22} \xrightarrow[\text{20}^{\circ}\text{, 4 h}]{\text{Bu}_4\text{N}^+\text{OH}^{-b}} ROH$

ROH = carbohydrate

(2) $\textbf{22} \xrightarrow[\text{20}^{\circ}\text{, }t_{1/2} = 3 \text{ h}]{\text{aq MeNH}_2^c} ROH$

ROH = uridine

An acetate ester can be selectively cleaved by NH_3/MeOH in the presence of a pivaloate ester.[c]

(3)   **22** $\xrightarrow[\text{20}°,\ 12\ \text{h, 58%}]{\text{0.5 }N\text{ NaOH/EtOH, H}_2\text{O}^d}$ ROH

ROH = nucleoside

(4)   **22** $\xrightarrow{\text{Li/NH}_3,\ \text{Et}_2\text{O}}$ $\xrightarrow{\text{NH}_4\text{Cl}}$ ROH, 70–85%[e]

(5)   **22** $\xrightarrow[\text{20}°]{\text{MeLi, Et}_2\text{O}^f}$ ROH

(6)   **22** $\xrightarrow[\text{20}°,\ 3\ \text{h, 94%}]{\text{KO-}t\text{-Bu/H}_2\text{O (8:2)}^g}$ ROH

This method (using "anhydrous hydroxide") also cleaves tertiary amides.[g]

[a] M. J. Robins, S. D. Hawrelak, T. Kanai, J.-M. Siefert, and R. Mengel, *J. Org. Chem.*, **44**, 1317 (1979).
[b] C. A. A. van Boeckel and J. H. van Boom, *Tetrahedron Lett.*, 3561 (1979).
[c] B. E. Griffin, M. Jarman, and C. B. Reese, *Tetrahedron*, **24**, 639 (1968).
[d] K. K. Ogilvie and D. J. Iwacha, *Tetrahedron Lett.*, 317 (1973).
[e] H. W. Pinnick and E. Fernandez, *J. Org. Chem.*, **44**, 2810 (1979).
[f] B. M. Trost, S. A. Godleski, and J. L. Belletire, *J. Org. Chem.*, **44**, 2052 (1979).
[g] P. G. Gassman and W. N. Schenk, *J. Org. Chem.*, **42**, 918 (1977).

23.   Adamantoate Ester: ROCO-1-adamantyl, 23

Compound **23** is formed selectively from a primary hydroxyl group (i.e., from the 5'-OH in a ribonucleoside) by reaction with adamantoyl chloride/Py (20°, 16 h). It is cleaved by alkaline hydrolysis (0.25 N NaOH, 20 min), but is stable to milder alkaline hydrolysis (e.g., NH$_3$/MeOH), conditions that cleave an acetate ester.[a]

[a] K. Gerzon and D. Kau, *J. Med. Chem.*, **10**, 189 (1967).

24.   Crotonate Ester: ROCOCH=CHCH$_3$, 24

25.   4-Methoxycrotonate Ester: ROCOCH=CHCH$_2$OCH$_3$, 25

Compounds **24** and **25**, prepared to protect a primary hydroxyl group in nucleosides, are cleaved by hydrazine (MeOH/Py, 2 h). Compound **25** (R = nucleoside) is 100-fold more reactive to hydrazinolysis and 2-fold less reactive to alkaline hydrolysis than the corresponding acetate.[a]

[a] R. Arentzen and C. B. Reese, *J. Chem. Soc., Chem. Commun.*, 270 (1977).

26. *(E)*-**2-Methyl-2-butenoate (Tigloate) Ester:** ROCO$-$C$=$C$-$H, **26**

$$\underset{\text{CH}_3\text{CH}_3}{||}$$

A tigloate ester, prepared in 80% yield to protect a hydroxy steroid is oxidized (OsO_4/HIO_4) to an α-keto ester that is cleaved by mild basic hydrolysis (pH 8.5, 12 h, 90% yield).[a]

[a] S. M. Kupchan, A. D. J. Balon, and E. Fujita, *J. Org. Chem.*, **27**, 3103 (1962).

27. Benzoate Ester: ROCOC$_6$H$_5$, 27

Formation

(1)   ROH $\xrightarrow[\text{18°, 2 days, } \rightarrow \text{ 60°, 1 h}]{\text{PhCOCl or (PhCO)}_2\text{O/Py}^a}$ ROCOPh

  ROH = carbohydrates

Regioselective benzoylation of methyl 4,6-*O*-benzylidene-α-D-galactopyranoside can be effected by phase transfer catalysis (e.g., PhCOCl/n-Bu$_4$N$^+$Cl$^-$, 40% NaOH, C$_6$H$_6$, 69% yield of the 2-benzoate; PhCOCl/n-Bu$_4$N$^+$Cl$^-$, 40% NaOH, HMPA, 62% yield of the 3-benzoate).[b]

(2)   ROH + $\xrightarrow[\text{20°, 15 min, 90\%}]{\text{Et}_3\text{N, DMF}^c}$ ROCOPh

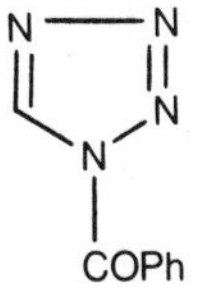

  ROH = nucleoside

(3)   ROH + PhCOCN $\xrightarrow[\text{5 min to 2 h, } > 80\%]{\text{Et}_3\text{N/CH}_3\text{CN}^d}$ ROCOPh

  ROH = steroids,[d] sugars[e]

Selective formation of benzoate esters, determined by steric factors, can be effected by reaction of the hydroxy compound with benzoyl cyanide. In hydroxy steroids the order of reactivity is C-21 $> -17\beta > -3\beta > -6\alpha, -3\alpha \gg -20\alpha > -20\beta > -7\beta \gg -6\beta > -22\beta > -22\alpha$.[d]

(4)   MeOH + PhCOCF(CF$_3$)$_2$ $\xrightarrow[\text{20°, 30 min, 90\%}]{\text{TMEDA}^f}$ ROCOPh

This reagent reacts with amines to form benzamides in high yields.[f]

Cleavage

Benzoate esters are more stable to hydrolysis than acetate esters.

(1) $ROCOPh \xrightarrow[\text{20°, 50 min, 90%}]{\text{1% NaOH/MeOH}^g} ROH$

(2) $ROCOPh \xrightarrow[\text{reflux, 20 h, 86%}]{\text{Et}_3\text{N/MeOH/H}_2\text{O (1:5:1)}^h} ROH$

(3)

$\xrightarrow[\text{20°, 7 days or 80°, 12 h, 80%}]{\text{H}_2\text{NNH}_2\text{/HOAc–Py (1:4)}^i}$

Regioselective cleavage of the 2′-benzoate has been effected by these conditions.[i]

(4)   A benzoate ester can be cleaved in 60–90% yield by electrolytic reduction at −2.3 V.[j]

---

[a] M. Gyr and T. Reichstein, *Helv. Chim. Acta,* **28,** 226 (1945); A. H. Haines, *Adv. Carbohydr. Chem. Biochem.,* **33,** 11 (1976).

[b] W. Szeja, *Synthesis,* 821 (1979).

[c] J. Stawinski, T. Hozumi, and S. A. Narang, *J. Chem. Soc., Chem. Commun.,* 243 (1976).

[d] M. Havel, J. Velek, J. Pospišek, and M. Souček, *Collect. Czech. Chem. Commun.,* **44,** 2443 (1979).

[e] A. Holý and M. Souček, *Tetrahedron Lett.,* 185 (1971).

[f] N. Ishikawa and S. Shin-ya, *Chem. Lett.,* 673 (1976).

[g] K. Mashimo and Y. Sato, *Tetrahedron,* **26,** 803 (1970).

[h] K. Tsuzuki, Y. Nakajima, T. Watanabe, M. Yanagiya, and T. Matsumoto, *Tetrahedron Lett.,* 989 (1978).

[i] Y. Ishido, N. Nakazaki, and N. Sakairi, *J. Chem. Soc., Perkin Trans. 1,* 2088 (1979).

[j] V. G. Mairanovsky, *Angew. Chem., Inter. Ed. Engl.,* **15,** 281 (1976).

---

28.   *o*-(Dibromomethyl)benzoate Ester: ROCOC₆H₄-*o*-CHBr₂, 28.

Compound **28,** prepared to protect nucleosides by reaction with the benzoyl chloride (CH$_3$CN, 65–90% yield), can be cleaved under nearly neutral conditions. This is a good example of a "protected protective group." The cleavage involves conversion of the —CHBr$_2$ group to —CHO by silver ion–assisted hydrolysis. The benzoate group, ortho to the —CHO group, now is rapidly hydrolyzed by neighboring group participation (the morpholine and hydroxide ion-catalyzed hydrolyses of methyl 2-formylbenzoate are particularly rapid).[a]

Cleavage[a]

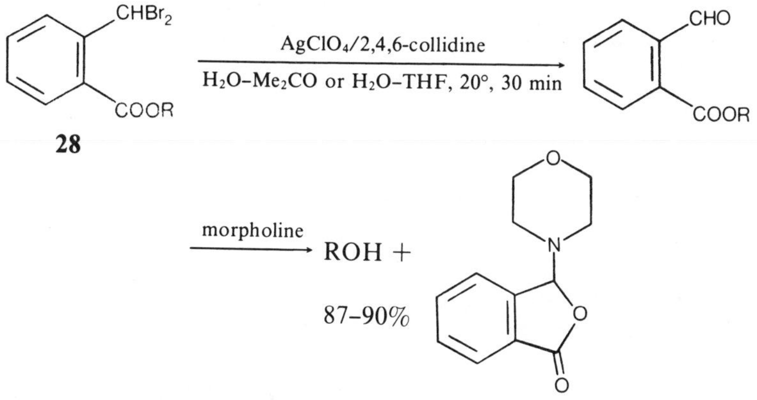

ROH = nucleoside

[a] J. B. Chattopadhyaya, C. B. Reese, and A. H. Todd, *J. Chem. Soc., Chem. Commun.*, 987 (1979).

29. *o*-(Methoxycarbonyl)benzoate Ester: ROCOC$_6$H$_4$-*o*-COOCH$_3$, 29

Compound **29** was prepared to protect the hydroxyl group in lactic acid during a synthesis of Val-Oxisoval-Val-Oxisoval-Val by reaction with the monomethyl ester monoacid chloride of phthalic acid (Et$_3$N, 20°, 12 h, 50–60% yield). It is cleaved by hydrazine (80°, 1 h, 40–60% yield).[a]

[a] G. Losse and H. Raue, *Chem. Ber.*, **98**, 1522 (1965).

30. *p*-Phenylbenzoate Ester: ROCOC$_6$H$_4$-*p*-C$_6$H$_5$, 30

Compound **30** was prepared to protect a hydroxyl group of a prostaglandin intermediate by reaction with the benzoyl chloride (Py, 25°, 1 h, 97% yield). It was a more crystalline, more readily separated derivative than 15 other esters that were investigated.[a]

[a] E. J. Corey, S. M. Albonico, U. Koelliker, T. K. Schaaf, and R. K. Varma, *J. Am. Chem. Soc.*, **93**, 1491 (1971).

31. 2,4,6-Trimethylbenzoate (Mesitoate) Ester: ROCOC$_6$H$_2$-2,4,6-(CH$_3$)$_3$, 31

Formation

(1) ROH + ClCOC$_6$H$_2$-2,4,6-Me$_3$ $\xrightarrow[\text{0°, 14 h} \rightarrow \text{23°, 1 h, 95\%}]{\text{Py, CHCl}_3{}^a}$ **31**

ROH = allylic alcohol

(2) $ROH + HOOCC_6H_2\text{-}2,4,6\text{-}Me_3 \xrightarrow[20°, \ 15 \ min]{(CF_3CO)_2O, \ C_6H_6{}^b}$ **31**

ROH = secondary alcohol

Cleavage

(1)   **31** $\xrightarrow[20°, \ 2 \ h]{LiAlH_4, \ Et_2O{}^b}$ ROH

(2)   **31** $\xrightarrow[20°, \ 24\text{--}72 \ h, \ 50\text{--}72\%]{\substack{\text{"anhydrous hydroxide"}{}^c \\ KO\text{-}t\text{-}Bu/H_2O \ (8:1)}}$ ROH

A mesitoate ester is stable to mild basic hydrolysis (2 N NaOH, 20°, 20 h; 12 N NaOH/EtOH, 50°, 15 min).[b]

[a] E. J. Corey, K. Achiwa, and J. A. Katzenellenbogen, *J. Am. Chem. Soc.*, **91**, 4318 (1969).

[b] I. J. Bolton, R. G. Harrison, B. Lythgoe, and R. S. Manwaring, *J. Chem. Soc. C*, 2944 (1971).

[c] P. G. Gassman and W. N. Schenk, *J. Org. Chem.*, **42**, 918 (1977).

32.   *p*-(P)-Benzoate Ester: ROCOC₆H₄-*p*-(P), 32

A primary hydroxyl group in a carbohydrate has been protected as a polymer-supported benzoate:

$ROH + ClCOC_6H_4\text{-}p\text{-}CH{=}CH_2 \xrightarrow[\text{(polymerization)}]{\text{styrene, AIBN, } 60°{}^a}$ **32**

(P) = styrene-divinylbenzene polymer

[a] R. D. Guthrie, A. D. Jenkins, and J. Stehlicek, *J. Chem. Soc. C*, 2690 (1971).

33.   α-Naphthoate Ester: ROCO-α-naphthyl, 33

Compound **33**, prepared to protect primary and secondary hydroxyl groups in a glucose by reaction with naphthoyl chloride (NaH, DMF, −10°, 9 h, 76% yield), is cleaved by sodium benzyloxide (PhCH₂ONa/PhCH₂OH, 37°, 2 h, 83% yield).[a]

[a] I. Watanabe, T. Tsuchiya, T. Takase, S. Umezawa, and H. Umezawa, *Bull. Chem. Soc. Jpn.*, **50**, 2369 (1977).

Carbonates

Carbonates, like esters, are cleaved by basic hydrolysis (e.g., see alkyl methyl or ethyl carbonate, compounds **34** and **35**). Advantage, however, can be taken of the properties of the second alkyl group to effect selective cleavage of a carbonate in the presence of an ester (e.g., cleavage of a 2,2,2-trichloroethyl carbonate by reaction with zinc, or an *S*-benzylthiocarbonate by oxidation).

34. Alkyl Methyl Carbonate: ROCOOCH₃, 34

Formation[a]

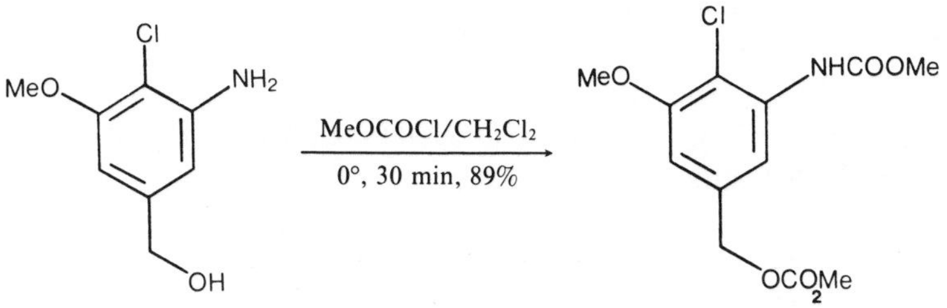

34a

Cleavage[a]

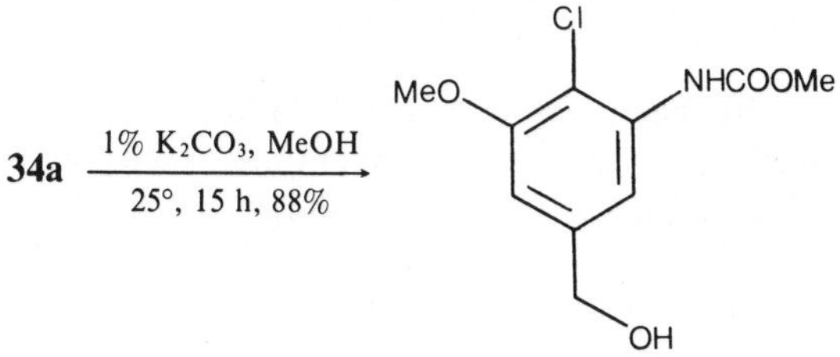

[a] A. I. Meyers, K. Tomioka, D. M. Roland, and D. Comins, *Tetrahedron Lett.,* 1375 (1978).

35. Alkyl Ethyl Carbonate: ROCOOC₂H₅, 35

An ethyl carbonate, prepared and cleaved by conditions similar to those described for a methyl carbonate, was used to protect a hydroxyl group in glucose.[a]

[a] F. Reber and T. Reichstein, *Helv. Chim. Acta,* **28,** 1164 (1945).

36. Alkyl 2,2,2-Trichloroethyl Carbonate: ROCOOCH₂CCl₃, 36

Compound **36** can be cleaved under mild conditions by β-elimination with zinc or by electrolytic reduction.

Formation[a]

$$\text{ROH} + \text{Cl}_3\text{CCH}_2\text{OCOCl} \xrightarrow[20°, 12\ \text{h}]{\text{Py}} \textbf{36}$$

Cleavage

$$(1)\quad \textbf{36} \xrightarrow[20°,\ 1\text{–}3\ h,\ 80\%]{Zn/HOAc^a} \text{ or } \xrightarrow[\text{reflux, short time}]{Zn/MeOH^a}$$

$$\text{or } \xrightarrow[20°,\ 3.5\ h,\ 100\%]{Zn\text{–}Cu/HOAc^b} \text{ROH}$$

$$(2)\quad \textbf{36} \xrightarrow[80\%]{-1.65\ V/MeOH,\ LiClO_4{}^c} \text{cholesterol}$$

A 2,2,2-tribromoethyl carbonate is cleaved by Zn–Cu/HOAc 10 times as fast as a trichloroethyl carbonate.[b]

[a] T. B. Windholz and D. B. R. Johnston, *Tetrahedron Lett.*, 2555 (1967).

[b] A. F. Cook, *J. Org. Chem.*, **33**, 3589 (1968).

[c] M. F. Semmelhack and G. E. Heinsohn, *J. Am. Chem. Soc.*, **94**, 5139 (1972).

37.   Alkyl Isobutyl Carbonate: $ROCOOCH_2CH(CH_3)_2$, 37

Compound **37** was prepared, by reaction with isobutyl chloroformate (Py, 20°, 3 days, 73% yield), to protect the 5'-OH group in thymidine. It was cleaved by acidic hydrolysis (80% HOAc, reflux, 15 min, 88% yield).[a]

[a] K. K. Ogilvie and R. L. Letsinger, *J. Org. Chem.*, **32**, 2365 (1967).

38.   Alkyl Vinyl Carbonate: $ROCOOCH{=}CH_2$, 38

Formation $(\rightarrow)$ / *Cleavage* $(\leftarrow)^a$

$$\text{ROH} \underset{\underset{\text{warm, 97\%}}{\xleftarrow[Na_2CO_3,\ H_2O\text{–dioxane}]{}}}{\xrightarrow[93\%]{CH_2{=}CHOCOCl/Py,\ CH_2Cl_2}} \textbf{38}$$

ROH = cholesterol

Phenols can be protected under similar conditions. Amines are converted by these conditions to carbamates that are stable to alkaline hydrolysis with sodium carbonate. Carbamates are cleaved by acidic hydrolysis (HBr/MeOH, CH_2Cl_2, 8 h), conditions that do not cleave alkyl or aryl vinyl carbonates.[a]

[a] R. A. Olofson and R. C. Schnur, *Tetrahedron Lett.*, 1571 (1977).

39. Alkyl Allyl Carbonate: $ROCOOCH_2CH{=}CH_2$, 39

Compound **39** can be cleaved under mild, aprotic conditions:

Formation **(→) / *Cleavage*** **(←)**[a]

$$\text{ROH} \underset{\overset{\text{Ni(CO)}_4\text{/TMEDA, DMF}}{\xleftarrow[55°,\ 4\ \text{h},\ 87\text{–}95\%]{}}}{\xrightarrow[0 \to 20°,\ 2\ \text{h},\ 90\%]{\text{CH}_2{=}\text{CHCH}_2\text{OCOCl, Py, THF}}} \mathbf{39}$$

ROH = primary, secondary alcohol

[a] E. J. Corey and J. W. Suggs, *J. Org. Chem.*, **38,** 3223 (1973).

40. Alkyl Cinnamyl Carbonate: $ROCOOCH_2CH{=}CHC_6H_5$, 40

A cinnamyl ester, used to protect a carboxyl group, is cleaved by the mild conditions shown below. The authors suggest that an alcohol (or an amine) can be protected as a cinnamyl carbonate, **40,** (or cinnamyl carbamate) and cleaved by the conditions, shown below, used to cleave an ester[a]:

Cleavage[a]

$$\text{RCO}_2\text{CH}_2\text{CH}{=}\text{CHPh} \xrightarrow[23°,\ 2\text{–}4\ \text{h}]{\text{Hg(OAc)}_2\text{/MeOH}} \xrightarrow[23°,\ 12\text{–}16\ \text{h}]{\text{KSCN, H}_2\text{O}} \text{RCOOH, 90\%}$$

[a] E. J. Corey and M. A. Tius, *Tetrahedron Lett.*, 2081 (1977).

41. Alkyl *p*-Nitrophenyl Carbonate: $ROCOOC_6H_4\text{-}p\text{-}NO_2$, 41

Formation **(→) / *Cleavage*** **(←)**[a]

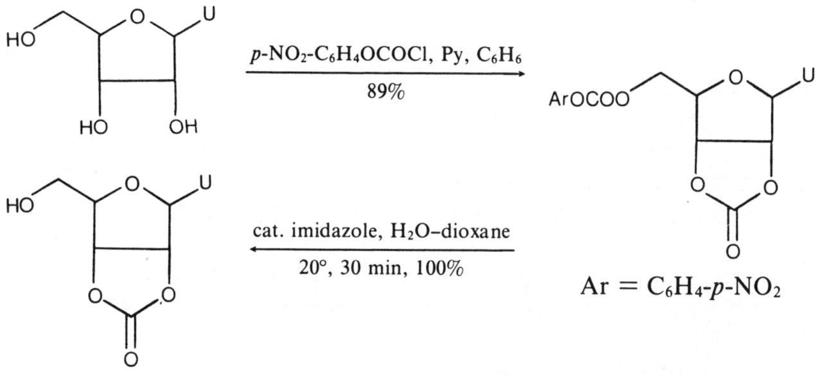

HO
O
U
HO
OH
p-NO2-C6H4OCOCl, Py, C6H6
89%
ArOCOO
O
U
O
O
O
HO
O
U
cat. imidazole, H2O–dioxane
20°, 30 min, 100%
O
O
O

Ar = $C_6H_4\text{-}p\text{-}NO_2$

Acetates, benzoates, and cyclic carbonates are stable to these hydrolysis conditions. [Cyclic carbonates are cleaved by more alkaline conditions (e.g., dil NaOH, 20°, 5 min, or aq Py, warm, 15 min, 100% yield).][a]

[a] R. L. Letsinger and K. K. Ogilvie, *J. Org. Chem.*, **32**, 296 (1967).

42. Alkyl Benzyl Carbonate: $ROCOOCH_2C_6H_5$, 42

A benzyl carbonate was prepared in 83% yield from the sodium alkoxide of glycerol and benzyl chloroformate (20°, 24 h).[a] It is cleaved by hydrogenolysis (H_2/Pd–C, EtOH, 20°, 2 h, 2 atm, 76% yield)[a] and electrolytic reduction (−2.7 V, $R_4N^+X^-$, DMF, 70% yield).[b] A benzyl carbonate was used to protect the hydroxyl group in lactic acid during a peptide synthesis.[c]

[a] B. F. Daubert and C. G. King, *J. Am. Chem. Soc.*, **61**, 3328 (1939).
[b] V. G. Mairanovsky, *Angew. Chem., Inter. Ed. Engl.*, **15**, 281 (1976).
[c] G. Losse and G. Bachmann, *Chem. Ber.*, **97**, 2671 (1964).

43. Alkyl *p*-Methoxybenzyl Carbonate: $ROCOOCH_2C_6H_4$-*p*-OCH_3, 43

44. Alkyl 3,4-Dimethoxybenzyl Carbonate: $ROCOOCH_2C_6H_3$-3,4-$(OCH_3)_2$, 44

Compounds **43** and **44,** prepared to protect the hydroxyl group in cholesterol, are cleaved by oxidation with triphenylmethyl fluoroborate, $Ph_3C^+BF_4^-$: **43,** 0°, 6 min, 90% yield; **44,** 0°, 15 min, 90% yield.[a]

[a] D. H. R. Barton, P. D. Magnus, G. Smith, G. Streckert, and D. Zurr, *J. Chem. Soc., Perkin Trans. 1*, 542 (1972).

45. Alkyl *o*-Nitrobenzyl Carbonate: $ROCOOCH_2C_6H_4$-*o*-NO_2, 45

46. Alkyl *p*-Nitrobenzyl Carbonate: $ROCOOCH_2C_6H_4$-*p*-NO_2, 46

Compounds **45** and **46** were prepared to protect a secondary hydroxyl group in a thienamycin precursor. Compound **45** was prepared from the chloroformate (DMAP, CH_2Cl_2, 0° → 20°, 3 h) and cleaved by irradiation, pH 7.[a] Compound **46** was prepared from the chloroformate (−78°, *n*-BuLi/THF, 85% yield) and cleaved by hydrogenolysis (H_2/Pd–C, dioxane–H_2O–EtOH–K_2HPO_4).[b] Compound **46** is also cleaved by electrolytic reduction (−1.1 V, $R_4N^+X^-$, ĐMF).[c]

[a] L. D. Cama and B. G. Christensen, *J. Am. Chem. Soc.*, **100**, 8006 (1978).
[b] D. B. R. Johnston, S. M. Schmitt, F. A. Bouffard, and B. G. Christensen, *J. Am. Chem. Soc.*, **100**, 313 (1978).
[c] V. G. Mairanovsky, *Angew. Chem., Inter. Ed., Engl.*, **15**, 281 (1976).

47. Alkyl *S*-Benzyl Thiocarbonate: $ROCOSCH_2C_6H_5$, 47

Formation (→) / Cleavage (←)[a]

$$\xrightarrow[\text{65-70\%}]{\text{PhCH}_2\text{SCOCl/Py}}$$

ROH   47

$$\xleftarrow[\text{20°, 4 days, 50-55\%}]{\text{H}_2\text{O}_2/\text{HOAc, KOAc, CHCl}_3}$$

ROH = carbohydrate

[a] J. J. Willard, *Can. J. Chem.,* **40**, 2035 (1962).

Miscellaneous Esters

48. Alkyl *N*-Phenylcarbamate: $ROCONHC_6H_5$, 48

Formation[a]

$$\text{ROH} \xrightarrow[\text{20°, 2-3 h, ~100\%}]{\text{PhN=C=O, Py}} \text{ROCONHPh}$$

ROH = nucleoside

Cleavage

(1) $\text{ROCONHPh} \xrightarrow[\text{reflux, 1.5 h, good yield}]{\text{NaOMe/MeOH}^b} \text{ROH}$

(2) $\text{ROCONHPh} \xrightarrow[\text{reflux, 3-4 h, 90\%}]{\text{LiAlH}_4/\text{THF or dioxane}^b} \text{ROH}$

Carbamates are more stable to hydrolysis than esters.

(3) $\text{ROCONHR}' \xrightarrow[\text{4-48 h, 25-80°}]{\text{Cl}_3\text{SiH/Et}_3\text{N, CH}_2\text{Cl}_2} \xrightarrow{\text{H}_2\text{O}} \text{ROH, 80-95\%}^c$

ROH = primary, secondary, tertiary, allylic, propargylic, or benzyl
alcohol

(4) Compound **48** is cleaved by passage through a trityl-cellulose column.[a]

[a] K. L. Agarwal and H. G. Khorana, *J. Am. Chem. Soc.,* **94**, 3578 (1972).

[b] H. O. Bouveng, *Acta Chem. Scand.,* **15**, 87, 96 (1961).

[c] W. H. Pirkle and J. R. Hauske, *J. Org. Chem.,* **42**, 2781 (1977).

49. Alkyl N-Imidazolylcarbamate: ROCO—N $\diagup\!\!=\!\!N$, 49

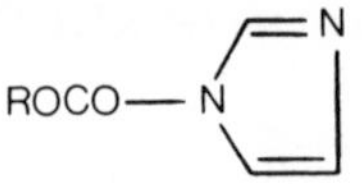

Formation (→) / Cleavage (←)[a]

$$
\text{ROH} \quad
\xrightarrow[\substack{20°, \ 20 \ h, \ 100\%}]{\substack{1,1'\text{-carbonyldiimidazole, } CH_2Cl_2}}
\quad \mathbf{49}
$$

$$
\xleftarrow[\substack{20°, \ 12 \ h}]{\substack{Na_2CO_3, \ H_2O, \ THF}}
$$

ROH = tylosin precursor

[a] A. A. Nagel and L. A. Vincent, *J. Org. Chem.*, **44**, 2050 (1979).

50. Borate Ester: (RO)₃B, 50

Compound **50** was prepared, by reaction with boric acid (C_6H_6, reflux), to protect a hydroxyl group during a synthesis of dihydro-β-santalol. It is readily cleaved by hydrolysis under acidic, basic, or neutral conditions.[a] Borate esters can also be prepared by the following method:

Formation[b]

$$
3 \ \text{ROH} + BH_3 \cdot SMe_2 \quad \xrightarrow[\substack{80-90\%}]{\substack{25°, \ 1 \ h}} \quad (RO)_3B
$$

ROH = primary, secondary, tertiary, and aromatic alcohols

[a] W. I. Fanta and W. F. Erman, *J. Org. Chem.*, **37**, 1624 (1972).
[b] C. A. Brown and S. Krishnamurthy, *J. Org. Chem.*, **43**, 2731 (1978).

51. Nitrate Ester: RONO₂, 51

A nitrate ester is stable to the mildly acidic conditions that cleave acetals and ketals, and to the mildly basic conditions that cleave esters. Nitrates have been prepared to protect carbohydrates[a] and steroids[b] by reaction with nitric acid (50–100% yield).[c] In general nitrate esters are cleaved by reduction (e.g., with $LiAlH_4$; H_2/Raney Ni or H_2/Pd; Zn/or Fe/HOAc; or Na_2S),[a] or by nucleophilic displacement (with H_2NNH_2).[a] Irradiation may provide a milder method of cleavage[d]:

Cleavage[d]

$$
\text{RONO}_2 \quad \xrightarrow[\substack{1 \ h, \ 92-100\%}]{\substack{h\nu, \ i\text{-PrOH}}} \quad \text{ROH}
$$

ROH = carbohydrate

[a] J. Honeyman and J. W. W. Morgan, *Adv. Carbohydr. Chem.*, **12**, 117 (1957).

[b] J. F. W. Keana, in *Steroid Reactions*, C. Djerassi, Ed., Holden-Day, San Francisco, 1963, pp. 75–76.

[c] R. Boschan, R. T. Merrow, and R. W. Van Dolah, *Chem. Rev.*, **55**, 485 (1955).

[d] R. W. Binkley and D. J. Koholic, *J. Org. Chem.*, **44**, 2047 (1979).

52.   Alkyl *N,N,N′,N′*-Tetramethylphosphorodiamidate: $ROPO[N(CH_3)_2]_2$, 52

Protection of a hydroxyl group as a phosphorodiamidate has been suggested.[a] Compounds such as **52** are stable to a variety of reagents (e.g., $MeLi/Et_2O$, 25°, 2 h; $LiAlH_4/Et_2O$, 25°, 2 h; 1 *N* KOH/EtOH, reflux, 15 h; and 0.2 *N* HCl/acetone, 25°, 2 h). Compound **52** (R = 3β-cholestanyl) is cleaved by reaction with *n*-butyllithium (5 eq *n*-BuLi in TMEDA, 25°, 30 min, 100% yield).[a]

Formation[a]

$$ROH + (Me_2N)_2POCl \xrightarrow[25°, \ 1-2 \ h, \ 90\%]{n\text{-BuLi/THF, TMEDA}} 52$$

ROH = primary, secondary, tertiary alcohol

[a] R. E. Ireland, D. C. Muchmore, and U. Hengartner, *J. Am. Chem. Soc.*, **94**, 5098 (1972).

53.   Alkyl 2,4-Dinitrophenylsulfenate: $ROSC_6H_3$-2,4-$(NO_2)_2$, 53

A nitrophenylsulfenate, cleaved by nucleophiles under very mild conditions, was developed as protection for a hydroxyl group during solid phase nucleotide synthesis.[a] Compound **53** is stable to acidic hydrolysis of acetonides prepared to protect 1,2-diols.[b]

Formation

$$ROH + ClSC_6H_3\text{-}2,4\text{-}(NO_2)_2 \xrightarrow[20°, \ 1 \ h, \ 70-85\%]{Py/DMF^a \ \text{or} \ CH_2Cl_2^b} 53$$

Cleavage

(1) $53 \xrightarrow[25°, \ 4 \ h, \ 63-80\%]{Nu^-/MeOH, \ H_2O^a} ROH$

$Nu^- = Na_2S_2O_3$, pH 8.9; NaCN, pH 8.9; Na_2S, pH 6.6; PhSH, pH 11.8.

(2) $53 \xrightarrow[54\%]{H_2/Raney \ Ni^b}$ or $\xrightarrow[5 \ h, \ 67\%]{Al/Hg(OAc)_2, \ MeOH^b} ROH$

(3) An alkyl *o*-nitrophenylsulfenate is cleaved by electrolytic reduction (-1.0 V, DMF, $R_4N^+X^-$).[c]

[a] R. L. Letsinger, J. Fontaine, V. Mahadevan, D. A. Schexnayder, and R. E. Leone, *J. Org. Chem.*, **29**, 2615 (1964).
[b] K. Takiura, S. Honda, and T. Endo, *Carbohydr. Res.*, **21**, 301 (1972).
[c] V. G. Mairanovsky, *Angew. Chem., Inter. Ed. Engl.*, **15**, 281 (1976).

PROTECTION FOR 1,2- AND 1,3-DIOLS

cis-1,2-Diols (e.g., in carbohydrates[a] and nucleosides[b]), and *cis*- and *trans*-1,3-diols can be protected as cyclic acetals and ketals (e.g., dioxolanes or dioxanes) or cyclic ortho esters that are cleaved by acidic hydrolysis, or as cyclic esters (e.g., carbonates and boronates) that are cleaved by basic hydrolysis.[b]

Cyclic acetals and ketals are formed by reaction of the diol and a carbonyl compound in the presence of an acid catalyst. They are formed in moderate yield by reaction of a diol with an enol ether (eq. 1).[c]

(1)  $HO(CH_2)_nOH + CH_2{=}CHOR \xrightarrow[\text{20°, 24 h, 50–60\%}]{KSF/CH_2Cl_2}$

$n = 2, 4,$ or 6; $R = Et$

enol ether = EtO-1-cyclohexenyl; dihydropyran

Depending on conditions either the kinetic or thermodynamic product can be isolated.[d]

Dioxanes and dioxolanes are stable to the alkaline conditions of *O*-alkylation or acylation, to reduction by lithium aluminum hydride or Na(Hg), and (with the exception of benzylidene acetals/ketals) to reduction by catalytic hydrogenation. Dioxanes and dioxolanes are stable to some oxidizing agents [e.g., CrO_3/Py; $NaIO_4$; $Pb(OAc)_4$, $<80°$; Ag_2O; $KMnO_4/OH^-$; $Al(O$-i-$Pr)_3$-acetone]. Benzylidene acetals react with O_3[e] or NBS[f] (eqs. 2 and 3).

(2)

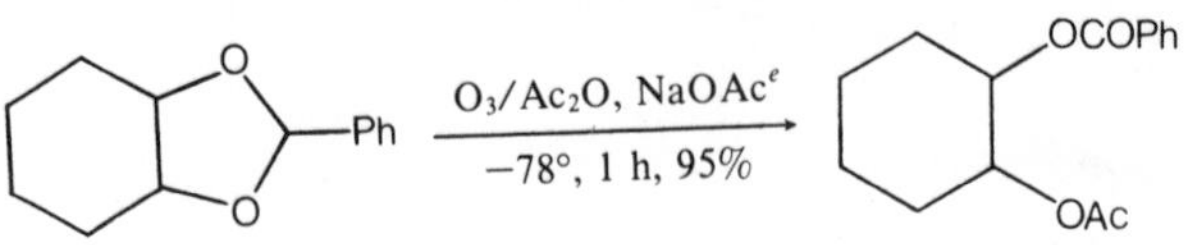

(3)

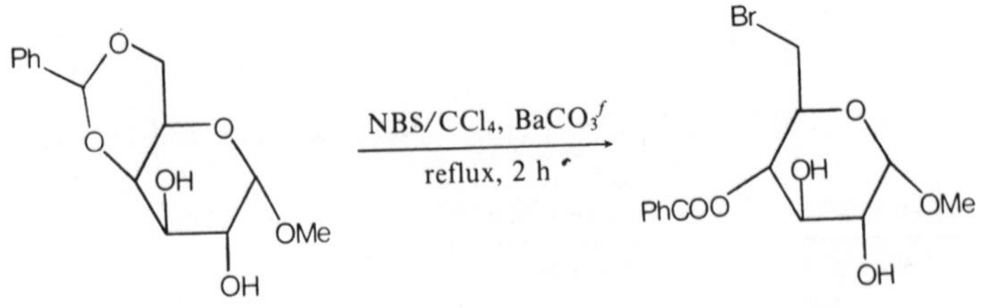

Kinetic studies of acetal/ketal formation, from cyclohexanone, and hydrolysis $(3 \times 10^{-3} N$ HCl/dioxane–H_2O, 20°) indicate the following orders of reactivity[g]:

Formation

$$HOCH_2C(CH_3)_2CH_2OH > HO(CH_2)_2OH > HO(CH_2)_3OH$$

Cleavage

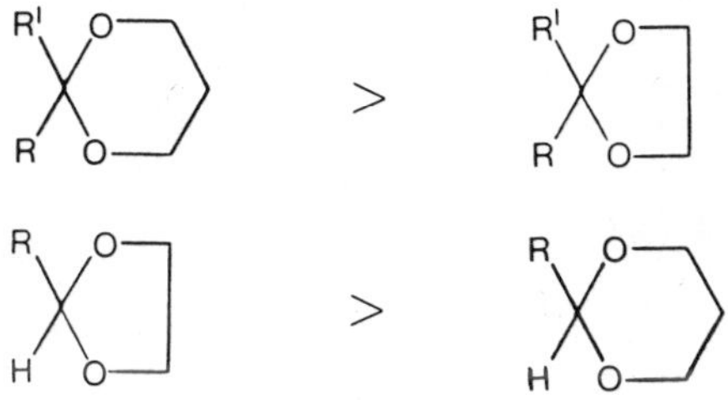

The relative rates of acid-catalyzed hydrolysis of some dioxolanes [dioxolane: aq HCl (1:1)] are 2,2-dimethyldioxolane: 2-methyldioxolane: dioxolane, 50,000:5,000:1.[h] Acidic hydrolysis of some cyclic acetals and ketals can lead to a mixture of products by migration of the dioxolane group (eq. 4).[d]

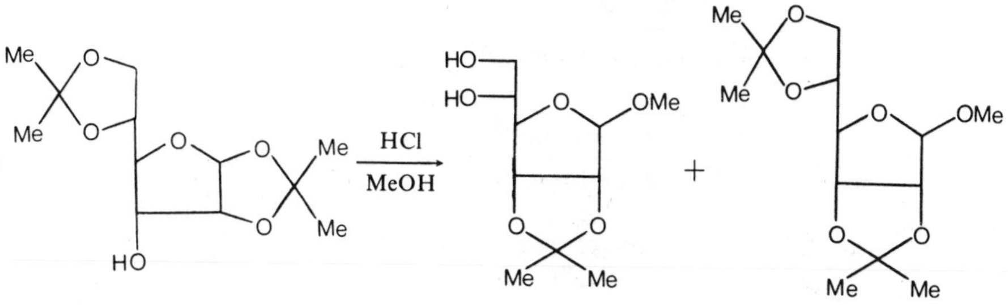

Cyclic ortho esters are more readily cleaved by acidic hydrolysis than are cyclic acetals or ketals. A number of substituted cyclic acetals and ketals have been developed to provide wide variation in ease of removal.

Cyclic carbonates are prepared from a diol and phosgene or a chloroformate in the presence of base. Cyclic boronates, prepared by reaction of the diol with a boronic acid [e.g., PhB(OH)₂] in the presence of pyridine, are readily cleaved by water.

The most useful protective groups for diols are listed in Reactivity Chart 3; conventions that are used in this section are described on p. xii.[i]

[a] H. G. Fletcher, Jr., *Methods Carbohydr. Chem.*, **II**, 307 (1963); T. G. Bonner, *Methods Carbohydr. Chem.*, **II**, 309, 314 (1963); O. Th. Schmidt, *Methods Carbohydr. Chem.*, **II**, 318 (1963); A. N. de Belder, *Adv. Carbohydr. Chem.*, **20**, 219–302 (1965).

[b] V. Amarnath and A. D. Broom, *Chem. Rev.*, **77**, 183 (1977); C. B. Reese, *Tetrahedron*, **34**, 3143 (1978).

[c] V. M. Thuy and P. Maitte, *Bull. Soc. Chim. Fr.*, II-264 (1979).

[d] D. M. Clode, *Chem. Rev.*, **79**, 491 (1979).

[e] P. Deslongchamps, C. Moreau, D. Fréhel, and R. Chênevert, *Can. J. Chem.*, **53**, 1204 (1975).

[f] S. Hanessian, *Carbohydr. Res.*, **2**, 86 (1966).

[g] M. S. Newman and R. J. Harper, *J. Am. Chem. Soc.*, **80**, 6350 (1958); S. W. Smith and M. S. Newman, *J. Am. Chem. Soc.*, **90**, 1249, 1253 (1968).

[h] P. Salomaa and A. Kankaanperä, *Acta Chem. Scand.*, **15**, 871 (1961).

[i] See also: C. B. Reese, "Protection of . . . Glycol Systems," in *Protective Groups in Organic Chemistry*, J. F. W. McOmie, Ed., Plenum, New York and London, 1973, pp. 120–135; H. M. Flowers, "Protection of the Hydroxyl Group," in *The Chemistry of the Hydroxyl Group*, S. Patai, Ed., Wiley-Interscience, New York, 1971, Vol. 10/2, pp. 1025, 1028–1035.

Cyclic Acetals and Ketals

1. Methylene Acetal: , **1**

Formation

(1) $C_6H_8(OH)_6$ $\xrightarrow[\text{50°, 4 days, 68\%}]{\text{40\% CH}_2\text{O, concd HCl}^a}$ **1a**

1a = trismethylenedioxy derivative

(2)   sugar $\xrightarrow[\text{90°, 1 h, good yield}]{\text{paraformaldehyde, H}_2\text{SO}_4\text{, HOAc}^b}$ **1b**

1b = bismethylenedioxy derivative

(3)   *cis*-cyclohexane-1,2-diol $\xrightarrow[\text{50°, 12 h, 62\%}]{\text{DMSO, NBS}^c}$ **1**

(4)   *cis*-cyclohexane-1,2-diol $\xrightarrow[\text{0–30°, 40 h, 46\%}]{\text{CH}_2\text{Br}_2\text{, NaH, DMF}^d}$ **1**

Cleavage

(1)   **1a** $\xrightarrow[\text{−80°, 30 min → 20°}]{\text{BCl}_3\text{/CH}_2\text{Cl}_2}$ $\xrightarrow[\text{NaOAc}]{\text{Ac}_2\text{O}}$ $C_6H_8(OAc)_6$, $61\%^a$

(2)   **1b** $\xrightarrow[\text{100°, 3 h}]{2\ N\ \text{HCl}^b}$ sugar

[a] T. G. Bonner, *Methods Carbohydr. Chem.,* **II,** 314 (1963).
[b] L. Hough, J. K. N. Jones, and M. S. Magson, *J. Chem. Soc.,* 1525 (1952).
[c] S. Hanessian, G. Y.-Chung, P. Lavallee, and A. G. Pernet, *J. Am. Chem. Soc.,* **94,** 8929 (1972).
[d] J. S. Brimacombe, A. B. Foster, B. D. Jones, and J. J. Willard, *J. Chem. Soc. C,* 2404 (1967).

2. Ethylidene Acetal: [structure: $-CH_3$], **2**

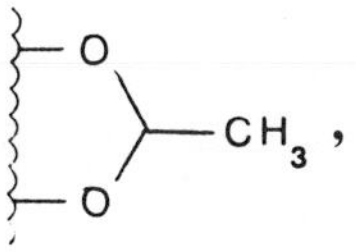

Formation

$$\text{sugar} \xrightarrow[\text{concd } H_2SO_4,\ 2\text{–}3\ h,\ 60\%]{\text{CH}_3\text{CHO, CH}_3\text{CH(OMe)}_2,\ \text{or paraldehyde}^a} \mathbf{2}$$

Cleavage

(1)   **2** $\xrightarrow[\text{reflux, 7 h}]{0.67\ N\ H_2SO_4,\ \text{aq Me}_2\text{CO}^a}$ sugar

(2)   **2** $\xrightarrow[20°,\ 5\ \text{min},\ 60\%]{\text{Ac}_2\text{O, cat. }H_2SO_4{}^a}$ [structure] OAc / OCH(OAc)CH$_3$

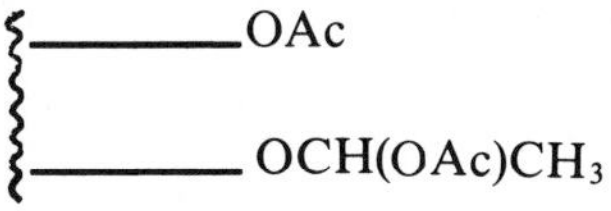

(3)   **2** $\xrightarrow[\text{reflux, 1.5 h}]{80\%\ \text{HOAc}^b}$ 1,2-diol

[a] T. G. Bonner, *Methods Carbohydr. Chem.,* **II,** 309 (1963).
[b] J. W. Van Cleve and C. E. Rist, *Carbohydr. Res.,* **4,** 82 (1967).

3. 1-*t*-Butylethylidene Ketal: [structure: CH_3, t-C_4H_9], **3**

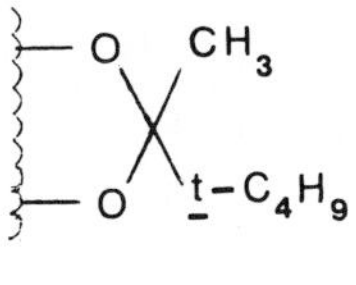

4. 1-Phenylethylidene Ketal: [structure: CH_3, C_6H_5], **4**

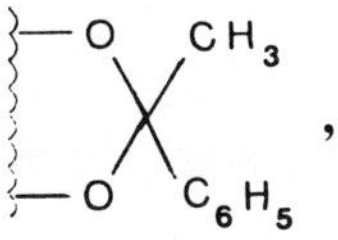

Compounds **3** and **4** were prepared selectively from the C_4–C_6 1,3-diol in glucose by an acid-catalyzed transketalization reaction [e.g., $Me_3CC(OMe)_2CH_3$, TsOH/DMF, 24 h, 79% yield; $PhC(OMe)_2Me$, TsOH/DMF, 24 h, 90% yield, respectively]. They are cleaved by acidic hydrolysis—**3**: HOAc, 20°, 90 min, 100% yield; **4**: HOAc, 20°, 3 days, 100% yield.[a]

[a] M. E. Evans, F. W. Parrish, and L. Long, Jr., *Carbohydr. Res.,* **3**, 453 (1967).

5. 2,2,2-Trichloroethylidene Acetal:

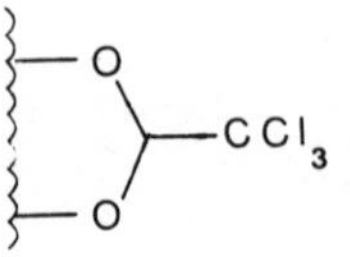

Trichloroacetaldehyde (chloral) reacts with glucose in the presence of sulfuric acid to form two mono- and four diacetals (e.g., compound **5**). Compound **5** is cleaved by reduction [H_2/Raney Ni, 50% NaOH/EtOH, 15 min].[a] Compound **5** can probably be cleaved by reaction with Zn/HOAc [cf. ROCH(R')OCH$_2$CCl$_3$ cleaved by Zn/HOAc, NaOAc, 20°, 3 h, 90% yield[b]].

[a] S. Forsén, B. Lindberg, and B.-G. Silvander, *Acta Chem. Scand.,* **19**, 359 (1965).
[b] R. U. Lemieux and H. Driguez, *J. Am. Chem. Soc.,* **97**, 4069 (1975).

6. Acetonide (Isopropylidene Ketal):

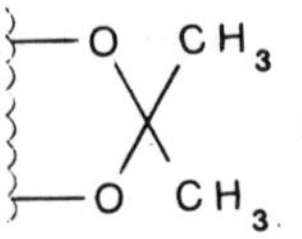

Formation

(1)   1,3-diol $\xrightarrow[\text{0°, 16 h, 75%}]{\text{CH}_3\text{C(OMe)=CH}_2,\ \text{dry HBr, CH}_2\text{Cl}_2{}^a}$ **6a**

1,3-diol = erythronolide precursor

(2)

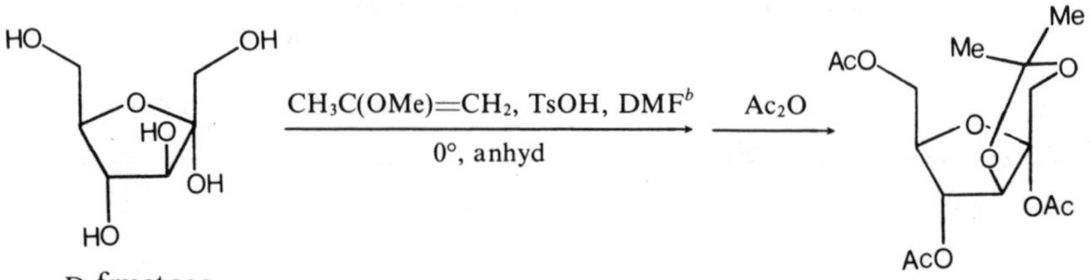

D-fructose   **6b,** 55%

Under these conditions 2-methoxypropene reacts to form the kinetically controlled 1,3-*O*-isopropylidene, **6b,** instead of the thermodynamically more stable 1,2-*O*-isopropylidene.[b]

(3)   sugar $\xrightarrow[\text{24 h}]{\text{Me}_2\text{C(OMe)}_2,\ \text{TsOH, DMF}^c}$ **6c + 6d**

79% 1.8%

sugar $=$ methyl α-D-glucopyranoside

$\quad$ **6c** $=$ methyl 4,6-O-isopropylidene-α-D-glucoside

$\quad$ **6d** $=$ methyl 2,3:4,6-di-O-isopropylidene-α-D-glucoside

(4) $\quad$ sugar $\xrightarrow[\text{36°, 5 h, 60–70\%}]{\text{anhyd Me}_2\text{CO, FeCl}_3{}^d}$ **6**

(5) $\quad$ nucleoside $\xrightarrow[\text{3–5 h, 90–100\%}]{\text{Me}_2\text{C(OMe)}_2\text{, cat.}^e}$ **6**

$\quad$ cat. $=$ di-p-nitrophenyl hydrogen phosphate

(6) $\quad$ nucleoside $\xrightarrow[\text{25°, 12 h, 90–100\%}]{\text{MeC(OEt)=CH}_2\text{, cat. HCl, DMF}^f}$ **6**

(7) $\quad$ diol $\xrightarrow[\text{10–30 min, 80–85\%}]{\text{MeC(OSiMe}_3\text{)=CH}_2\text{, concd HCl or Me}_3\text{SiCl}^g}$ **6**

$\quad$ diol $=$ *cis-* or *trans-c*-C_6H_{10}-1,2-$(OH)_2$;

$\qquad$ *trans-c*-C_7H_{12}-1,2-$(OH)_2$; acyclic diols

(8) $\quad$ The classical method of formation of an acetonide is by reaction of a diol with acetone and an acid catalyst.[h,i]

Cleavage

A variety of acid-catalyzed hydrolysis conditions have been required to cleave an acetonide:

$\quad$ i. $\quad$ **6a,** 1 N HCl/THF (1:1), 20°[a]

$\quad$ ii. $\quad$ **6,** 2 N HCl, 80°, 6 h[j]

$\quad$ iii. $\quad$ **6,** 60–80% HOAc, 25°, 2 h, 92% yield of *cis*-1,2-diol[k]

$\quad$ iv. $\quad$ **6,** 80% HOAc, reflux, 30 min, 78% yield of *trans*-1,2-diol[k]

$\quad$ v. $\quad$ **6,** TsOH/MeOH, 25°, 5 h[l]

$\quad$ vi. $\quad$ **6,** TsOH/MeOH, 25°, 1 day; a 2′,3′-ribonucleoside acetonide was not cleaved[m]

$\quad$ vii. $\quad$ **6,** Dowex 50-W(H⁺), H_2O, 70°, excellent yield[n]

$\quad$ viii. $\quad$ **6,** BCl₃, 25°, 2 min, 100% yield[o]

$\quad$ ix. $\quad$ **6,** Br₂, Et₂O[p]

[a] E. J. Corey, S. Kim, S. Yoo, K. C. Nicolaou, L. S. Melvin, Jr., D. J. Brunelle, J. R. Falck, E. J. Trybulski, R. Lett, and P. W. Sheldrake, *J. Am. Chem. Soc.*, **100**, 4620 (1978).

[b] E. Fanton, J. Gelas, and D. Horton, *J. Chem. Soc., Chem. Commun.*, 21 (1980).

[c] M. E. Evans, F. W. Parrish, and L. Long, Jr., *Carbohydr. Res.*, **3**, 453 (1967).

[d] P. P. Singh, M. M. Gharia, F. Dasgupta, and H. C. Srivastava, *Tetrahedron Lett.*, 439 (1977).

[e] A. Hampton, *J. Am. Chem. Soc.*, **83**, 3640 (1961).

[f] S. Chládek and J. Smrt, *Collect. Czech. Chem. Commun.*, **28**, 1301 (1963).

[g] G. L. Larson and A. Hernandez, *J. Org. Chem.*, **38**, 3935 (1973).

[h] O. Th. Schmidt, *Methods Carbohydr. Chem.*, **II**, 318 (1963).

[i] A. N. de Belder, *Adv. Carbohydr. Chem.*, **20**, 219 (1965).

[j] T. Ohgi, T. Kondo, and T. Goto, *Tetrahedron Lett.*, 4051 (1977).

[k] M. L. Lewbart and J. J. Schneider, *J. Org. Chem.*, **34**, 3505 (1969).

[l] A. Ichihara, M. Ubukata, and S. Sakamura, *Tetrahedron Lett.*, 3473 (1977).

[m] J. Kimura and O. Mitsunobu, *Bull. Chem. Soc. Jpn.*, **51**, 1903 (1978).

[n] P.-T. Ho, *Tetrahedron Lett.*, 1623 (1978).

[o] T. J. Tewson and M. J. Welch, *J. Org. Chem.*, **43**, 1090 (1978).

[p] A. N. de Belder, *Adv. Carbohydr. Chem.*, **20**, 219–302 (1963), see p. 235.

7. Butylidene Acetal: , **7**

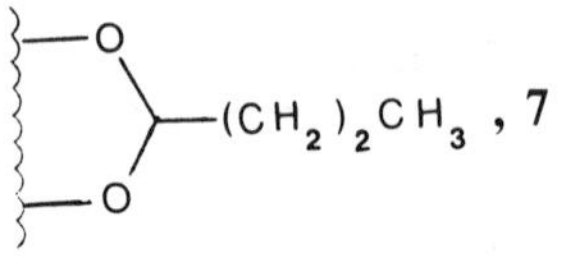

Compound **7**, prepared from a 1,2-diol by reaction with butyraldehyde (cat. concd HCl/DMF, 24 h), is cleaved by acidic hydrolysis (cat. concd HCl/CF$_3$CO$_2$H, EtOH, H$_2$O, 65°, 5 h, ~100% yield).[a]

[a] L. Yüceer, *Carbohydr. Res.*, **56**, 87 (1977).

8. Cyclopentylidene Ketal: 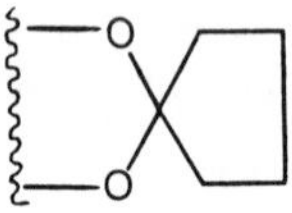   , **8**

9. Cyclohexylidene Ketal: 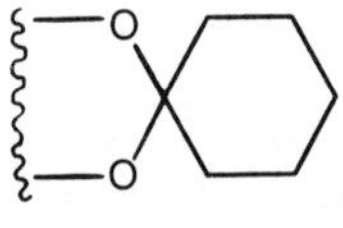   , **9**

10. Cycloheptylidene Ketal: 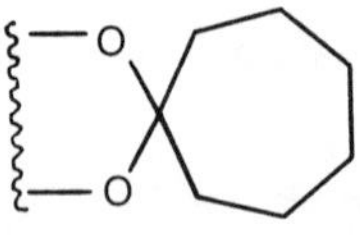   , **10**

Compounds **8, 9,** and **10** can be prepared by an acid-catalyzed reaction of a diol and the cycloalkanone in the presence of ethyl orthoformate and mesitylenesulfonic acid.[a]

The relative ease of acid-catalyzed hydrolysis [0.53 *M* H$_2$SO$_4$/H$_2$O–PrOH

(65:35), 20°] for compounds **8, 10, 6,** and **9** is $C_5 \sim C_7 > C_3$ (acetonide) $\gg C_6$ (e.g., $t_{1/2}$ for 1,2-*O*-alkylidene-α-D-glucopyranoses of C_5, C_7, C_3 acetonide, and C_6 derivatives are 8, 10, 20, and 124 h, respectively).[a]

Compound **9** can also be prepared from a diol and 1-(trimethylsiloxy)cyclohexene (concd HCl, 20°, 10–30 min, 70–75% yield)[b] and cleaved by acidic hydrolysis (10% HCl/Et₂O, 25°, 5 min[b]; CF₃CO₂H/H₂O, 20°, 6 min to 2 h, 65–85% yield[c]).

[a] W. A. R. van Heeswijk, J. B. Goedhart, and J. F. G. Vliegenthart, *Carbohydr. Res.,* **58,** 337 (1977).
[b] G. L. Larson and A. Hernandez, *J. Org. Chem.,* **38,** 3935 (1973).
[c] S. L. Cook and J. A. Secrist, *J. Am. Chem. Soc.,* **101,** 1554 (1979).

11.   Benzylidene Acetal:

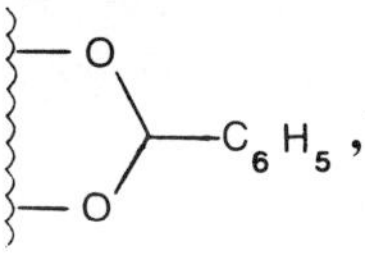

Formation

(1)   sugar $\xrightarrow[\text{28°, 4 h}]{\text{PhCHO/ZnCl}_2{}^{a}}$ **11**

(2)   glucose osazone $\xrightarrow[\text{25°, 4 h}]{\text{PhCHO/DMSO, concd H}_2\text{SO}_4{}^{b}}$ **11**

In DMSO a dioxane derivative of the sugar, formed kinetically, is converted to the thermodynamically more stable dioxolane.[b]

(3)   diol $+$

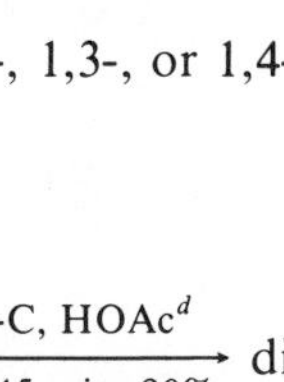

$\xrightarrow[\text{25°, 16 h, 45–82%}]{\text{K}_2\text{CO}_3 \text{ or Py, CH}_2\text{Cl}_2{}^{c}}$ **11**

diol = 1,2-, 1,3-, or 1,4-compound

Cleavage

(1)   **11** $\xrightarrow[\text{25°, 30–45 min, 90%}]{\text{H}_2/\text{Pd-C, HOAc}^{d}}$ diol

(2)   **11** $\xrightarrow[\text{85%}]{\text{Na/NH}_3{}^{e}}$ *N*-acetylglucosamine

(3)   Compound **11** is cleaved by acidic hydrolysis (e.g., 0.01 N H$_2$SO$_4$, 100°, 3 h, 92% yield[f]; 80% HOAc, 25°, $t_{1/2}$ for uridine = 60 h[g]), conditions that do not cleave a methylenedioxy group.[f]

(4)   **11** $\xrightarrow{\text{-2.9 V, R}_4\text{N}^+\text{X}^-,\ \text{DMF}^h}$ sugar

(5)   **11** $\xrightarrow[25°,\ 8\ h,\ 80\%]{\text{Ph}_3\text{C}^+\text{BF}_4^-,\ \text{CH}_3\text{CN}^i}$ 1 : 1 mixture of diol monobenzoates

An acetonide is stable to these cleavage conditions.[i]

(6)   A benzylidene acetal is cleaved in 100% yield by boron trichloride, a reagent that cleaves a variety of groups including acetonides.[j]

[a] H. G. Fletcher, Jr., *Methods Carbohydr. Chem.*, **II**, 307 (1963).
[b] R. M. Carman and J. J. Kibby, *Aust. J. Chem.*, **29**, 1761 (1976).
[c] R. M. Munavu and H. H. Szmant, *Tetrahedron Lett.*, 4543 (1975).
[d] W. H. Hartung and R. Simonoff, *Org. React.*, **7**, 263–326; see pp. 271, 284, 302 (1953).
[e] M. Zaoral, J. Ježek, R. Straka, and K. Masek, *Collect. Czech. Chem. Commun.*, **43**, 1797 (1978).
[f] R. M. Hann, N. K. Richtmyer, H. W. Diehl, and C. S. Hudson, *J. Am. Chem. Soc.*, **72**, 561 (1950).
[g] M. Smith, D. H. Rammler, I. H. Goldberg, and H. G. Khorana, *J. Am. Chem. Soc.*, **84**, 430 (1962).
[h] V. G. Mairanovsky, *Angew, Chem., Inter. Ed. Engl.*, **15**, 281 (1976).
[i] S. Hanessian and A. P. A. Staub, *Tetrahedron Lett.*, 3551 (1973).
[j] T. G. Bonner, E. J. Bourne, and S. McNally, *J. Chem. Soc.*, 2929 (1960).

12. *p*-Methoxybenzylidene Acetal: $\displaystyle \rangle\!\!\!< ^{O}_{O} \rangle\!\!- C_6H_4\text{-}\underline{p}\text{-OCH}_3$, **12**

13. 2,4-Dimethoxybenzylidene Acetal: $\displaystyle \rangle\!\!\!< ^{O}_{O} \rangle\!\!- C_6H_3\text{-}2,4\text{-}(OCH_3)_2$, **13**

14. *p*-Dimethylaminobenzylidene Acetal: $\displaystyle \rangle\!\!\!< ^{O}_{O} \rangle\!\!- C_6H_4\text{-}\underline{p}\text{-N}(CH_3)_2$, **14**

Compounds **12**[a] and **13**[b] were prepared, by reaction with the methoxybenzaldehyde (acid cat., 70–95% yield), to protect the 2′,3′-diol system in nucleosides. Compounds **12** and **13** are cleaved by very mild acidic hydrolysis: **12**, 80% HOAc,

25°, 10 h, 100% cleaved (10 times faster than an O-benzylidene acetal[a]); **13**, 80% HOAc, 25° or 90% CF_3CO_2H, 30°, 2–10 min, good yield.[b]

Compound **14**,[b] like compounds **12** and **13**, was prepared (by reaction with 4-N,N-dimethylaminobenzaldehyde, acid cat., 25°, 24 h, 55–80% yield) to protect the 2′,3′-diol system in nucleosides. Compound **14** is cleaved in good yield by acidic hydrolysis (e.g., 80% HOAc, 25° or 90% CF_3CO_2H, 30°, 2–10 min).[b]

[a] M. Smith, D. H. Rammler, I. H. Goldberg, and H. G. Khorana, *J. Am. Chem. Soc.*, **84**, 430 (1962).
[b] F. Cramer, W. Saenger, K.-H. Scheit, and J. Tennigkeit, *Liebigs Ann. Chem.*, **679**, 156 (1964).

15. *o*-Nitrobenzylidene Acetal:

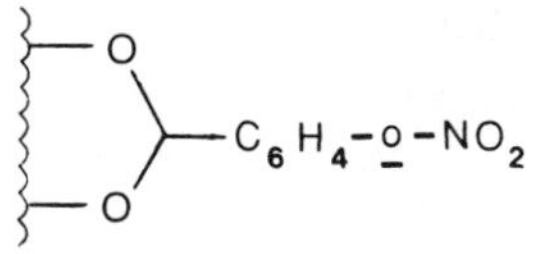

Compound **15**, prepared to protect a *cis*-1,2-diol in a sugar, has been cleaved in 88% yield by irradiation. Cleavage of a *m*- or *p*-nitro derivative by irradiation could not be effected.[a]

[a] P. M. Collins and N. N. Oparaeche, *J. Chem. Soc., Chem. Commun.*, 532 (1972).

16. *p*-Ⓟ-Benzylidene Acetal:

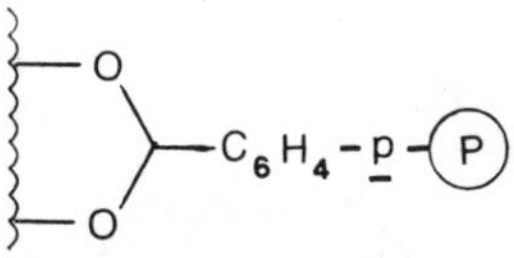

Compound **16** was prepared by reaction with a polymer-bound benzaldehyde (Ⓟ = divinylbenzene-styrene copolymer)/TsOH, 89% yield, from the C_4,C_6-hydroxyl groups in sucrose, and cleaved by acidic hydrolysis (CF_3CO_2H-dioxane, 70–80% yield).[a]

[a] J.M.J.M. Fréchet and G. Pellé, *J. Chem. Soc., Chem. Commun.*, 225 (1975).

17. **Phenanthrylidene Derivative:**

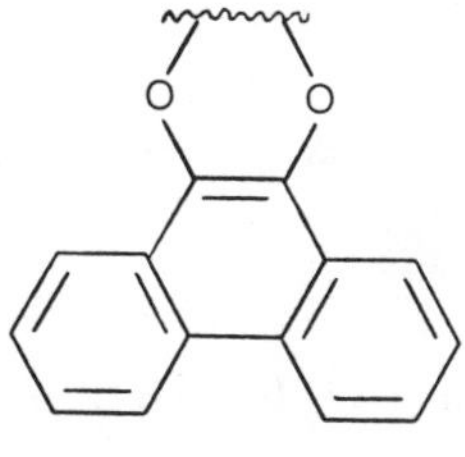

Compound **17** was prepared to protect a *cis*-1,2-diol in a sugar by reaction with 9,10-phenanthraquinone under irradiation (20°, C_6H_6, 15 h, ~50% yield). It is cleaved by oxidation with ozone ($-10°$, 5 h) to the ester, followed by basic hydrolysis (NaOMe/CHCl$_3$, $0°$, 12 h, 75% yield).[a]

[a] B. Helferich and E. von Gross, *Chem. Ber.,* **85,** 531 (1952); B. Helferich, E. N. Mulcahy, and H. Ziegler, *Chem. Ber.,* **87,** 233 (1954); B. Helferich and M. Gindy, *Chem. Ber.,* **87,** 1488 (1954).

Cyclic Ortho Esters

A variety of cyclic ortho esters,[a,b] including cyclic orthoformates, have been developed to protect *cis*-1,2-diols in nucleosides. Cyclic ortho esters are more readily cleaved by acidic hydrolysis (e.g., by a phosphate buffer, pH 4.5–7.5, or by 0.005–0.05 *M* HCl)[c] than are acetonides.

[a] C. B. Reese, *Tetrahedron,* **34,** 3143 (1978).

[b] V. Amarnath and A. D. Broom, *Chem. Rev.,* **77,** 183 (1977).

[c] M. Ahmad, R. G. Bergstrom, M. J. Cashen, A. J. Kresge, R. A. McClelland, and M. F. Powell, *J. Am. Chem. Soc.,* **99,** 4827 (1977).

18.   Methoxymethylene Acetal: **, 18**

19.   Ethoxymethylene Acetal: **, 19**

Compounds **18**[a] and **19**[b] were prepared to protect *cis*-1,2-diols in nucleosides by reaction with trimethyl or triethyl orthoformate (acid catalyst, 77%, 45–80% yields, respectively). Compounds **18** and **19** are cleaved by mild acidic hydrolysis: **18,** 0.01 *N* HCl, $20°$, 20 min, followed by basic hydrolysis of the resulting monoformates[a]; **19,** 0.01 *N* HCl, $20°$, 10 min, or oxalic acid, warm, followed by basic hydrolysis.[b]

[a] B. E. Griffin, M. Jarman, C. B. Reese, and J. E. Sulston, *Tetrahedron,* **23,** 2301 (1967).

[b] J. Žemlička, *Chem. Ind.* (*London*), 581 (1964); F. Eckstein and F. Cramer, *Chem. Ber.,* **98,** 995 (1965).

20.   Dimethoxymethylene Ortho Ester: **, 20**

Compound **20** was prepared, by reaction with tetramethyl orthocarbonate (TsOH/dioxane, 42–82% yield), to protect the *cis*-1,2-diol system in nucleosides.

It is readily cleaved by acidic hydrolysis ($0.01\ N$ HCl, $20°$, $t_{1/2} = 10$ min; 98% HCOOH) to a cyclic carbonate that can then be cleaved by basic hydrolysis.[a]

[a] G. R. Niaz and C. B. Reese, *J. Chem. Soc., Chem. Commun.,* 552 (1969).

21. 1-Methoxyethylidene Ortho Ester: , **21**

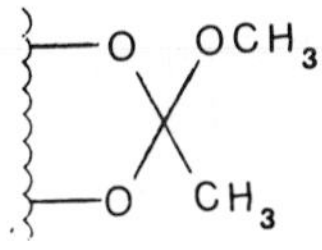

Compound **21** was prepared by acid-catalyzed transketalization [MeC(OMe)₃, TsOH, 55% yield] to protect the 1,2-diol system in nucleosides. It is very readily cleaved by acidic hydrolysis [1% HOAc (pH 3), $20°$, 30 sec, 100% yield] to a mixture of mono 2'- and 3'-acetates.[a]

[a] C. B. Reese and J. E. Sulston, *Proc. Chem. Soc.,* 214 (1964).

22. 1,2-Dimethoxyethylidene Ortho Ester: , **22**

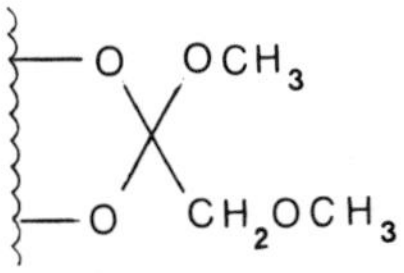

Compound **22** was prepared by acid-catalyzed transketalization [MeOCH₂C-(OMe)₃, mesitylenesulfonic acid, DMF, $20°$, 3–7 h, satisfactory yields] to protect 1,2-diols in nucleosides. It is cleaved by acidic hydrolysis (98% HCOOH, $25°$, 30 min, 69% yield) to a mono methoxyacetate.[a]

[a] J. H. van Boom, G. R. Owen, J. Preston, T. Ravindranathan, and C. B. Reese, *J. Chem. Soc. C,* 3230 (1971).

23. α-Methoxybenzylidene Ortho Ester: , **23**

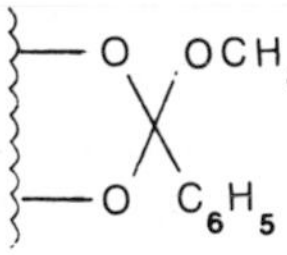

Compound **23** was prepared and cleaved by conditions similar to those described for compound **21**. Acidic hydrolysis gave a mixture of mono 2'- and 3'-benzoates.[a]

[a] C. B. Reese and J. E. Sulston, *Proc. Chem. Soc.,* 214 (1964).

24. 1-(*N,N*-Dimethylamino)ethylidene Derivative: , **24**

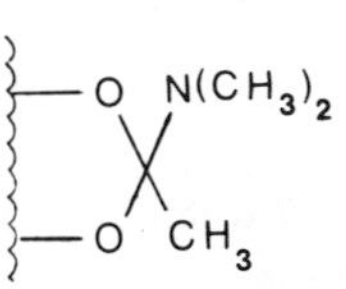

25.   α-(N,N-Dimethylamino)benzylidene Derivative:

, 25

Compounds **24**[a] and **25**[a] were prepared in high yields from cis-1,2-diols and $CH_3C(OMe)_2NMe_2$ or $PhC(OMe)_2NMe_2$, respectively. They are readily cleaved by hydrolysis (3% $MeOH–H_2O$, 20° or reflux, 100% yield of diol; dil HOAc, 1:1 mixture of mono 2- and 3-acetates or benzoates).

[a] S. Hanessian and E. Moralioglu, *Can. J. Chem.*, **50**, 233 (1972).

26.   1,3-(1,1,3,3-Tetraisopropyldisiloxanylidene) Derivative:

, 26

The author wished to protect, simultaneously, the 3'- and 5'-hydroxyl groups in nucleosides. 1,3-Dichloro-1,1,3,3-tetraisopropyldisiloxane [$ClSi(i-Pr)_2OSi(i-Pr)_2Cl$] reacts first with the primary 5'-hydroxyl group, then intramolecularly with the 3'-hydroxyl group to form compound **26** in 70–80% yield. The reagent will react with a 2',3'-diol system in a nucleoside. Compound **26** is stable to 0.3 M TsOH/dioxane, 10% $CF_3CO_2H/CHCl_3$, 5 M NH_3/dioxane–H_2O, and Et_3N or pyridine. It is cleaved by a variety of conditions: n-$Bu_4N^+F^-$, THF, 10 min; n-Bu_3NHF^-, 2 h; 0.2 M HCl/dioxane–H_2O, 24 h; 0.2 M HCl/MeOH, 9 h; and 0.2 M NaOH/dioxane–H_2O, 1 week.

Selective cleavage at the 3'-position was effected by reaction with 0.2 M NaOH/dioxane–H_2O, 20°, 95% yield or n-Bu_3NHF^-/THF, 20°, 60% yield. Attempted cleavage at the 5'-position (with 0.2 M HCl/dioxane–H_2O) gave a mixture of products: 2'-OTIPDSi[x]: 3'-OTIPDSi[x]: 5'-OTIPDSi[x] (3:5:2), [TIPDSi[x] = $HOSi(i-Pr_2)_2OSi(i-Pr)_2-$].[a]

[a] W. T. Markiewicz, *J. Chem. Res. Synop.*, 24 (1979).

27.   Stannoxane Derivative:

, 27

Compound **27** was prepared from a 1,2- or 1,3-diol by reaction with dibutyltin oxide. Reaction of compound **27** with 1 eq of benzoyl chloride or toluenesulfonyl chloride converts one of the hydroxyl groups to a benzoate or toluenesulfonate.[a]

[a] A. Shanzer, *Tetrahedron Lett.,* **21**, 221 (1980).

28. Cyclic Carbonates: , **28**

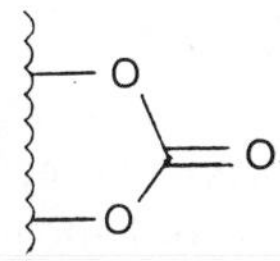

Cyclic carbonates,[a,b] prepared from 1,2- and 1,3-diols and phosgene or a chloroformate, are stable to the acid conditions that hydrolyze an acetonide (H_2SO_4, MeOH, 45°, 5 h).[c] They are readily cleaved by basic hydrolysis.

Formation

(1)   diol + Cl_2CO $\xrightarrow[\text{1 h}]{\text{Py, 20°}[d]}$ **28**

(2)   1,2-diol + $ClCO_2C_6H_4\text{-}p\text{-}NO_2$ $\xrightarrow[\text{20°, 5 days, 72\%}]{\text{Py}[e]}$ **28**

(3)   diol + *N,N'*-carbonyldiimidazole $\xrightarrow[\text{12 h to 4 days, 90\%}]{C_6H_6, \text{ heat}[f]}$ **28**

Cleavage

(1)   **28** $\xrightarrow[\text{70°}]{Ba(OH)_2[g]}$ 1,2-diol

(2)   **28** $\xrightarrow[\text{reflux, 15 min, 100\%}]{\text{aq Py}[e]}$ uridine

(3)   **28** $\xrightarrow[\text{25°, 5 min, 100\%}]{0.5 \text{ M NaOH/50\% aq dioxane}[e]}$ uridine

[a] L. Hough, J. E. Priddle, and R. S. Theobald, *Adv. Carbohydr. Chem.,* **15**, 91–158 (1960).
[b] V. Amarnath and A. D. Broom, *Chem. Rev.,* **77**, 183 (1977).
[c] W. N. Haworth and C. R. Porter, *J. Chem. Soc.,* 2796 (1929).
[d] W. N. Haworth and C. R. Porter, *J. Chem. Soc.,* 151 (1930).
[e] R. L. Letsinger and K. K. Ogilvie, *J. Org. Chem.,* **32**, 296 (1967).
[f] J. P. Kutney and A. H. Ratcliffe, *Synth. Commun.,* **5**, 47 (1975).
[g] W. G. Overend, M. Stacey, and L. F. Wiggins, *J. Chem. Soc.,* 1358 (1949).

29. Cyclic Boronates: , **29**

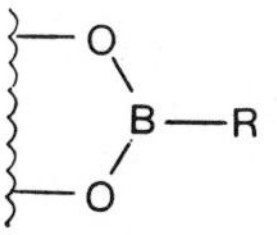

30.   Phenyl Boronate: 29, R = C_6H_5

31.   p-(P)-Phenyl Boronate: 29, R = p-(P)-C_6H_4

Cyclic boronates have been used to a limited extent (because they are hydrolyzed very readily) to protect diols in carbohydrates[a] and nucleosides. Cyclic boronates that have been isolated from aqueous solutions may have been insoluble rather than stable.[a] Ethyl boronates have been studied extensively,[b] but are liquids and provide less satisfactory protection.

Formation

$$\text{diol} + \text{R-B(OH)}_2 \xrightarrow{\quad C_6H_6{}^c \text{ or Py}^d \quad} \text{30 or 31}$$

30, R = Ph, 90% yield[c]

31, R = p-(P)-$C_6H_4{}^d$

Cleavage

(1)   **30** $\xrightarrow{\quad \text{HO(CH}_2)_3\text{OH, acetone}^a \quad}$ diol + (CH$_2$)$_3$ B—Ph (cyclic)

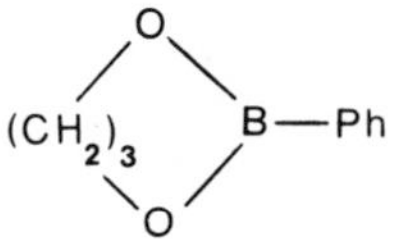

(2)   **31** $\xrightarrow[\text{30 min, 83\%}]{\quad \text{Me}_2\text{CO–H}_2\text{O (4:1)}^d \quad}$ diol

[a]   R. J. Ferrier, *Adv. Carbohydr. Chem. Biochem.,* **35**, 31–80 (1978).

[b]   W. V. Dahlhoff and R. Köster, *J. Org. Chem.,* **41**, 2316 (1976), and references cited therein.

[c]   R. J. Ferrier, *Methods Carbohydr. Chem.,* **VI**, 419–426 (1972).

[d]   J. M. J. Fréchet, L. J. Nuyens, and E. Seymour, *J. Am. Chem. Soc.,* **101**, 432 (1979).

3

Protection for Phenols and Catechols

*Included in Reactivity Chart 4.

---

*Included in Reactivity Chart 4.

A phenolic hydroxyl group is present in many compounds of biological interest (e.g., tyrosine, thyroxine, estrone, codeine, Terramycin; the catechol group is present in adrenalin). O-Protection is often required for phenols, which react readily either at oxygen or carbon or both, with oxidizing agents, electrophiles, and, as the phenoxide ion, with even mild alkylating and acylating agents. A phenol, like an alcohol, can be protected either as an ether or an ester. Since the same groups can often be used to protect both aliphatic and aromatic hydroxyl groups, the reader should consult the chapter on protection of alcohols for groups that may be suitable for protection of a phenol. A catechol can be protected in the presence of a phenol as a cyclic acetal or ketal or cyclic ester. In general aromatic ethers and esters are more readily cleaved than the corresponding aliphatic compounds.

   The more important phenol and catechol protective groups are included in Reactivity Chart 4.[a]

[a] See also: E. Haslam, "Protection of Phenols and Catechols," in *Protective Groups in Organic Chemistry,* J. F. W. McOmie, Ed., Plenum, New York and London, 1973, pp. 145–182.

PROTECTION FOR PHENOLS
ETHERS

Simple *n*-alkyl ethers, formed in basic solution from a phenol and a halide or sulfate, are very stable. Formerly, drastic conditions (e.g., refluxing HBr) were required for their cleavage. More recently several types of ethers have been investigated for the protection of phenols, generally involving milder methods for deblocking (e.g., via nucleophilic displacement, hydrogenolysis of benzyl ethers, and mild acid hydrolysis of acetal-type ethers).

1. Methyl Ether: $ArOCH_3$, 1

Two general methods used to prepare methyl ethers are as follows:

Formation

(1a)

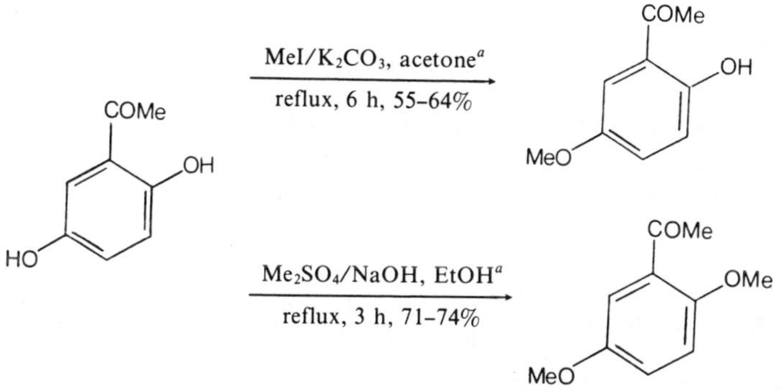

$$\xrightarrow[\text{reflux, 6 h, 55–64\%}]{\text{MeI/K}_2\text{CO}_3\text{, acetone}^a}$$

$$\xrightarrow[\text{reflux, 3 h, 71–74\%}]{\text{Me}_2\text{SO}_4\text{/NaOH, EtOH}^a}$$

(1b) $ArOH + RX$ or $R_2'SO_4$ $\xrightarrow[\substack{\text{PhCH}_2\overset{+}{\text{N}}\text{-}n\text{-Bu}_3\text{Br}^- \\ 25°, \ 2\text{–}13 \ \text{h}, \ 75\text{–}95\%}]{\text{NaOH/CH}_2\text{Cl}_2, \ \text{H}_2\text{O}^b}$ $ArOR$ or $ArOR'$

Ar = simple; 2- or 2,6-substituted^{b,c}

R = Me, CH_2=$CHCH_2$—, CH_2CHCH_2—, n-Bu, c-C_5H_{11},
$\overset{\displaystyle \diagdown \diagup}{\text{O}}$

PhCH$_2$, —CH$_2$COOEt^b

R′ = Me, Et^b

(1c) Methyl, ethyl, and benzyl ethers have been prepared in the presence of tetraethylammonium fluoride as a Lewis base (alkyl halide, DME, 20°, 3 h, 60–85% yields).^{c}

(2) p-NO$_2$—C$_6$H$_4$—ONa + MeN(NO)CONH$_2$ $\xrightarrow[0 \to 25°, \ 6 \ \text{h}]{\text{DME}}$

[p-NO$_2$—C$_6$H$_4$—O$^-$ + CH$_2$N$_2$] $\to$ p-NO$_2$—C$_6$H$_4$—OMe, >90%^{d}

Cleavage

Aryl methyl ethers are readily accessible compounds, stable to many common reagents. A number of mild methods have been developed to effect their cleavage, in place of the original drastic conditions of strong acid. Aryl methyl ethers are cleaved by reaction with iodotrimethylsilane (eq. 1), *vide infra,* with strong nucleophiles (eqs. 2–6), with Lewis acids (eqs. 7 and 8), by fusion with pyridine hydrochloride (eq. 9), by a Grignard reagent (eq. 10), and by refluxing with hydrobromic acid (eq. 11).

$$(1)\quad AroMe + Me_3SiI \xrightarrow[25-50°,\ 12-140\ h]{CHCl_3^e} AroH$$

$$Ar = Ph,\ 25°,\ 48\ h,\ 100\%$$

$$Ar = Ph;\ p\text{-}C_6H_5\text{—}C_6H_4;\ o\text{-},\ m\text{-}\ or\ p\text{-}Br\text{—}C_6H_4;$$

$$o\text{-},\ m\text{-},\ or\ p\text{-}NH_2\text{—}C_6H_4;\ m\text{-}CH_3\text{—}C_6H_4$$

Iodotrimethylsilane in quinoline (180°, 70 min) selectively cleaves an aryl methyl group, in 72% yield, in the presence of a methylenedioxy group.[f]

$$(2)\quad AroMe + EtS^-Na^+ \xrightarrow[reflux,\ 3\ h]{DMF} AroH,\ 94\text{–}98\%^g$$

Potassium thiophenoxide has been used to cleave an aryl methyl ether without causing migration of a double bond.[h]

Sodium benzylselenide (PhCH$_2$SeNa) and sodium thiocresolate (p-CH$_3$—C$_6$H$_4$SNa) cleave a dimethoxyaryl compound regioselectively, reportedly due to steric factors in the former case[i] and to electronic factors in the latter case.[j]

(3) Sodium sulfide in *N*-methylpyrrolidone (140°, 2–4 h) cleaves aryl methyl ethers in 78–85% yield.[k]

(4)

Lithium diphenylphosphide selectively cleaves an aryl methyl ether in the presence of an aryl ethyl ether.[l]

It also cleaves a phenyl benzyl ether and phenyl allyl ether to phenol in 88% and 78% yield, respectively.[m]

$$(5)\quad AroMe + NaCN \xrightarrow[125-180°,\ 5-48\ h]{DMSO^n} ArCH,\ 65\text{–}90\%$$

This cleavage reaction is successful for aromatic systems containing ketones, amides, and carboxylic acids; mixtures are obtained from nitro-substituted aromatic compounds; there is no reaction with 5-methoxyindole (180°, 48 h).[n]

(6) $ArOMe + LiI \xrightarrow[\text{reflux, 10 h}]{\text{collidine}^o} ArOH, \sim \text{quant}$

Aryl ethyl ethers are cleaved more slowly; dialkyl ethers are stable to these conditions.[o]

(7) $\xrightarrow[25°, \, <1 \text{ h}, \, 94\%]{\text{AlBr}_3/\text{EtSH}^p} C_{3,17}$-dihydroxy compound

A methylenedioxy group, used to protect a catechol, is cleaved under similar conditions in satisfactory yields; methyl and ethyl esters are stable (0–20°, 2 h).[p]

Regioselective cleavage of dimethoxyaryl derivatives with methanesulfonic acid/methionine has been reported.[q]

(8)   3,3'-dimethoxybiphenyl $\xrightarrow[-80 \rightarrow 20°, \, 12 \text{ h}]{\text{BBr}_3/\text{CH}_2\text{Cl}_2{}^r}$ 3,3'-dihydroxybiphenyl, 77–86%

Methylenedioxy groups and diphenyl ethers are stable to these cleavage conditions.[r]

Either an aryl methyl ether or a methylenedioxy group can be cleaved with boron trichloride under various conditions.[s]

Boron triiodide rapidly cleaves methyl ethers of o-, m-, or p-substituted aromatic aldehydes (0°, 25°; 0.5–5 min; 40–86% yield).[t]

Boron tribromide is reported to be more effective than iodotrimethylsilane for cleaving aryl methyl ethers.[u]

(9)   codeine $\xrightarrow[220°, \, 6 \text{ min}]{\text{Py} \cdot \text{HCl}}$ morphine, 34%[v]

(10) $ArOCH_3 \xrightarrow[155-165°, \, 15 \text{ min}, \, 80\%]{\text{xs MeMgI}^w} ArOH$

(11) $p\text{-MeO}-C_6H_4(CH_2)_7C_6H_4\text{-}p\text{-OMe} \xrightarrow[\text{reflux, 30 min, 85\%}]{48\% \text{ HBr/HOAc}^x}$

$p\text{-HO}-C_6H_4(CH_2)_7C_6H_4\text{-}p\text{-OH}$

This aryl methyl ether was stable to sulfuric acid (60°, 30 min; reflux, 30 min).[x]

[a]   G. N. Vyas and N. M. Shah, *Org. Synth., Coll. Vol. IV*, 836 (1963).

[b]   A. McKillop, J.-C. Fiaud, and R. P. Hug, *Tetrahedron*, **30**, 1379 (1974).

[c]   J. M. Miller, K. H. So, and J. H. Clark, *Can. J. Chem.*, **57**, 1887 (1979).

[d]   S. M. Hecht and J. W. Kozarich, *Tetrahedron Lett.*, 1397 (1973).

[e] M. E. Jung and M. A. Lyster, *J. Org. Chem.*, **42**, 3761 (1977).

[f] J. Minamikawa and A. Brossi, *Tetrahedron Lett.*, 3085 (1978).

[g] G. I. Feutrill and R. N. Mirrington, *Tetrahedron Lett.*, 1327 (1970); *Aust. J. Chem.*, **25**, 1719, 1731 (1972).

[h] J. W. Wildes, N. H. Martin, C. G. Pitt, and M. E. Wall, *J. Org. Chem.*, **36**, 721 (1971).

[i] R. Ahmad, J. M. Saá, and M. P. Cava, *J. Org. Chem.*, **42**, 1228 (1977).

[j] C. Hansson and B. Wickberg, *Synthesis*, 191 (1976).

[k] M. S. Newman, V. Sankaran, and D. R. Olson, *J. Am. Chem. Soc.*, **98**, 3237 (1976).

[l] R. E. Ireland and D. M. Walba, *Org. Synth.*, **56**, 44 (1977).

[m] F. G. Mann and M. J. Pragnell, *Chem. Ind. (London)*, 1386 (1964).

[n] J. R. McCarthy, J. L. Moore, and R. J. Cregge, *Tetrahedron Lett.*, 5183 (1978).

[o] I. T. Harrison, *J. Chem. Soc., Chem. Commun.*, 616 (1969).

[p] M. Node, K. Nishide, K. Fuji, and E. Fujita, *J. Org. Chem.*, **45**, 4275 (1980).

[q] N. Fujii, H. Irie, and H. Yajima, *J. Chem. Soc., Perkin Trans. 1*, 2288 (1977).

[r] J. F. W. McOmie and D. E. West, *Org. Synth., Coll. Vol. V*, 412 (1973).

[s] M. Gerecke, R. Borer, and A. Brossi, *Helv. Chim. Acta*, **59**, 2551 (1976).

[t] J. M. Lansinger and R. C. Ronald, *Synth. Commun.*, **9**, 341 (1979).

[u] E. H. Vickery, L. F. Pahler, and E. J. Eisenbraun, *J. Org. Chem.*, **44**, 4444 (1979).

[v] M. Gates and G. Tschudi, *J. Am. Chem. Soc.*, **78**, 1380 (1956).

[w] R. Mechoulam and Y. Gaoni, *J. Am. Chem. Soc.*, **87**, 3273 (1965).

[x] I. Kawasaki, K. Matsuda, and T. Kaneko, *Bull. Chem. Soc. Jpn.*, **44**, 1986 (1971).

2. Methoxymethyl Ether (MOM Ether): $ArOCH_2OCH_3$, 2

Formation

(1) $ArOH + ClCH_2OMe$ $\xrightarrow[\text{20°, 20 min, 80–95\%}]{\text{CH}_2\text{Cl}_2,\ \text{NaOH–H}_2\text{O, Adogen (phase transfer cat.)}^a}$ **2**

(2) $ArOK + ClCH_2OMe$ $\xrightarrow[\text{80\%}]{\text{CH}_3\text{CN, 18-crown-6}^b}$ **2**

(3) $ArOH + MeOCH_2OMe$ $\xrightarrow[\text{reflux, 12 h, 60–80\%}]{\text{TsOH, CH}_2\text{Cl}_2,\ \text{mol sieves, N}_2{}^c}$ **2**

This method of formation[c] avoids the use of the carcinogen chloromethyl methyl ether.

Cleavage

(1)   **2** $\xrightarrow[\text{25°, 12 h, quant}]{\text{HCl, }i\text{-PrOH, THF}^c}$ ArOH

(2)   **2** $\xrightarrow[\text{90°, 40 h, high yield}]{2\ N\ \text{HOAc}^d}$ ArOH

The group has been used in a synthesis of 13-desoxydelphonine from *o*-cresol, a synthesis that required the group to be stable to many reagents.[e]

[a] F. R. van Heerden, J. J. van Zyl, G. J. H. Rall, E. V. Brandt, and D. G. Roux, *Tetrahedron Lett.*, 661 (1978).

[b] G. J. H. Rall, M. E. Oberholzer, D. Ferreira, and D. G. Roux, *Tetrahedron Lett.*, 1033 (1976).

[c] J. P. Yardley and H. Fletcher, 3rd, *Synthesis*, 244 (1976).

[d] M. A. A.-Rahman, H. W. Elliott, R. Binks, W. Küng, and H. Rapoport, *J. Med. Chem.*, **9**, 1 (1966).

[e] K. Wiesner, *Pure Appl. Chem.*, **51**, 689 (1979).

3. Methoxyethoxymethyl Ether (MEM Ether): $ArOCH_2OCH_2CH_2OCH_3$, 3

A 2-methoxyethoxymethyl ether was used to protect one phenol group during a total synthesis of gibberellic acid.[a]

Formation[a]

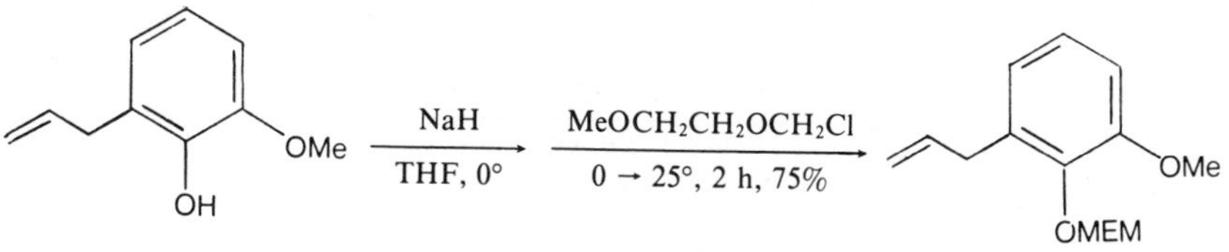

Cleavage[a]

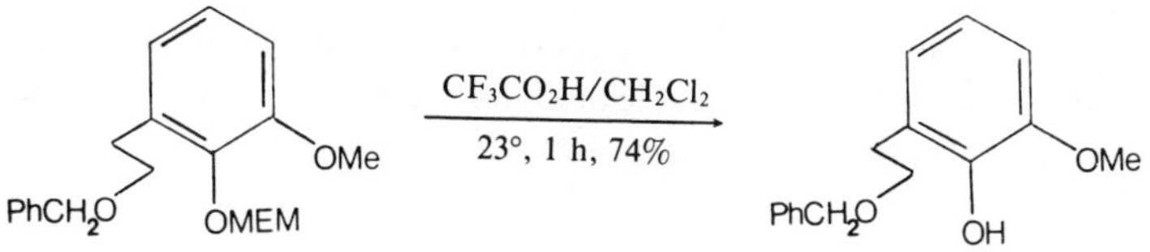

[a] E. J. Corey, R. L. Danheiser, S. Chandrasekaran, P. Siret, G. E. Keck, and J.-L. Gras, *J. Am. Chem. Soc.*, **100**, 8031 (1978).

4. Methylthiomethyl Ether (MTM Ether): $ArOCH_2SCH_3$, 4

Formation (→) / Cleavage (←)[a]

$$\text{ArOH} \underset{\text{HgCl}_2/\text{CH}_3\text{CN–H}_2\text{O, reflux, 10 h, 90–95\%}}{\overset{\text{NaOH, ClCH}_2\text{SMe, HMPA, 25°, 16 h, 91–94\%}}{\rightleftharpoons}} \textbf{4}$$

Aryl methylthiomethyl ethers are stable to the conditions used to hydrolyze primary alkyl MTM ethers (e.g., $HgCl_2/CH_3CN-H_2O$, 25°, 6 h). They are moderately stable to acidic conditions (95% recovered from $HOAc/THF-H_2O$, 25°, 4 h).[a]

[a] R. A. Holton and R. G. Davis, *Tetrahedron Lett.*, 533 (1977).

5. Tetrahydropyranyl Ether: ArO-2-tetrahydropyranyl, 5

Compound **5**, prepared from a phenol and dihydropyran (HCl/EtOAc, 25°, 24 h), is cleaved by aqueous oxalic acid (MeOH, 50–90°, 1–2 h).[a]

[a] H. N. Grant, V. Prelog, and R. P. A. Sneeden, *Helv. Chim. Acta,* **46,** 415 (1963).

6. Phenacyl Ether: $ArOCH_2COC_6H_5$, 6

Formation (→) / Cleavage (←)[a]

$$\xrightarrow[\text{reflux, 1–2 h, 85–95\%}]{\text{BrCH}_2\text{COPh, K}_2\text{CO}_3\text{, acetone}}$$

ArOH   **6**

$$\xleftarrow[\text{25°, 1 h, 88–96\%}]{\text{Zn/HOAc}}$$

Phenacyl and *p*-bromophenacyl ethers of phenols are stable to 1% ethanolic alkali (reflux, 2 h), and to 5 N sulfuric acid in ethanol–water. Compound **6**, prepared from β-naphthol, is cleaved in 82% yield by 5% ethanolic alkali (reflux, 2 h).

[a] J. B. Hendrickson and C. Kandall, *Tetrahedron Lett.,* 343 (1970).

Cyclopropylmethyl Ether: $ArOCH_2$-*c*-C_3H_5, A

For a particular phenol, the authors required a protective group that would be stable to reduction (by complex metals, catalytic hydrogenation, and Birch conditions) and that could be easily and selectively removed.[a]

Formation (→) / Cleavage (←)[a]

$$\xrightarrow[\text{0°, 30 min}]{\text{KO-}t\text{-Bu, DMF}} \quad \xrightarrow[\text{20°, 20 min → 40°, 6 h}]{c\text{-C}_3\text{H}_5\text{CH}_2\text{Br}}$$

ArOH   **A,** 80%

$$\xleftarrow[\text{reflux, 2 h, 94\%}]{\text{aq HCl, MeOH}}$$

[a] W. Nagata, K. Okada, H. Itazaki, and S. Uyeo, *Chem. Pharm. Bull.,* **23,** 2878 (1975).

7.   Allyl Ether: ArOCH$_2$CH=CH$_2$, 7

Allyl ethers can be prepared by reaction of a phenol and the allyl bromide in the presence of base. Several reagents have been used to effect their cleavage:

Cleavage

(1)   7 $\xrightarrow[\text{60–80°, 6 h, > 95\%}]{\text{Pd/C, TsOH, H}_2\text{O or MeOH}^a}$ ArOH

(2)   7 $\xrightarrow[\text{reflux, 1 h, 40–57\%}]{\text{SeO}_2/\text{HOAc, dioxane}^b}$ ArOH + uncleaved **7**

(3)   7 $\xrightarrow[\text{90°, 1 h}]{\text{Ph}_3\text{P/Pd(OAc)}_2\text{, HCOOH}}$ [ArOCH=CHCH$_3$] → ArOH^c

(4)   7 $\xrightarrow[\text{reflux, 10 h, 62\%}]{\text{NaAlH}_2(\text{OCH}_2\text{CH}_2\text{OCH}_3)_2\text{, PhCH}_3{}^d}$ ArOH

An aryl allyl ether is selectively cleaved by this reagent (which also cleaves aryl benzyl ethers) in the presence of an *N*-allylamide.^{d}

[a] R. Boss and R. Scheffold, *Angew. Chem., Inter. Ed., Engl.,* **15,** 558 (1976).

[b] K. Kariyone and H. Yazawa, *Tetrahedron Lett.,* 2885 (1970).

[c] H. Hey and H.-J. Arpe, *Angew. Chem., Inter. Ed., Engl.,* **12,** 928 (1973).

[d] T. Kametani, S.-P. Huang, M. Ihara, and K. Fukumoto, *J. Org. Chem.,* **41,** 2545 (1976).

Isopropyl Ether: ArOCH(CH$_3$)$_2$, B

An isopropyl ether was developed as a phenol protective group that would be more stable to Lewis acids than an aryl benzyl ether.^{a}

Formation (→) / Cleavage (←)^a

$$\text{ArOH} \quad \xrightarrow[\text{20°, 19 h}]{\text{Me}_2\text{CHBr/K}_2\text{CO}_3\text{–DMF, acetone}} \quad \textbf{B}$$

$$\text{ArOH} \quad \xleftarrow[\substack{\text{or TiCl}_4/\text{CH}_2\text{Cl}_2\text{, 0°, slower} \\ \text{(no reaction with SnCl}_4)}]{\text{BCl}_3/\text{CH}_2\text{Cl}_2\text{, 0°, rapid}} \quad \textbf{B}$$

[a] T. Sala and M. V. Sargent, *J. Chem. Soc., Perkin Trans. 1,* 2593 (1979).

8.   Cyclohexyl Ether: ArO-*c*-C$_6$H$_{11}$, 8

Formation[a]

$$p\text{-HOC}_6\text{H}_4\text{CH}_2\text{CHCOOMe} \xrightarrow[\text{reflux, 24 h, 60\%}]{\text{cyclohexene, BF}_3 \cdot \text{Et}_2\text{O, CH}_2\text{Cl}_2} \textbf{8a}$$
$$|$$
$$\text{NHCOCF}_3$$

Cleavage[a]

$$\textbf{8a} \xrightarrow[\text{0°, 30 min, 100\%}]{\text{HF}} \text{Tyr-OMe}$$

$$\textbf{8a} \xrightarrow[\text{25°, 2 h, 99\%}]{\text{5.3 } N \text{ HBr/HOAc}} \text{Tyr-OMe}$$

An ether that would not undergo rearrangement to a 3-alkyl derivative during acid-catalyzed removal of —NH protective groups was required to protect the phenol group in tyrosine. Four compounds were investigated: *O*-cyclohexyl- (**8a**), *O*-isobornyl- (**8b**), *O*-[1-(5-pentamethylcyclopentadienyl)ethyl]- (**8c**), and *O*-isopropyltyrosine (**8d**).

Compounds **8b** and **8c** do not undergo rearrangement, but are very labile in trifluoroacetic acid (100% cleaved in 5 min). The cyclohexyl (**8a**) and isopropyl (**8d**) derivatives are more stable to acid, but undergo some rearrangement. The cyclohexyl group combines minimal rearrangement with ready removal.[a]

[a] M. Engelhard and R. B. Merrifield, *J. Am. Chem. Soc.*, **100**, 3559 (1978).

9.   *t*-Butyl Ether: ArOC(CH$_3$)$_3$, 9

An aryl *t*-butyl ether can be prepared by an acid-catalyzed reaction.[a] (Most aryl ethers are formed under basic conditions.)

Formation[a]

$$p\text{-HOC}_6\text{H}_4\text{CH}_2\text{CHCOOH} \xrightarrow[\text{25°, 6–10 h, 93\%}]{\text{Me}_2\text{C}=\text{CH}_2, \text{ cat. concd H}_2\text{SO}_4, \text{ CH}_2\text{Cl}_2} \textbf{9a}$$
$$|$$
$$\text{NHCOOCH}_2\text{Ph}$$

Cleavage[a]

$$p\text{-Me}_3\text{COC}_6\text{H}_4\text{CH}_2\text{CHCOOCMe}_3 \xrightarrow[\text{100\%}]{\text{H}_2/\text{Pd–C}} \xrightarrow[\text{25°, 16 h, 81\%}]{\text{anhyd CF}_3\text{COOH}} \text{Tyr-OH}$$
$$|$$
$$\text{NHCOOCH}_2\text{Ph}$$

9a

A *t*-butyl ether can be prepared in pyridine by reaction of a phenol with a *t*-butyl halide (20–30°, few h, 65–90% yield).[b]

[a] H. C. Beyerman and J. S. Bontekoe, *Recl. Trav. Chim. Pays-Bas,* **81**, 691 (1962).
[b] H. Masada and Y. Oishi, *Chem. Lett.,* 57 (1978).

10. Benzyl Ether: ArOCH₂C₆H₅, 10

Aryl benzyl ethers can be prepared from a phenol and phenyldiazomethane or by treating an alkaline solution of the phenol with a benzyl halide. The benzyl group is readily removed by hydrogenolysis and by acid-catalyzed hydrolysis.

Cleavage

(1)

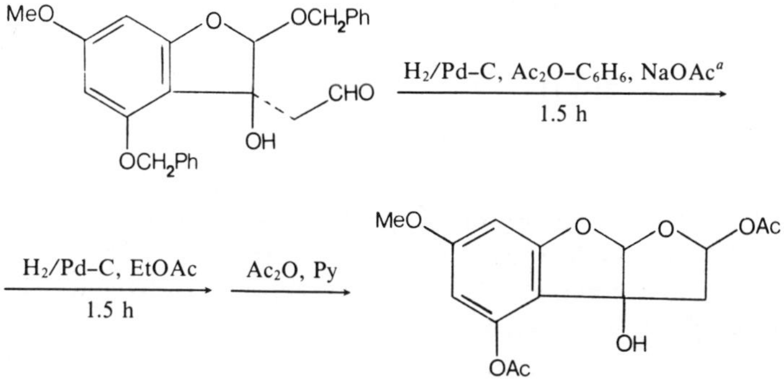

H_2/Pd–C, Ac_2O–C_6H_6, NaOAc[a]
1.5 h

H_2/Pd–C, EtOAc Ac_2O, Py
1.5 h

Catalytic hydrogenation in acetic anhydride–benzene removes the aromatic benzyl ether and forms a monoacetate; hydrogenation in ethyl acetate removes the aliphatic benzyl ether to give, after acetylation, the diacetate.[a]

(2) *p*-PhCH₂O—C₆H₄CH₂CHCOOH $\xrightarrow[\text{25°, 1.5 h, 95–100\%}]{\text{Pd–C, 1,4-cyclohexadiene}^b}$ BOC-Tyr-OH

  |
  NHCOOCMe₃

Palladium black, a more reactive catalyst than Pd–C, must be used to cleave the more stable aliphatic benzyl ethers.[b]

(3)

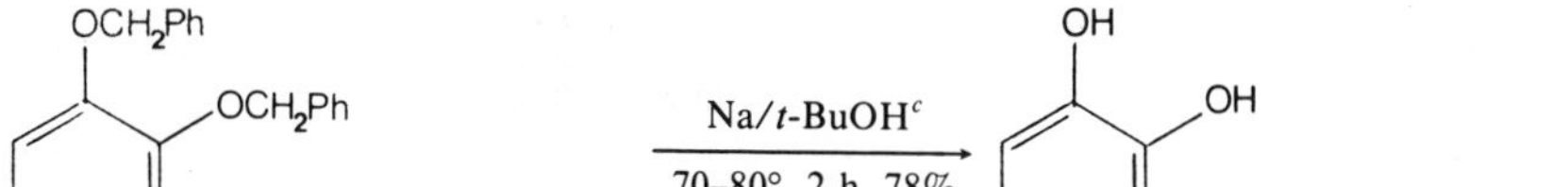

$\xrightarrow[\text{70–80°, 2 h, 78\%}]{\text{Na}/t\text{-BuOH}^c}$

In this example sodium in butyl alcohol cleaves two aryl benzyl ethers and reduces a double bond that is conjugated with an aromatic ring; nonconjugated double bonds are stable.[c]

(4) $\quad$ ArOCH$_2$Ph $\xrightarrow[\text{25°, 40 min, 80–90\%}]{\text{BF}_3\cdot\text{Et}_2\text{O, EtSH}^d}$ AroH

Addition of sodium sulfate prevents hydrolysis of a dithioacetal group present in the compound; replacement of ethanethiol with ethanedithiol prevents cleavage of a dithiolane group.[d]

(5) $\quad$ p-PhCH$_2$O—C$_6$H$_4$CH$_2$CHCOOH $\xrightarrow[\text{CF}_3\text{COOH, 0°, 30 min, 100\%}]{\text{CF}_3\text{OSO}_2\text{F or CH}_3\text{OSO}_2\text{F/PhSMe}^e}$ Tyr-OH
$\qquad\qquad\qquad\qquad\quad |$
$\qquad\qquad\qquad\qquad\ \text{NH}_2$

Thioanisole suppresses acid-catalyzed rearrangement of the benzyl group to form 3-benzyltyrosine.[e]

The more acid-stable 2,6-dichlorobenzyl ether is cleaved in a similar manner.[e]

(6) $\quad$ p-PhCH$_2$O—C$_6$H$_4$CH$_2$CHCOOMe $\xrightarrow[\text{25–50°, 100\%}]{\text{Me}_3\text{SiI, CH}_3\text{CN}^f}$ Tyr-OMe or Tyr-OH
$\qquad\qquad\qquad\qquad\quad\ |$
$\qquad\qquad\qquad\qquad\ \text{NHCOOCMe}_3$

Selective removal of protective groups is possible with this reagent since a carbamate, >NCOOCMe$_3$, is cleaved in 6 min at 25°; an aryl benzyl ether is cleaved in 100% yield, with no formation of 3-benzyltyrosine, in 1 h at 50°, at which time a methyl ester begins to be cleaved.[f]

[a] G. Büchi and S. M. Weinreb, *J. Am. Chem. Soc.*, **93**, 746 (1971).

[b] A. M. Felix, E. P. Heimer, T. J. Lambros, C. Tzougraki, and J. Meienhofer, *J. Org. Chem.*, **43**, 4194 (1978).

[c] B. Loev and C. R. Dawson, *J. Am. Chem. Soc.*, **78**, 6095 (1956).

[d] K. Fuji, K. Ichikawa, M. Node, and E. Fujita, *J. Org. Chem.*, **44**, 1661 (1979).

[e] Y. Kiso, H. Isawa, K. Kitagawa, and T. Akita, *Chem. Pharm. Bull*, **26**, 2562 (1978).

[f] R. S. Lott, V. S. Chauhan, and C. H. Stammer, *J. Chem. Soc., Chem. Commun.*, 495 (1979).

11. $\quad$ *o*-Nitrobenzyl Ether: ArOCH$_2$C$_6$H$_4$-*o*-NO$_2$, 11

Formation (→) / Cleavage (←)[a]

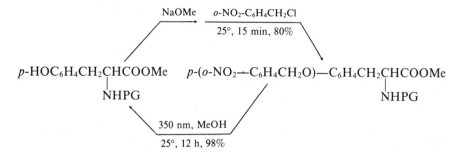

p-HOC$_6$H$_4$CH$_2$CHCOOMe $\qquad$ p-(o-NO$_2$—C$_6$H$_4$CH$_2$O)—C$_6$H$_4$CH$_2$CHCOOMe
$\qquad\qquad |$ $\qquad\qquad\qquad\qquad\qquad\qquad\qquad\qquad\qquad\qquad\quad |$
$\qquad\quad$ NHPG $\qquad\qquad\qquad\qquad\qquad\qquad\qquad\qquad\qquad\qquad\qquad$ NHPG

An *o*-nitrobenzyl ether can be cleaved by photolysis. In tyrosine this avoids the use of acid-catalyzed cleavage and the attendant conversion to 3-benzyltyrosine.[a] (Note that this unwanted conversion may also be suppressed by the addition of thioanisole, e.g., **Cleavage** of benzyl ethers, **10**, eq. 5.)

[a] B. Amit, E. Hazum, M. Fridkin, and A. Patchornik, *Int. J. Pept. Protein Res.*, **9**, 91 (1977).

12. 9-Anthrylmethyl Ether: ArOCH₂-9-anthryl, 12

Formation (→) / Cleavage (←)[a]

$$\text{ArONa} \quad \xrightarrow[\text{DMF, 25°}]{\text{9-anthrylmethyl chloride}} \quad \mathbf{12}$$

$$\mathbf{12} \quad \xrightarrow[\text{25°, 20 min, 85–99\%}]{\text{CH}_3\text{S}^-\text{Na}^+,\ \text{DMF}} \quad$$

Aryl 9-anthrylmethyl ethers are also cleaved by CF_3CO_2H/CH_2Cl_2 (0°, 10 min, 100% yield); they are stable to CF_3CO_2H/dioxane (25°, 1 h).[a]

[a] N. Kornblum and A. Scott, *J. Am. Chem. Soc.*, **96**, 590 (1974).

13. 4-Picolyl Ether: ArOCH₂-4-pyridyl, 13

Formation (→)[a] */ Cleavage (←)*[a,b]

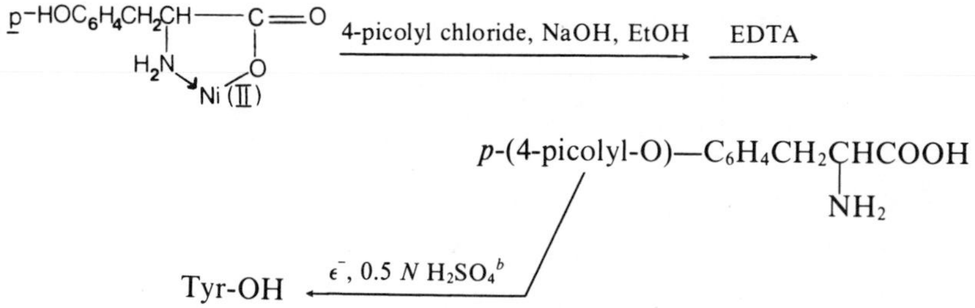

EDTA = ethylenediaminetetraacetic acid

An aryl 4-picolyl ether is stable to trifluoroacetic acid, used to cleave an *N-t*-butoxycarbonyl group.[a]

[a] A. Gosden, D. Stevenson, and G. T. Young, *J. Chem. Soc., Chem. Commun.*, 1123 (1972).
[b] P. M. Scopes, K. B. Walshaw, M. Welford, and G. T. Young, *J. Chem. Soc.*, 782 (1965).

Silyl Ethers

Aryl and alkyl trimethylsilyl ethers can often be cleaved by refluxing in aqueous methanol, an advantage for acid- or base-sensitive substrates. The ethers are stable to Grignard and Wittig reactions, and to reduction with lithium aluminum hydride at $-15°$. Aryl t-butyldimethylsilyl ethers require acid- or fluoride ion-catalyzed hydrolysis for removal.

14. Trimethylsilyl Ether: ArOSi(CH₃)₃, 14

Formation

(1) $\text{ArOH} \xrightarrow[\text{30-35°, 12 h}]{\text{Me}_3\text{SiCl/Py}^a}$ **14,** satisfactory yield

(2)

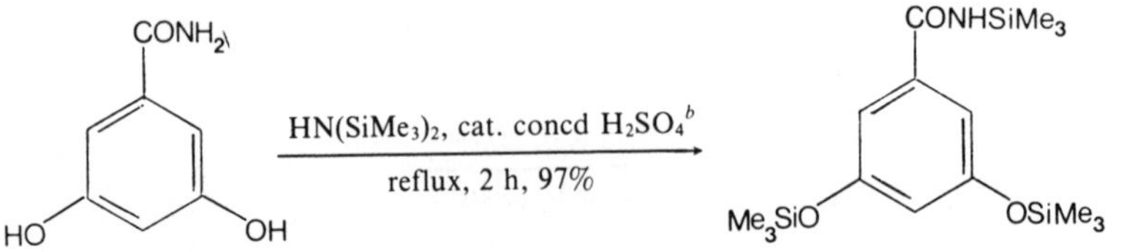

Cleavage[b]

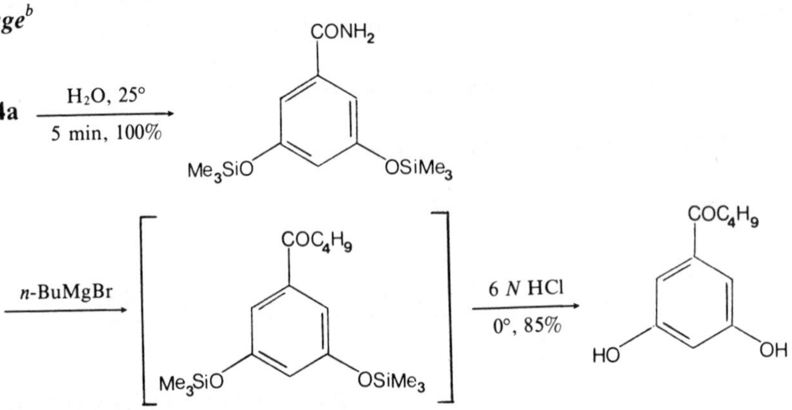

[a]   Cl. Moreau, F. Roessac, and J. M. Conia, *Tetrahedron Lett.*, 3527 (1970).

[b]   S. A. Barker and R. L. Settine, *Org. Prep. Proced. Int.*, **11**, 87 (1979).

15. *t*-Butyldimethylsilyl Ether: ArOSi(CH₃)₂C(CH₃)₃, 15

Formation $(\rightarrow)$ / *Cleavage* $(\leftarrow)^a$

$$\text{ArOH} \quad \overset{\overset{\textstyle t\text{-BuMe}_2\text{SiCl, DMF}}{\xrightarrow{\hspace{3cm}}}}{\underset{\underset{\textstyle \text{imidazole, 25°, 3 h, 96\%}}{}}{}} \quad \textbf{15}$$

$$\underset{\text{THF, 25°, 2 days, 77\%}}{\xleftarrow{\text{0.1 } M \text{ HF, 0.1 } M \text{ NaF, pH 5}}}$$

In this substrate[a] a mixture of products resulted from attempted cleavage of the *t*-butyldimethylsilyl ether with tetra-*n*-butylammonium fluoride, the reagent generally used.[b]

[a] P. M. Kendall, J. V. Johnson, and C. E. Cook, *J. Org. Chem.*, **44**, 1421 (1979).
[b] E. J. Corey and A. Venkateswarlu, *J. Am. Chem. Soc.*, **94**, 6190 (1972).

ESTERS

Aryl esters, prepared from the phenol and an acid chloride or anhydride in the presence of base, are readily cleaved by saponification. 9-Fluorenecarboxylates and 9-xanthenecarboxylates are also cleaved by photolysis. To permit selective removal, a number of carbonate esters have been investigated: aryl benzyl carbonates can be cleaved by hydrogenolysis; aryl 2,2,2-trichloroethyl carbonates by Zn/THF–H$_2$O.

16.   Aryl Acetate: ArOCOCH$_3$, 16

Formation[a]

$$ArOH + CH_3COCl \xrightarrow[25°, \ 30 \ min, \ 90\%]{NaOH–dioxane, \ Bu_4N^+HSO_4^-} 16$$

Phase transfer catalysis with tetra-*n*-butylammonium hydrogensulfate effects acylation of sterically hindered phenols and selective acylation of a phenol in the presence of an aliphatic secondary alcohol.

Cleavage

(1)   ArOAc $\xrightarrow[25°, \ 0.75 \ h, \ 94\%]{NaHCO_3/aq \ MeOH}$ ArOH[b]

(2)   **16a**

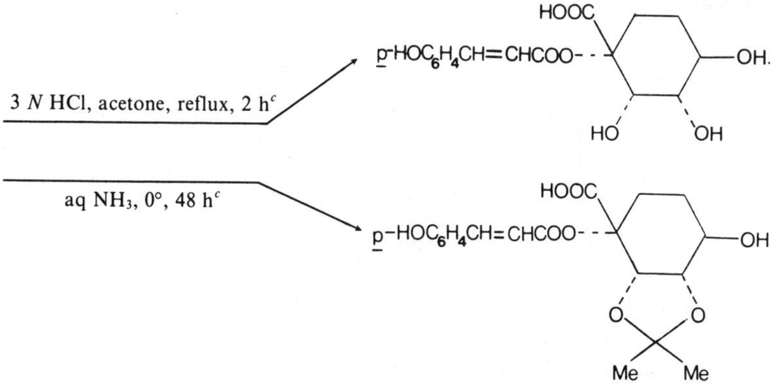

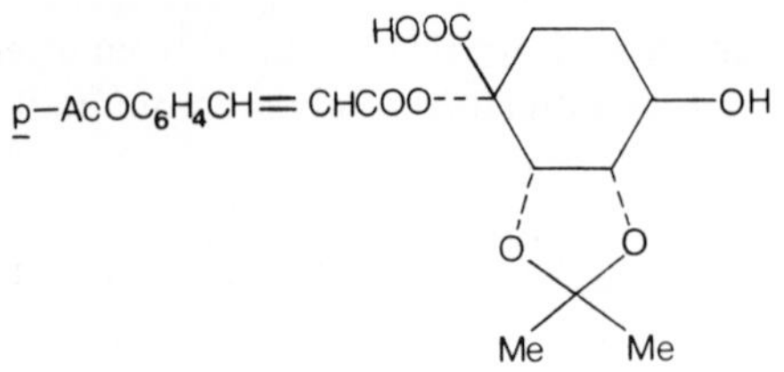

16a

An aryl acetate is readily cleaved (2 *N* NaOH, EtOH, 10 → 25°, 30 min, 82% yield) in the presence of an *N*-acetyl group.[d]

(3)

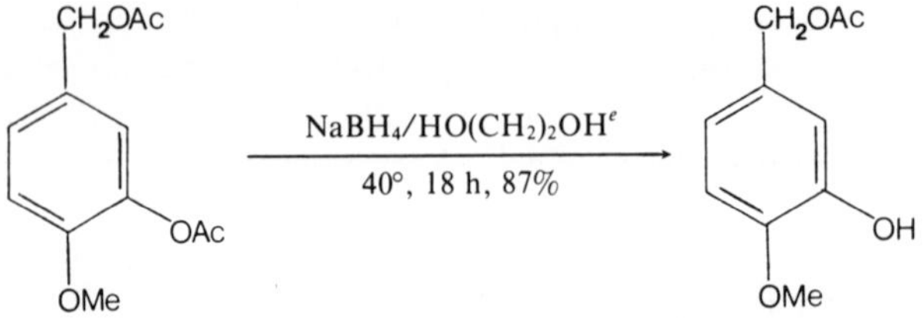

NaBH₄/HO(CH₂)₂OH[e]

40°, 18 h, 87%

(4) Lithium aluminum hydride can be used to effect efficient ester cleavage if no other functional group is present that can be attacked by this strong reducing agent.[f]

[a] V. O. Illi, *Tetrahedron Lett.,* 2431 (1979).
[b] For example, see G. Büchi and S. M. Weinreb, *J. Am. Chem. Soc.,* **93**, 746 (1971).
[c] E. Haslam, G. K. Makinson, M. O. Naumann, and J. Cunningham, *J. Chem. Soc.,* 2137 (1964).
[d] N. Komoto, Y. Enomoto, M. Miyagaki, Y. Tanaka, K. Nitanai, and H. Umezawa, *Agric. Biol. Chem.,* **43**, 555 (1979).
[e] J. Quick and J. K. Crelling, *J. Org. Chem.,* **43**, 155, (1978).
[f] H. Mayer, P. Schudel, R. Rüegg, and O. Isler, *Helv. Chim. Acta,* **46**, 650 (1963).

17. Aryl Pivaloate: ArOCOC(CH₃)₃, 17

Formation (→) / Cleavage (←)[a]

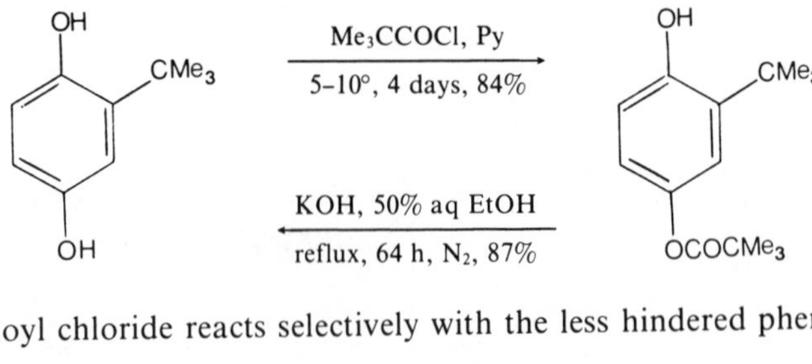

Me₃CCOCl, Py

5–10°, 4 days, 84%

KOH, 50% aq EtOH

reflux, 64 h, N₂, 87%

Pivaloyl chloride reacts selectively with the less hindered phenol group.

[a] L. K. T. Lam and K. Farhat, *Org. Prep. Proced. Int.,* **10**, 79 (1978).

18.   Aryl Benzoate: ArOCOC₆H₅, 18

Aryl benzoates, stable to alkylation conditions using K_2CO_3/Me_2SO_4,[a] are cleaved by more basic hydrolysis (KOH).[a] They are stable to anhydrous hydrogen chloride,[b] but are cleaved by hydrochloric acid.[c]

Formation

(1) $ArOH + [(ClCO)_2 + Me_2NCHO$ $\xrightarrow{PhCOOH}$ $Me_2\overset{+}{N}{=}CHOCOPh]$ $\xrightarrow[20°, 2\ h]{Py}$

   Cl^-

   $ArOCOPh,\ 90\%$[d]

(2) $ArOH +$ [pyridinium structure: N-Me, 2-SCOPh, Cl⁻] $\xrightarrow[\text{aq NaOH, 80\%}^e]{\text{aq NaHCO}_3 \text{ or}}$ $ArOCOPh$

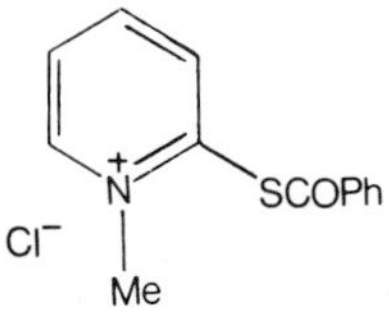

   This reagent forms aryl benzoates under aqueous conditions. (It also acylates amines and carboxylic acids.)[e]

(3)   Monoesterification of a symmetrical dihydroxy aromatic compound can be effected by reaction with polymer-bound benzoyl chloride to give compound 18a, which can be alkylated with diazomethane to form, after basic hydrolysis, a monomethyl ether[f]:

   hydroquinone $\xrightarrow[\text{reflux, 15 h}]{\text{P}\text{-C}_6\text{H}_4\text{COCl/Py, C}_6\text{H}_6{}^f}$ $p\text{-HOC}_6\text{H}_4\text{OCOC}_6\text{H}_4\text{-}p\text{-}\text{P}$

   18a

   Ⓟ- = 2% cross-linked divinylbenzene-styrene copolymer

Cleavage

(1)   **18a** $\xrightarrow[25°, 20\ h, \text{ or } 60°, 3\ h]{0.5\ M\ \text{NaOH, dioxane–H}_2\text{O}^f}$ $ArOH$, quant

(2)   Under anhydrous conditions, cesium carbonate or bicarbonate quantitatively cleaves an aryl dibenzoate or diacetate to the monoester; yields are considerably lower with potassium carbonate.[g]

   $C_6H_4\text{-1,3-(OCOPh)}_2$ $\xrightarrow[\text{reflux, 24 h, } >95\%]{Cs_2CO_3/DME^g}$ $3\text{-PhCO}_2\text{-C}_6\text{H}_4\text{OH}$

[a] M. Gates, *J. Am. Chem. Soc.*, **72**, 228 (1950).

[b] D. D. Pratt and R. Robinson, *J. Chem. Soc.*, 1577 (1922).

[c] A. Robertson and R. Robinson, *J. Chem. Soc.*, 1710 (1927).

[d] P. A. Stadler, *Helv. Chim. Acta*, **61**, 1675 (1978).

[e] M. Yamada, Y. Watabe, T. Sakakibara, and R. Sudoh, *J. Chem. Soc., Chem. Commun.*, 179 (1979).

[f] C. C. Leznoff and D. M. Dixit, *Can. J. Chem.*, **55**, 3351 (1977).

[g] H. E. Zaugg, *J. Org. Chem.*, **41**, 3419 (1976).

19.   Aryl 9-Fluorenecarboxylate:

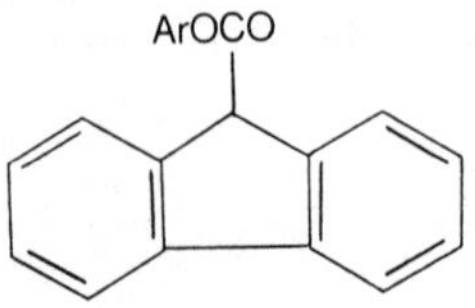

, 19

Formation (→) / Cleavage (←)[a]

$$
\text{ArOH} \quad \xrightarrow[\text{25°, 1 h, 65\%}]{\text{9-fluorenecarbonyl chloride, Py, C}_6\text{H}_6} \quad \textbf{19}
$$

$$
\xleftarrow[\text{reflux, 4 h, 60\%}]{h\nu,\ \text{Et}_2\text{O}}
$$

ArOH = β-naphthol

Aryl xanthenecarboxylates, **a**, were prepared and cleaved in the same way.[a]

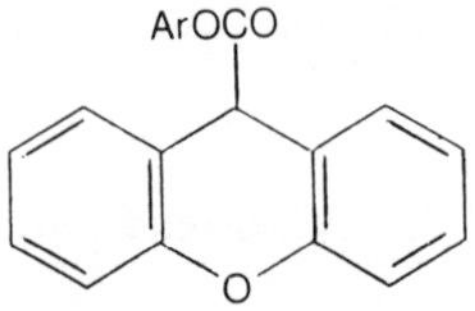

a

[a] D. H. R. Barton, Y. L. Chow, A. Cox, and G. W. Kirby, *J. Chem. Soc.*, 3571 (1965).

Carbonates

20.   Aryl Methyl Carbonate: ArOCOOCH₃, 20

In an early synthesis a methyl carbonate, prepared by reaction of a phenol with methyl chloroformate, was cleaved selectively in the presence of a phenyl ester[a]:

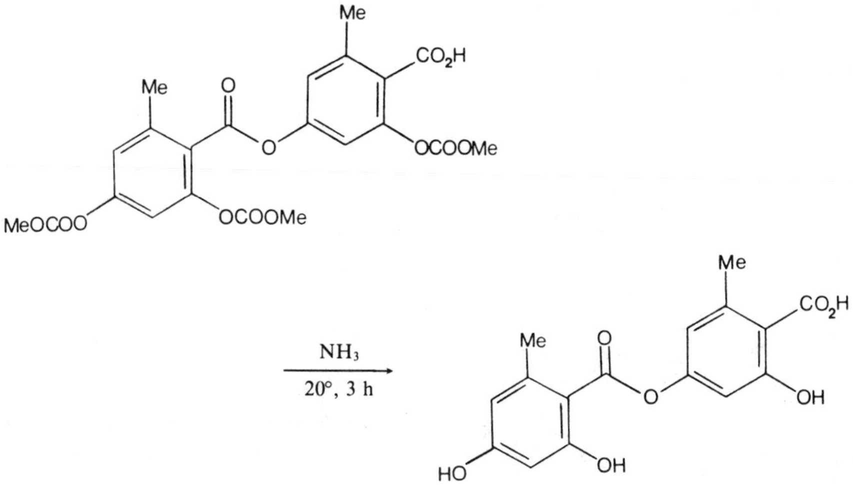

More recently an ethyl carbonate was cleaved by refluxing in acetic acid for 6 h.[b]

[a] E. Fischer and H. O. L. Fischer, *Ber.,* **46**, 1138 (1913).
[b] E. Haslam, R. D. Haworth, and G. K. Makinson, *J. Chem. Soc.,* 5153 (1961).

21.   Aryl 2,2,2-Trichloroethyl Carbonate: $ArOCOOCH_2CCl_3$, 21

Formation[a]

$$ArOH \xrightarrow[\text{25°, 12 h}]{Cl_3CCH_2OCOCl,\ Py\ or\ aq\ NaOH} 21$$

Cleavage

(1) $21 \xrightarrow[\text{25°, 1–3 h}]{Zn/HOAc}$ or $\xrightarrow[\text{heat, few min}]{Zn/CH_3OH} ArOH$[a]

(2) $21 \xrightarrow[\text{25°, 4 h}]{Zn/THF–H_2O,\ pH\ 4.2^b} ArOH$

The authors suggest that selective cleavage should be possible by this method since at pH 4.2, 25°, 2,2,2-trichloroethyl esters are cleaved in 10 min, 2,2,2-trichloroethyl carbamates are cleaved in 30 min, and the 2,2,2-trichloroethyl carbonate of estrone, formed in 87% yield from estrone and the acid chloride, is cleaved in 4 h (97% yield).[b]

[a] T. B. Windholz and D. B. R. Johnston, *Tetrahedron Lett.,* 2555 (1967).
[b] G. Just and K. Grozinger, *Synthesis,* 457 (1976).

22.   Aryl Vinyl Carbonate: ArOCOOCH=CH$_2$, 22

Formation ($\rightarrow$) / Cleavage ($\leftarrow$)^a

$$\xrightarrow[95\%]{CH_2=CHOCOCl,\ Py}$$

ArOH   22

$$\xleftarrow[96\%]{Na_2CO_3,\ warm\ aq\ dioxane}$$

ArOH = phenol, 2-naphthol

Selective protection of an aryl —OH or an amine —NH group is possible by reaction of the compound with vinyl chloroformate. Vinyl carbamates (RR'NCO$_2$CH=CH$_2$) are stable to the basic conditions (Na$_2$CO$_3$) used to cleave vinyl carbonates, **22**. Conversely, vinyl carbonates are stable to the acidic conditions (HBr/CH$_3$OH/CH$_2$Cl$_2$) used to cleave vinyl carbamates. Vinyl carbonates are cleaved by more acidic conditions: 2 N anhyd HCl/dioxane, 25°, 3 h, 10% yield; HBF$_4$, 25°, 12 h, 30% yield; 2 N HCl/CH$_3$OH–H$_2$O(4:1), 60°, 8 h, 100% yield.^{a}

^{a} R. A. Olofson and R. C. Schnur, *Tetrahedron Lett.*, 1571 (1977).

23.   Aryl Benzyl Carbonate: ArOCOOCH$_2$C$_6$H$_5$, 23

Formation ($\rightarrow$) / Cleavage ($\leftarrow$)^a

$$\xrightarrow{PhCH_2OCOCl,\ Py,\ CH_2Cl_2,\ THF}$$

ArOH   23

$$\xleftarrow[20°]{H_2/Pd\text{–}C,\ EtOH}$$

o-Bromobenzyl carbonates have been developed for use in solid-phase peptide synthesis. An aryl *o*-bromobenzyl carbonate is stable to acidic cleavage (CF$_3$CO$_2$H) of a *t*-butyl carbamate; a benzyl carbonate is cleaved. The *o*-bromo derivative is quantitatively cleaved with hydrogen fluoride (0°, 10 min).^{b}

^{a} M. Kuhn and A. von Wartburg, *Helv. Chim. Acta,* **52,** 948 (1969).
^{b} D. Yamashiro and C. H. Li, *J. Org. Chem.,* **38,** 591 (1973).

24. Aryl Carbamate: ArOCONHR, 24

Formation[a]

$$\text{Tyr-OMe} \xrightarrow[\text{60°, 2 h, 65–85\%}]{\text{R'NCO}} \textbf{24a}$$

R' = *i*-Bu, Ph

Cleavage[a]

(1) **24a** $\xrightarrow[\text{20°, 2 h, 78\%}]{2\,N\,\text{NaOH}}$ Tyr-OH

(2) **24a** $\xrightarrow[\text{20°, 3 h, 59–87\%}]{\text{H}_2\text{NNH}_2\cdot\text{H}_2\text{O, DMF}}$ Tyr-NHNH$_2$

[a] G. Jäger, R. Geiger, and W. Siedel, *Chem. Ber.,* **101,** 2762 (1968).

Sulfonates

An aryl methane- or toluenesulfonate ester is stable to reduction with lithium aluminum hydride, to the acidic conditions used for nitration of an aromatic ring (HNO$_3$/HOAc),[a] and to the high temperatures (200–250°) of an Ullman reaction. Aryl sulfonate esters, formed by reaction of a phenol with a sulfonyl chloride in pyridine or aqueous sodium hydroxide, are cleaved by warming in aqueous sodium hydroxide.[b]

[a] E. M. Kampouris, *J. Chem. Soc.,* 2651 (1965).
[b] F. G. Bordwell and P. J. Boutan, *J. Am. Chem. Soc.,* **79,** 717 (1957).

25. Aryl Methanesulfonate: ArOSO$_2$CH$_3$, 25

In a synthesis of decinine a phenol was protected as a methanesulfonate that was stable during an Ullman coupling reaction and during condensation, catalyzed by calcium hydroxide, of an amine with an aldehyde. It was cleaved by warm sodium hydroxide solution.[a]

An aryl methanesulfonate was cleaved to a phenol by phenyllithium or phenylmagnesium bromide[b]; it was reduced to an aromatic hydrocarbon by sodium in liquid ammonia.[c]

[a] I. Lantos and B. Loev, *Tetrahedron Lett.,* 2011 (1975).
[b] J. E. Baldwin, D. H. R. Barton, I. Dainis, and J. L. C. Pereira, *J. Chem. Soc. C,* 2283 (1968).
[c] G. W. Kenner and N. R. Williams, *J. Chem. Soc.,* 522 (1955).

26.   Aryl Toluenesulfonate: ArOSO₂C₆H₄-*p*-CH₃, 26

Formation[a]

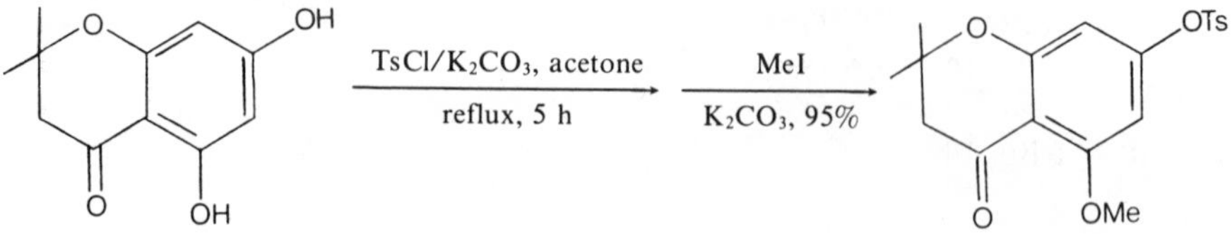

Cleavage[a]

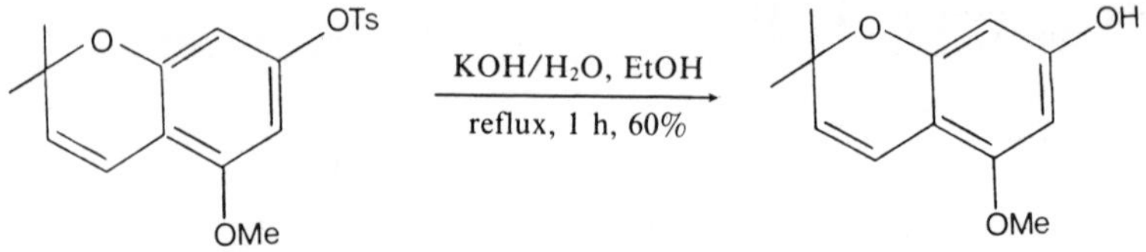

An aryl toluenesulfonate is stable to lithium aluminum hydride (Et₂O, reflux, 4 h) and to *p*-toluenesulfonic acid (C₆H₅CH₃, reflux, 15 min).[a]

o-Aminophenol can be selectively protected as a sulfonate or a sulfonamide[b]:

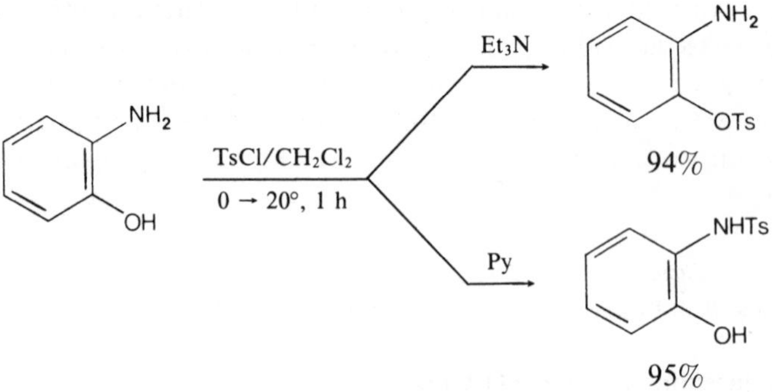

[a] M. L. Wolfrom, E. W. Koos, and H. B. Bhat, *J. Org. Chem.*, **32**, 1058 (1967).
[b] K. Kurita, *Chem. Ind.* (*London*), 345 (1974).

PROTECTION FOR CATECHOLS
(1,2-DIHYDROXYBENZENES)

Catechols can be protected as diethers or diesters by methods that have been described to protect phenols. However formation of cyclic acetals and ketals (e.g., methylenedioxy, acetonide, cyclohexylidenedioxy, and diphenylmethylenedioxy derivatives) or cyclic esters (e.g., borates or carbonates) selectively protects the two adjacent hydroxyl groups in the presence of isolated phenol groups.

CYCLIC ACETALS AND KETALS

27. Methylene Acetal: , **27**

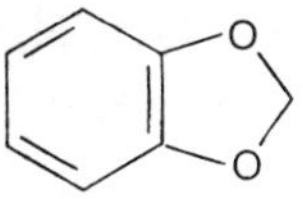

The methylenedioxy group, often present in natural products, is stable to many reagents. Efficient methods for both formation and removal of the group are now available.

Formation

(1) a catechol $\xrightarrow[\text{Adogen, reflux, 3 h, 76–86\%}]{\text{CH}_2\text{Br}_2/\text{NaOH}/\text{H}_2\text{O}^a}$ **27**

   Adogen $= \text{R}_3\text{N}^+\text{CH}_3\text{Cl}^-$, a phase transfer catalyst

   $\text{R} = \text{C}_8\text{–C}_{10}$ straight chain alkyl groups

   Earlier methods required anhydrous conditions and aprotic solvents.[a]

(2) a catechol $\xrightarrow[\text{110}°, \text{1.5 h, 70–98\%}]{\text{CH}_2\text{X}_2, \text{DMF, KF or CsF}^b}$ **27**

   $\text{X} = \text{Cl, Br}$

   Strong base is not necessary.

Cleavage

(1) **27** $\xrightarrow[\text{0}°, \text{0.5–1 h, 73–78\%}]{\text{AlBr}_3, \text{EtSH}^c}$ a catechol

   Aluminum bromide cleaves aryl and alkyl methyl ethers in high yield; methyl esters are stable.[c]
   Selective cleavage of an aryl methylenedioxy group, or an aryl methyl ester, by boron trichloride has been investigated.[d]

(2)

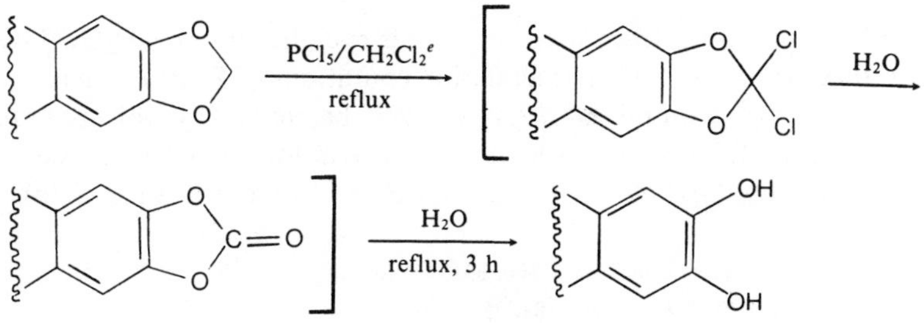

   61%

(3)   A 4-nitro-1,2-methylenedioxybenzene has been cleaved to a catechol with 2 N NaOH, 90°, 30 min[f]; a similar compound substituted with a 4-nitro or 4-formyl group has been cleaved by NaOCH₃/DMSO, 150°, 2.5 min (13–74% catechol, 6–60% recovered starting material).[g]

[a] A. P. Bashall and J. F. Collins, *Tetrahedron Lett.*, 3489 (1975).
[b] J. H. Clark, H. L. Holland, and J. M. Miller, *Tetrahedron Lett.*, 3361 (1976).
[c] M. Node, K. Nishide, M. Sai, K. Ichikawa, K. Fuji, and E. Fujita, *Chem. Lett.*, 97 (1979).
[d] M. Gerecke, R. Borer, and A. Brossi, *Helv. Chim. Acta*, **59**, 2551 (1976).
[e] G. L. Trammell, *Tetrahedron Lett.*, 1525 (1978).
[f] E. Haslam and R. D. Haworth, *J. Chem. Soc.*, 827 (1955).
[g] S. Kobayashi, M. Kihara, and Y. Yamahara, *Chem. Pharm. Bull.*, **26**, 3113 (1978).

28.   Acetonide Derivative: , **28**

Formation (→) / Cleavage (←)[a]

a catechol   **28**

An acetonide has also been cleaved by refluxing in acetic acid–water (100°, 18 h, 90% yield of catechol).[b]

[a] K. Ogura and G.-i. Tsuchihashi, *Tetrahedron Lett.*, 3151 (1971).
[b] E. J. Corey and S. D. Hurt, *Tetrahedron Lett.*, 3923 (1977).

29.   Cyclohexylidene Ketal: , **29**

Compound **29**, prepared from a catechol and cyclohexanone (Al₂O₃/TsOH, CH₂Cl₂, reflux, 36 h),[a] is stable to metalation conditions (RX/BuLi) that cleave aryl methyl ethers.[b] Compound **29** is cleaved by acidic hydrolysis (concd HCl/EtOH, reflux, 1.5 h → 20°, 12 h); it is stable to milder acidic hydrolysis that cleaves tetrahydropyranyl ethers (1 N HCl/EtOH, reflux, 5 h, 91% yield).[c]

[a] G. Schill and E. Logemann, *Chem. Ber.*, **106**, 2910 (1973).
[b] G. Schill and K. Murjahn, *Chem. Ber.*, **104**, 3587 (1971).
[c] J. Boeckmann and G. Schill, *Chem. Ber.*, **110**, 703 (1977).

30. Diphenylmethylene Ketal:

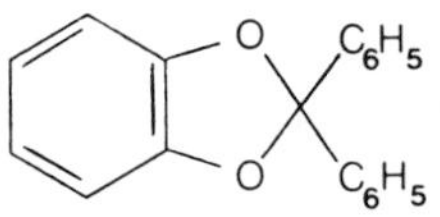

, **30**

Formation (→) / Cleavage (←)

$$\text{Ph}_2\text{CCl}_2,\ \text{Py, acetone}^a$$

a catechol   **30**

$$\xleftarrow{\text{H}_2/\text{Pd–C, THF}^b}$$

Compound **30** has also been prepared from a 1,2,3-trihydroxybenzene (Ph$_2$CCl$_2$, 160°, 5 min, 80% yield) and cleaved by acidic hydrolysis (HOAc, reflux, 7 h).[c]

[a] W. Bradley, R. Robinson, and G. Schwarzenbach, *J. Chem. Soc.,* 793 (1930).

[b] E. Haslam, R. D. Haworth, S. D. Mills, H. J. Rogers, R. Armitage, and T. Searle, *J. Chem. Soc.,* 1836 (1961).

[c] L. Jurd, *J. Am. Chem. Soc.,* **81,** 4606 (1959).

CYCLIC ESTERS

31. Cyclic Borate:

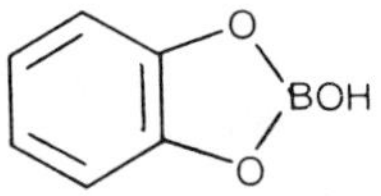

, **31**

A cyclic borate can be used to protect a catechol group during base-catalyzed alkylation or acylation of an isolated phenol group; the borate ester is then readily hydrolyzed by dilute acid.[a]

Formation[a]

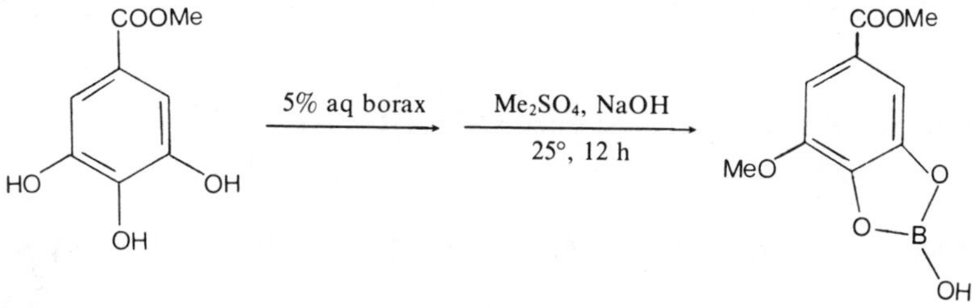

Cleavage[a]

$$\mathbf{a} \xrightarrow[\text{good yield}]{H_2SO_4}$$

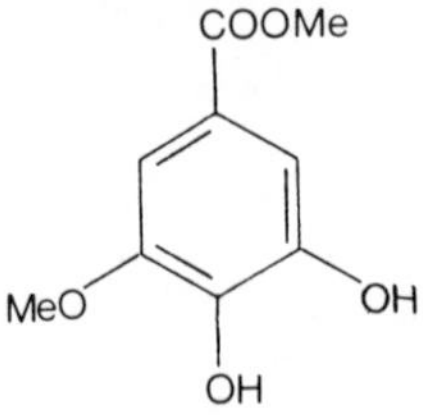

[a] R. R. Scheline, *Acta Chem. Scand.*, **20**, 1182 (1966).

32. Cyclic Carbonate: , **32**

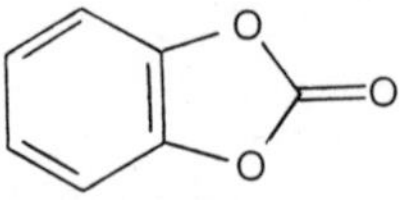

Cyclic carbonates have been used to a limited extent only (since they are readily hydrolyzed) to protect the catechol group in a polyhydroxy benzene.

Formation[a]

1,2,3-trihydroxybenzene $\xrightarrow[\text{NaOH}]{\text{ClCOCl}}$ or $\xrightarrow[\text{heat}]{(PhO)_2CO}$

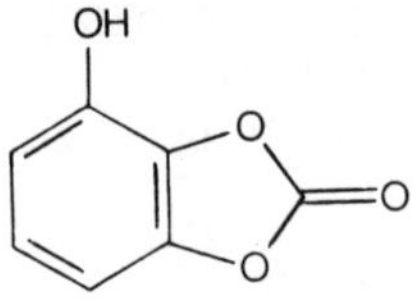

32a

Cleavage

(1) **32a** $\xrightarrow{CH_2N_2}$ $\xrightarrow[\text{reflux, 30 min}]{H_2O^b}$ 2,3-dihydroxyanisole

(2) **32** $\xrightarrow[\text{reflux, 3 h}]{\text{aq Me}_2SO_4, \text{ NaOH}^c}$ 1,2-dimethoxybenzene

[a] A. Einhorn, J. Cobliner, and H. Pfeiffer, *Ber.*, **37**, 100 (1904).
[b] H. Hillemann, *Ber.*, **71**, 34 (1938).
[c] W. Baker, J. A. Godsell, J. F. W. McOmie, and T. L. V. Ulbricht, *J. Chem. Soc.*, 4058 (1953).

Protection for 2-Hydroxybenzenethiols:

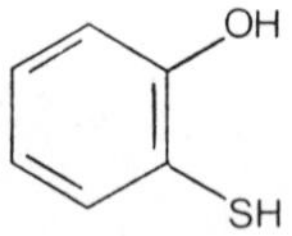

Two derivatives have been prepared which may prove useful as protective groups for 2-hydroxybenzenethiols.

Formation

(1) 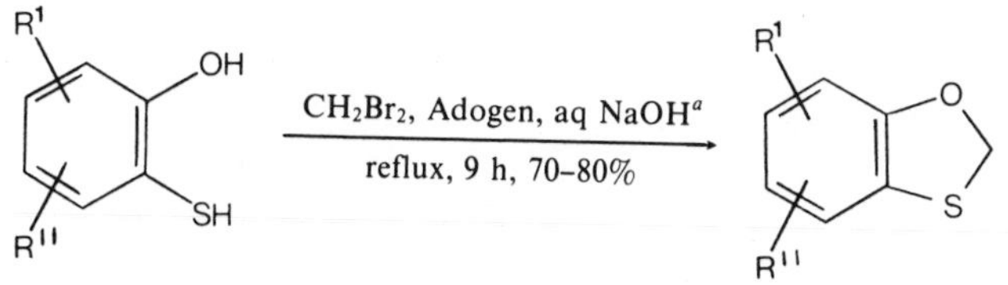
R¹
OH
SH
R''
CH2Br2, Adogen, aq NaOH[a]
reflux, 9 h, 70–80%
R¹
O
S
R''

R′, R″ = H, Me, Cl

Adogen = $Me\overset{+}{N}R_3Cl^-$, phase transfer catalyst

R = C_8–C_{10} straight chain alkyl groups

(2) 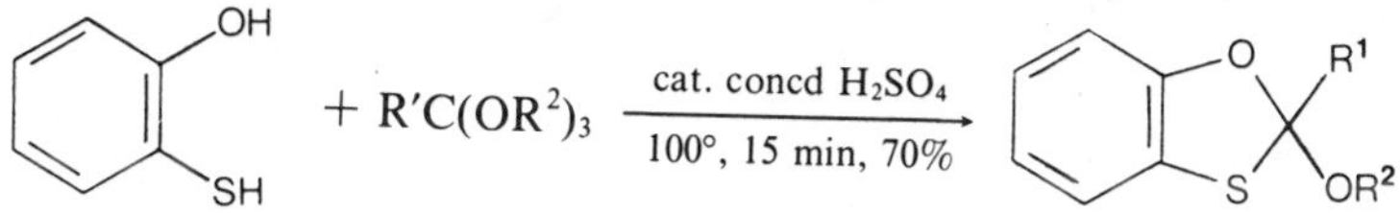
OH
SH
+ R′C(OR2)3
cat. concd H2SO4
100°, 15 min, 70%
O
S
R¹
OR²

$\xrightarrow[\text{reflux, 24 h, 60–75\%}]{R^3MgX/Et_2O^b}$ 2-HO-$C_6H_4SC(OR^2)R^1R^3$

R^1 = H, Me, Ph; R^2 = Me, Et; R^3 = Me, Et, Ph

[a] S. Cabiddu, A. Maccioni, and M. Secci, *Synthesis,* 797 (1976).

[b] S. Cabiddu, S. Melis, L. Bonsignore, and M. T. Cocco, *Synthesis,* 660 (1975).

4

Protection for The Carbonyl Group

*Included in Reactivity Chart 5.

114

*Included in Reactivity Chart 5.

During a synthetic sequence a carbonyl group may have to be protected against attack by various reagents such as strong or moderately strong nucleophiles including organometallic reagents; acidic, basic, catalytic, or hydride reducing agents; and some oxidants. Because of the order of reactivity of the carbonyl group [e.g., aldehydes (aliphatic > aromatic) > acyclic ketones and cyclohexanones $>$ cyclopentanones $>$ α,β-unsaturated ketones or α,α-disubstituted ketones $>>$ aromatic ketones], it may be possible to protect a reactive carbonyl group selectively in the presence of a less reactive one. In keto steroids the order of reactivity to ketalization is C_3 or Δ^4-$C_3 > C_{17} > C_{12} > C_{20} > C_{17,21\text{-}(OH)_2}C_{20} > C_{11}.$[a]

The most useful protective groups are the acyclic and cyclic acetals or ketals, and the acyclic or cyclic thio acetals or ketals. The protective group is introduced by treating the carbonyl compound in the presence of acid with an alcohol, diol, thiol, or dithiol. Cyclic and acyclic acetals and ketals are stable to aqueous and nonaqueous bases, to nucleophiles including organometallic reagents, and to hydride reduction. A 1,3-dithiane or 1,3-dithiolane, prepared to protect an aldehyde, is converted by strong base to an anion. The oxygen derivatives are stable to neutral and basic catalytic reduction, and to reduction by sodium in ammonia. Although the sulfur analogs poison hydrogenation catalysts, they can be cleaved by Raney Ni and by sodium/ammonia. The oxygen derivatives are stable to most oxidants; the sulfur derivatives are cleaved by a wide range of oxidants. The oxygen, but not the sulfur, analogs are readily cleaved by acidic hydrolysis. Sulfur derivatives are cleaved under neutral conditions by mercury(II), silver(I), or copper(II) salts; oxygen analogs are stable to these conditions. The properties of oxygen and sulfur derivatives are combined in the cyclic 1,3-oxathianes and 1,3-oxathiolanes.

The carbonyl group forms a number of other very stable derivatives. They are less used as protective groups because of the greater difficulty involved in their removal. Such derivatives include cyanohydrins, hydrazones, imines, oximes, and semicarbazones. Enol ethers are used to protect one carbonyl group in a 1,2- or 1,3-dicarbonyl compound.

Derivatives of carbonyl compounds that have been used as protective groups in synthetic schemes are described in this chapter; the more important protective groups are listed in Reactivity Chart 5. Conventions that are used in this chapter are explained on p. xii.[b,c]

---

[a] H. J. E. Loewenthal, *Tetrahedron,* **6,** 269 (1959).

[b] See also: H. J. E. Loewenthal, "Protection of Aldehydes and Ketones," in *Protective Groups in Organic Chemistry,* J. F. W. McOmie, Ed., Plenum, New York and London, 1973, pp. 323–402.

[c] J. F. W. Keana, in *Steroid Reactions,* C. Djerassi, Ed., Holden-Day, San Francisco, 1963, pp. 1–66, 83–87.

ACETALS AND KETALS

Acyclic Acetals and Ketals

Methods similar to those used to form and cleave dimethyl acetal and ketal derivatives can be used for other dialkyl acetals and ketals.

1. Dimethyl Acetals and Ketals: RR'C(OCH₃)₂, 1

Formation (→) / Cleavage (←)

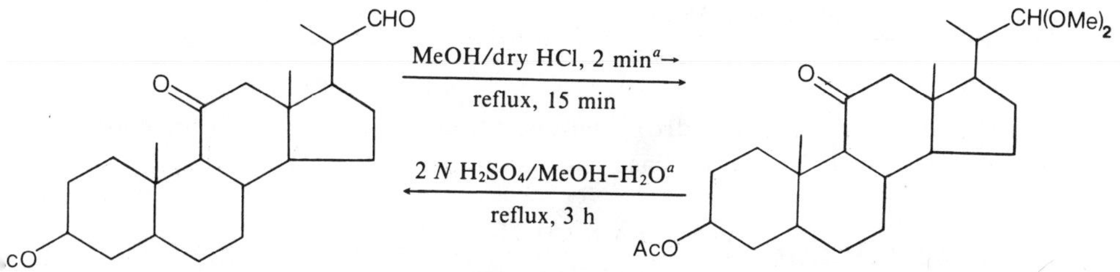

(2) C_4-C_8 cycloalkanone $\xrightarrow[25°, \text{3 days}]{\text{MeOH/(MeO)}_4\text{Si, dry HCl}^b}$ 1

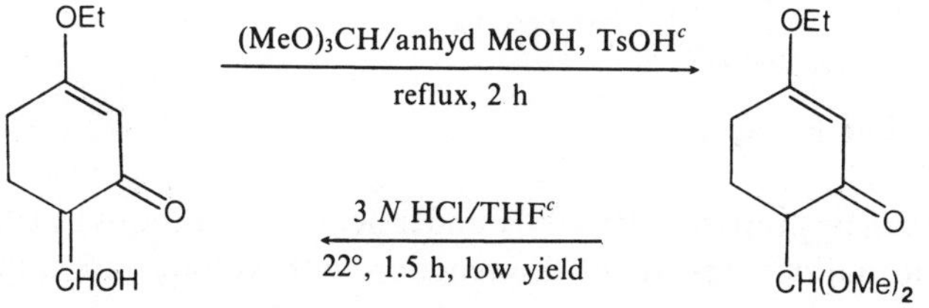

Diethyl ketals of cyclohexanones have been prepared under similar conditions (EtOH, TsOH, 0–23°, 15 min to 6 h, 80–95% yield) in the presence of molecular sieves to force the equilibrium by adsorbing water.[d]

(4)   cyclohexanone $\xrightarrow[\substack{\text{acidic ion-exchange resin}^e \\ -28°, 86\%; 24°, 7-46\%}]{\text{MeOH}}$ 1

(5)   RR'CO + (MeO)₃CH $\xrightarrow[\text{5 min to 15 h}]{\text{Montmorillonite Clay K-10}^f}$ 1, >90%

R = H, alkyl, Ph

R' = H, (—CH₂—)₅, (—CH₂—)₁₁, Ph, PhCH₂, PhCH=CH

Montmorillonite Clay = activated $Al_2O_3/SiO_2/H_2O$

Diethyl ketals have been prepared in satisfactory yield by reaction of the carbonyl compound and ethanol in the presence of Montmorillonite Clay.[g]

(6)

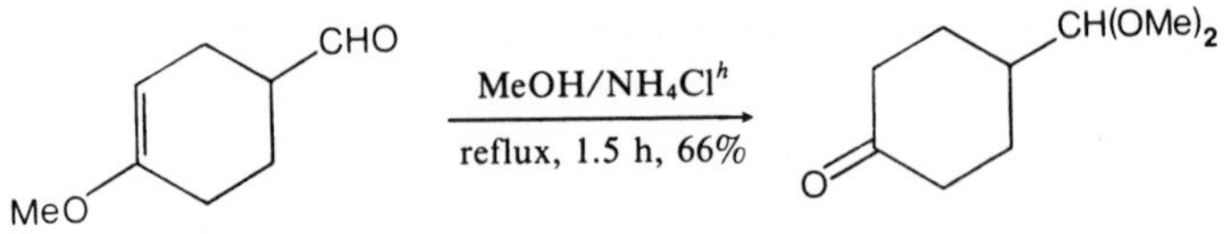

$$\text{(structure with CHO)} \xrightarrow[\text{reflux, 1.5 h, 66\%}]{\text{MeOH/NH}_4\text{Cl}^h} \text{(structure with CH(OMe)}_2\text{)}$$

(7) $\text{RCHO} \xrightarrow[\text{25}°, \text{ 15 min, 75–85\%}]{\text{MeOH, PhSO}_2\text{NHOH}^i} \text{RCH(OMe)}_2$

In the absence of base N-hydroxybenzenesulfonamide catalyzes the formation of dimethyl acetals. In strong base hydroxamic acids are formed.[i]

(8) $m\text{-NO}_2\text{-C}_6\text{H}_4\text{CHO} \xrightarrow[\text{reflux, 30 min, 85\%}]{\text{Me}_2\text{SO}_4/2 \ N \ \text{NaOH, MeOH–H}_2\text{O}^j} \mathbf{1}$

Dialkyl acetals and ketals of some substrates can be formed under alkaline conditions.[j]

(9) $\text{RCHO} \xrightarrow[\text{25}°, \text{ 10 min, 80–100\%}]{\text{MeOH/LaCl}_3*/(\text{MeO})_3\text{CH}^k} \text{RCH(OMe)}_2$

R = alkyl, Ph, PhCH=CH, . . .

Dimethyl acetals can be prepared efficiently under neutral conditions by catalysis with lanthanoid halides; results of the reaction with ketones are unpredictable.[k]

Cleavage

Examples are shown below of the classical method of acidic hydrolysis used to cleave acetals and ketals; some newer reagents are also described.

(1)

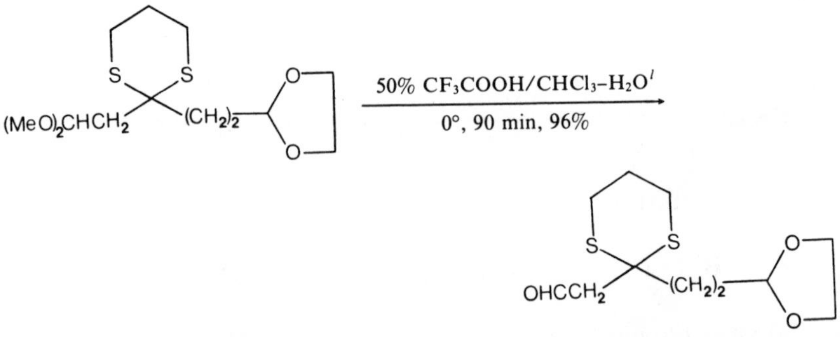

$$\text{(MeO)}_2\text{CHCH}_2\text{-(dithiane)-(CH}_2)_2\text{-(dioxolane)} \xrightarrow[\text{0}°, \text{ 90 min, 96\%}]{\text{50\% CF}_3\text{COOH/CHCl}_3\text{–H}_2\text{O}^l} \text{OHCCH}_2\text{-(dithiane)-(CH}_2)_2\text{-(dioxolane)}$$

Note that a dimethyl acetal is selectively hydrolyzed by trifluoroacetic acid in the presence of a 1,3-dioxolane and a 1,3-dithiane.[l]

---

*Or CeCl_3, NdCl_3, ErCl_3, YbCl_3.

(2)

(3) $RR'C(OMe)_2 + SiO_2$ $\xrightarrow[\text{0.5 - 24 h, 90–95\%}]{\text{H}_2\text{O or 10\% (HOOC)}_2 \text{ or 15\% H}_2\text{SO}_4{}^n}$ **1**

  R = cyclopropenyl

(4) $RR'C(OMe)_2$ $\xrightarrow[\text{25°, 15 min, 85–95\%}]{\text{Me}_3\text{SiI/CH}_2\text{Cl}_2{}^o}$ $RR'CO + MeOSiMe_3 + MeI$

Under these nonaqueous cleavage conditions 1,3-dithiolanes, alkyl and tri-methylsilyl enol ethers, and enol acetates are stable. 1,3-Dioxolanes give complex mixtures. Alcohols, epoxides, triphenylmethyl, *t*-butyl, and benzyl ethers and es-ters are reactive. Most other ethers and esters, amines, amides, ketones, olefins, acetylenes, and halides are expected to be stable.[o]

[a] A. F. B. Cameron, J. S. Hunt, J. F. Oughton, P. A. Wilkinson, and B. M. Wilson, *J. Chem. Soc.*, 3864 (1953).

[b] W. W. Zajac and K. J. Byrne, *J. Org. Chem.*, **35**, 3375 (1970).

[c] E. Wenkert and T. E. Goodwin, *Synth. Commun.*, **7**, 409 (1977).

[d] D. P. Roelofsen, E. R. J. Wils, and H. Van Bekkum, *Recl. Trav. Chim. Pays-Bas*, **90**, 1141 (1971).

[e] N. B. Lorette, W. L. Howard, and J. H. Brown, Jr., *J. Org. Chem.*, **24**, 1731 (1959).

[f] E. C. Taylor and C.-S. Chiang, *Synthesis*, 467 (1977).

[g] V. M. Thuy and P. Maitte, *Bull. Soc. Chim. Fr.*, 2558 (1975).

[h] J. I. DeGraw, L. Goodman, and B. R. Baker, *J. Org. Chem.*, **26**, 1156 (1961).

[i] A. Hassner, R. Wiederkehr, and A. J. Kascheres, *J. Org. Chem.*, **35**, 1962 (1970).

[j] E. Schmitz, *Chem. Ber.*, **91**, 410 (1958).

[k] A. L. Gemal and J.-L. Luche, *J. Org. Chem.*, **44**, 4187 (1979).

[l] R. A. Ellison, E. R. Lukenbach, and C.-W. Chiu, *Tetrahedron Lett.*, 499 (1975).

[m] E. W. Colvin, R. A. Raphael, and J. S. Roberts, *J. Chem. Soc., Chem. Commun.*, 858 (1971).

[n] F. Huet, A. Lechevallier, M. Pellet, and J. M. Conia, *Synthesis*, 63 (1978).

[o] M. E. Jung, W. A. Andrus, and P. L. Ornstein, *Tetrahedron Lett.*, 4175 (1977).

2.   Diethyl Acetals and Ketals: $RR'C(OC_2H_5)_2$, 2

See also discussion of Acyclic Acetals and Ketals.

Formation

(1)   cyclohexanone $\xrightarrow[\text{0–5°, 3 h, 100\%}]{\text{(EtO)}_3\text{CH/Amberlyst-15}{}^a}$ **2**

Amberlyst-15, an acidic ion-exchange resin, is an efficient catalyst for acetal and ketal formation under nonaqueous conditions.

(2) $RR'CO \xrightarrow[\text{3-6 h, 65-85\%}]{\text{(EtO)}_3\text{CH/graphite bisulfate}^b}$ **2**

In one example an ethyl enol ether was formed.

[a] S. A. Patwardhan and S. Dev, *Synthesis,* 348 (1974).

[b] J. P. Alazard, H. B. Kagan, and R. Setton, *Bull. Soc. Chim. Fr.,* 499 (1977).

3. Bis(2,2,2-trichloroethyl) Acetals and Ketals: RR'C(OCH₂CCl₃)₂, 3

Formation[a]

$RR'C(OR'')_2$ + TsOH/C₆H₆, reflux

1.5 eq Cl₃CCH₂OH → $RR'C(OR'')OCH_2CCl_3$ 50–80%

4 eq Cl₃CCH₂OH → **3**, 45–75%

R″ = Me or Et

Cleavage[a]

3 $\xrightarrow[\text{reflux, 3-12 h}]{\text{Zn/EtOAc or THF}}$ RR'CO, 40–100%

It is more efficient to make compound **3** by an exchange reaction with a dimethyl or diethyl acetal or ketal than directly from the carbonyl compound. Compound **3** is cleaved by Zn/EtOAc under nonacidic, aprotic conditions, as well as by acid hydrolysis.[a]

[a] J. L. Isidor and R. M. Carlson, *J. Org. Chem.,* **38**, 554 (1973).

4. Dibenzyl Acetals and Ketals: RR'C(OCH₂C₆H₅)₂, 4

Formation

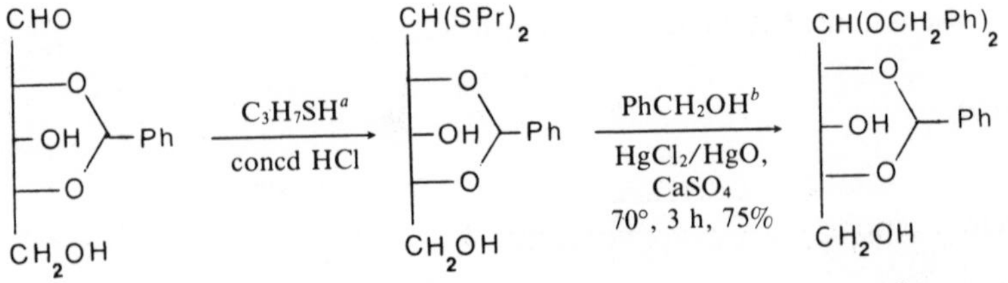

4a

Cleavage[b]

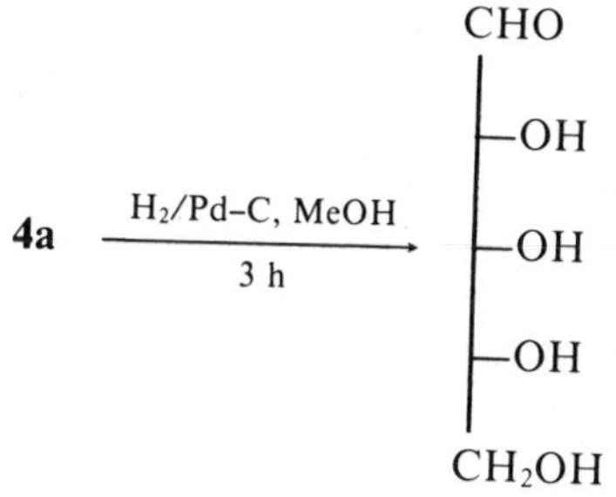

$$\mathbf{4a} \xrightarrow[\text{3 h}]{\text{H}_2/\text{Pd–C, MeOH}} \begin{array}{c} \text{CHO} \\ | \\ \text{—OH} \\ | \\ \text{—OH} \\ | \\ \text{—OH} \\ | \\ \text{CH}_2\text{OH} \end{array}$$

[a] H. Zinner, *Chem. Ber.*, **83**, 275 (1950).

[b] J. H. Jordaan and W. J. Serfontein, *J. Org. Chem.*, **28**, 1395 (1963).

Diacetyl Ketal: RR′C(OCOCH₃)₂, A

Formation[a]

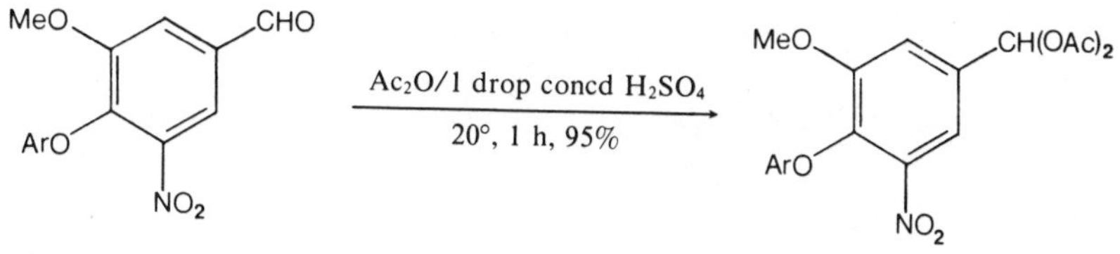

Ac₂O/1 drop concd H₂SO₄

20°, 1 h, 95%

Cleavage[a]

NaNO₂/HOAc–H₂SO₄

0°, 4 h, 72%; NaH₂PO₃·H₂O

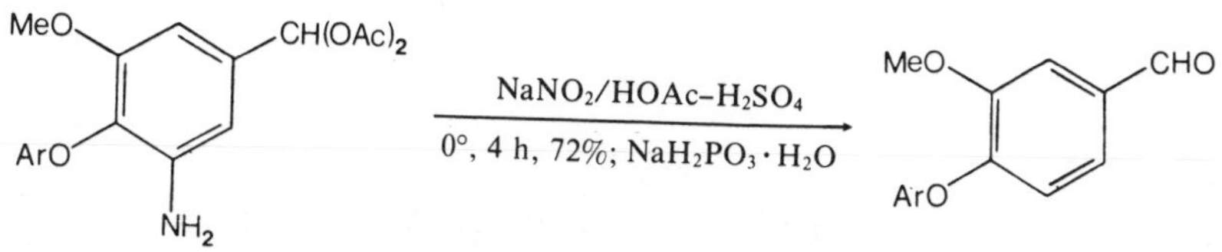

[a] M. Tomita, T. Kikuchi, K. Bessho, T. Hori, and Y. Inubushi, *Chem. Pharm. Bull.*, **11**, 1484 (1963).

Cyclic Acetals and Ketals

Kinetic studies of acetal/ketal formation from cyclohexanone and hydrolysis (3×10^{-3} N HCl/dioxane–H₂O, 20°) indicate the following orders of reactivity[a]:

(1) **Formation:** HOCH₂C(CH₃)₂CH₂OH > HO(CH₂)₂OH > HO(CH₂)₃OH

(2) 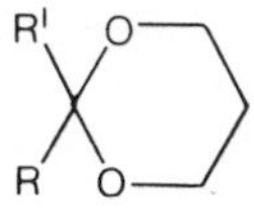 is hydrolyzed faster than

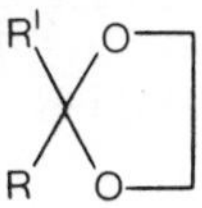

(3) 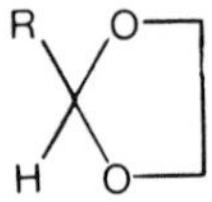 [structure] is hydrolyzed faster than [structure]

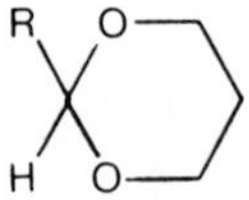

A review[b] discusses the condensation of aldehydes and ketones with glycerol to give 1,3-dioxanes and 1,3-dioxolanes.

[a] M. S. Newman and R. J. Harper, *J. Am. Chem. Soc.,* **80,** 6350 (1958); S. W. Smith and M. S. Newman, *J. Am. Chem. Soc.,* **90,** 1249, 1253 (1968).

[c] A. J. Showler and P. A. Darley, *Chem. Rev.,* **67,** 427–440 (1967).

5.   1,3-Dioxanes: $RR'C \begin{smallmatrix} O \\ \\ O \end{smallmatrix} (CH_2)_3$, **5**

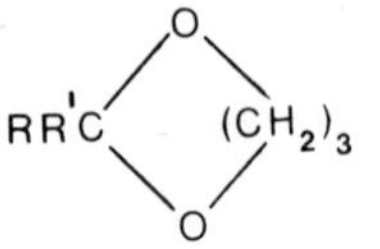

Formation

(1)

$$\xrightarrow[\text{reflux, 35 min, 61\%}]{\text{HO(CH}_2)_3\text{OH/TsOH, C}_6\text{H}_6{}^a}$$

MeO

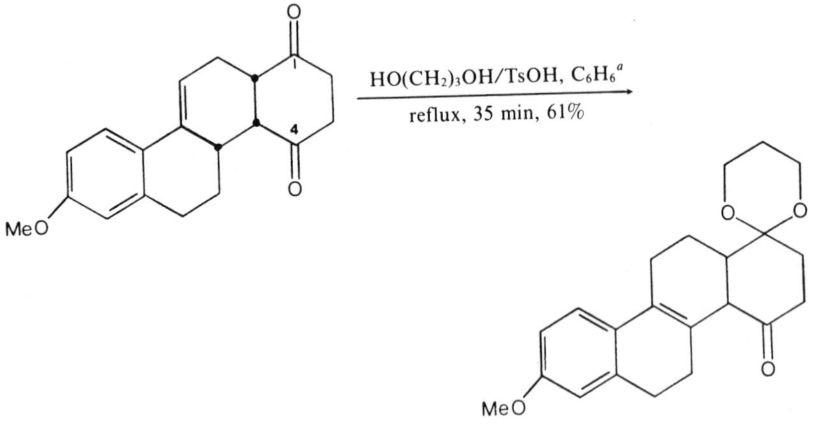

Selective monoprotection of the C_1-carbonyl group was more successful with 1,3-propanediol than with ethylene glycol.[a]

(2)   RR'CO $\xrightarrow[\text{Amberlyst-15, 5 min}]{\text{HO(CH}_2)_3\text{OH/THF}^b}$ **5,** 50–70%

1,3-Dioxanes and 1,3-dioxolanes are readily prepared by passing a mixture of ketone and 1,3-propanediol or ethylene glycol, respectively, through a column packed with Amberlyst-15, an acid ion-exchange resin.[b]

Cleavage

Conditions used to cleave 1,3-dioxolanes, **8,** should also be considered for methods to cleave 1,3-dioxanes.

(1) 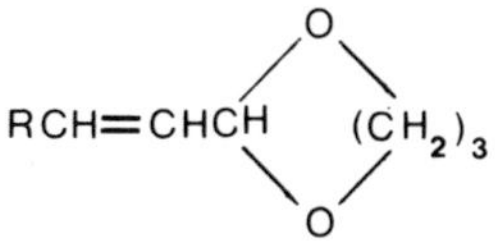 $RCH=CHCH$ $\underset{O}{\overset{O}{\diamond}}$ $(CH_2)_3$ $\xrightarrow[\text{20°, 20 min, 78\%}]{\text{2.5\% aq HCl/acetone}^c}$ $RCH=CHCHO$

[a] J. E. Cole, W. S. Johnson, P. A. Robins, and J. Walker, *J. Chem. Soc.*, 244 (1962).

[b] A. E. Dann, J. B. Davis, and M. J. Nagler, *J. Chem. Soc., Perkin Trans. 1*, 158 (1979).

[c] S. P. Tanis and K. Nakanishi, *J. Am. Chem. Soc.*, **101**, 4398 (1979).

6.   5-Methylene-1,3-dioxane:

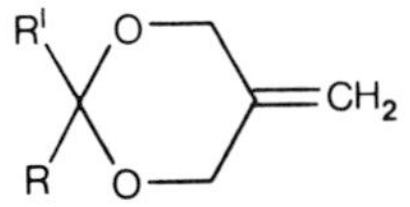

$\underset{R}{\overset{R^I}{\diamond}}\underset{O}{\overset{O}{\diamond}}=CH_2$, **6**

Formation[a]

$RR'CO$ $\xrightarrow[\text{reflux, 90\%}]{(HOCH_2)_2C=CH_2, \text{ TsOH/C}_6\text{H}_6}$ **6**

Cleavage[a]

6 $\xrightarrow[\text{reflux, 3 h, 96\%}]{\text{cat. RhCl(PPh}_3)_3, \text{ aq EtOH}}$ $\underset{R}{\overset{R^I}{\diamond}}\underset{O}{\overset{O}{\diamond}}$ $\xrightarrow[\text{HgCl}_2/\text{HgO}]{H_3O^+ \text{ or}}$ $RR'CO$
98%

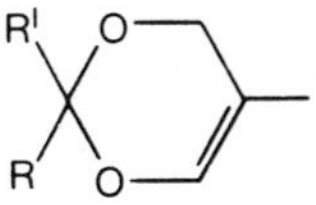

6 $\xrightarrow[\text{0°, 2 min}]{\text{Ph}_3\text{C}^+\text{BF}_4^-, \text{ CH}_2\text{Cl}_2}$ $\xrightarrow{H_2O}$ $RR'CO, 86\%$

6 $\xrightarrow{\text{OsO}_4/\text{NaIO}_4}$ $\underset{R}{\overset{R^I}{\diamond}}\underset{O}{\overset{O}{\diamond}}=O$ $\xrightarrow[\text{25°, 4 h}]{\text{Al/Hg, aq THF}}$ $RR'CO, 80\%$

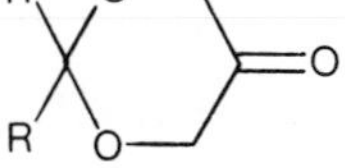

6 $\xrightarrow[\text{25°, 14 h}]{m\text{-Cl-C}_6\text{H}_4\text{CO}_3\text{H}, \text{ CH}_2\text{Cl}_2}$ $\underset{R}{\overset{R^I}{\diamond}}\underset{O}{\overset{O}{\diamond}}\overset{O}{\diamond}$ $\xrightarrow[\text{0°, 5 min}]{\text{BF}_3\cdot\text{Et}_2\text{O}}$

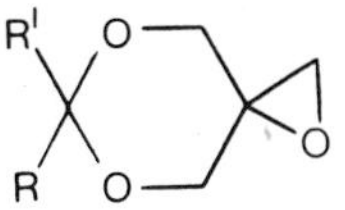

$\left[\underset{R}{\overset{R^I}{\diamond}}\underset{O}{\overset{O}{\diamond}}-CHO\right]$ $\xrightarrow{Py}$ $\xrightarrow{H_2O}$ $RR'CO, 80\%$

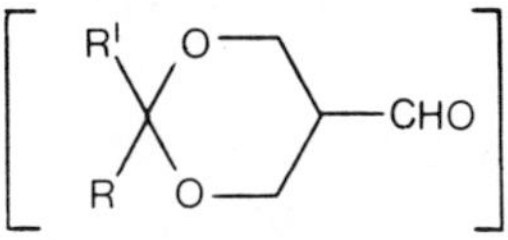

[a] E. J. Corey and J. W. Suggs, *Tetrahedron Lett.*, 3775 (1975).

7.   5,5-Dibromo-1,3-dioxane:

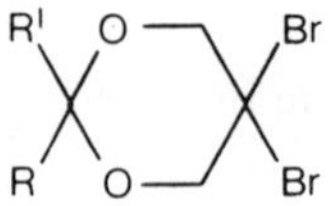

, **7**

Formation (→) / Cleavage (←)[a]

$$\underset{\text{heat, several hours, 84–94\%}}{\overset{\text{(HOCH}_2)_2\text{CBr}_2,\ \text{cat. TsOH, C}_6\text{H}_6}{\longrightarrow}}$$

RR′CO   **7**

$$\underset{25°,\ 1\ \text{h},\ \sim 90\%}{\overset{\text{Zn-Ag, THF/HOAc}}{\longleftarrow}}$$

[a] E. J. Corey, E. J. Trybulski, and J. W. Suggs, *Tetrahedron Lett.,* 4577 (1976).

8.   1,3-Dioxolanes: RR′C (CH$_2$)$_2$, **8**

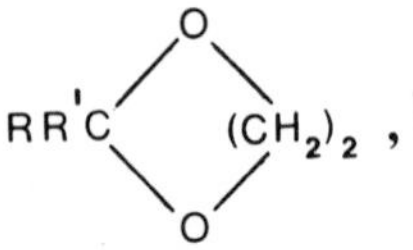

1,3-Dioxolanes are among the most widely used protective groups for carbonyl compounds, and a great variety of techniques have been developed for their formation:

Formation

HO(CH$_2$)$_2$OH/TsOH, C$_6$H$_6$, reflux, 75–85% yield[a]
HO(CH$_2$)$_2$OH/TsOH, (EtO)$_3$CH, 25°, 65% yield[b]
HO(CH$_2$)$_2$OH/BF$_3$·Et$_2$O, HOAc, 35–40°, 15 min, 90% yield[c]
HO(CH$_2$)$_2$OH/HCl, 25°, 12 h, 55–90% yield[d]

For acid-sensitive compounds

HO(CH$_2$)$_2$OH/oxalic acid, CH$_3$CN, 25°, 95% yield[e]
HO(CH$_2$)$_2$OH/adipic acid, C$_6$H$_6$, reflux, 17–24 h, 10–85% yield[f]
HO(CH$_2$)$_2$OH/SeO$_2$, CHCl$_3$, 28°, 4 h, 60% yield[g]
HO(CH$_2$)$_2$OH/C$_5$H$_5$N̟H Cl⁻, C$_6$H$_6$, reflux, 6 h, 85% yield[h]
HO(CH$_2$)$_2$OH/C$_5$H$_5$N̟H⁻OTs, C$_6$H$_6$, reflux, 1–3 h, 90–95% yield[i]
HO(CH$_2$)$_n$OH (n = 2, 3)/MeOC̟HNMe$_2$ MeSO$_4$⁻, 0–25°, 2 h, 40–95% yield[j]
HO(CH$_2$)$_n$OH (n = 2, 3)/column packed with an acid ion-exchange resin, 5 min, 50–90% yield[k]

Acid-catalyzed exchange dioxolanation

2-Methoxy-1,3-dioxolane/TsOH, C_6H_6, 40–50°, 4 h, 85% yield[l]

2-Ethyl-2-methyl-1,3-dioxolane/TsOH, reflux, 75% yield[m]

2-Dimethylamino-1,3-dioxolane/cat. HOAc, CH_2Cl_2, 83% yield[n]

Diethylene orthocarbonate, $C(-OCH_2CH_2O-)_2$,/TsOH or wet $BF_3 \cdot Et_2O$, $CHCl_3$, 20°, 70–95% yield[o]

2-Dimethylamino-1,3-dioxolane protects a reactive ketone under mild conditions: it reacts selectively with a C_3-keto steroid in the presence of a Δ^4-3-keto steroid. C_{12}- and C_{20}- keto steroids do not react.[n]

Diethylene orthocarbonate is useful for the protection of o-hydroxybenzaldehydes.[o]

1,3-Dioxolanes have been prepared from a carbonyl compound and an epoxide (e.g., ketone/$SnCl_4$, CCl_4, 20°, 4 h, 53% yield[p] or aldehyde/Et_4NBr^-, 125–220°, 2–4 h, 20–85% yield[q]). Perhalo ketones can be protected by reaction with ethylene chlorohydrin under basic conditions (K_2CO_3, pentane, 25°, 2 h, 85% yield[r] or NaOH, EtOH–H_2O, 95% yield[s]).

[a] R. A. Daignault and E. L. Eliel, *Org. Synth., Collect. Vol. V,* 303 (1973).

[b] F. F. Caserio and J. D. Roberts, *J. Am. Chem. Soc.,* **80,** 5837 (1958).

[c] L. F. Fieser and R. Stevenson, *J. Am. Chem. Soc.,* **76,** 1728 (1954).

[d] E. G. Howard and R. V. Lindsey, *J. Am. Chem. Soc.,* **82,** 158 (1960).

[e] N. H. Anderson and H.-S. Uh, *Synth. Commun.,* **3,** 125 (1973).

[f] J. J. Brown, R. H. Lenhard, and S. Bernstein, *J. Am. Chem. Soc.,* **86,** 2183 (1964).

[g] E. P. Oliveto, H. Q. Smith, C. Gerold, L. Weber, R. Rausser, and E. B. Hershberg, *J. Am. Chem. Soc.,* **77,** 2224 (1955).

[h] F. T. Bond, J. E. Stemke, and D. W. Powell, *Synth. Commun.,* **5,** 427 (1975).

[i] R. Sterzycki, *Synthesis,* 724 (1979).

[j] W. Kantlehner and H.-D. Gutbrod, *Liebigs Ann. Chem.,* 1362 (1979).

[k] A. E. Dann, J. B. Davis, and M. J. Nagler, *J. Chem. Soc., Perkin Trans. 1,* 158 (1979).

[l] B. Glatz, G. Helmchen, H. Muxfeldt, H. Porcher, R. Prewo, J. Senn, J. J. Stezowski, R. J. Stojda, and D. R. White, *J. Am. Chem. Soc.,* **101,** 2171 (1979).

[m] H. J. Dauben, B. Löken, and H. J. Ringold, *J. Am. Chem. Soc.,* **76,** 1359 (1954).

[n] H. Vorbrueggen, *Steroids,* **1,** 45 (1963).

[o] D. H. R. Barton, C. C. Dawes, and P. D. Magnus, *J. Chem. Soc., Chem. Commun.,* 432 (1975).

[p] J. L. E. Erickson and F. E. Collins, *J. Org. Chem.,* **30,** 1050 (1965).

[q] F. Nerdel, J. Buddrus, G. Scherowsky, D. Klamann, and M. Fligge, *Liebigs Ann. Chem.,* **710,** 85 (1967).

[r] H. E. Simmons and D. W. Wiley, *J. Am. Chem. Soc.,* **82,** 2288 (1960).

[s] R. J. Stedman, L. D. Davis, and L. S. Miller, *Tetrahedron Lett.,* 4915 (1967).

Cleavage

1,3-Dioxolanes, **8,** can be cleaved by acid-catalyzed exchange dioxolanation, acid-catalyzed hydrolysis, or oxidation. Some representative examples are shown below.

Acid-Catalyzed Exchange Dioxolanation

(1) 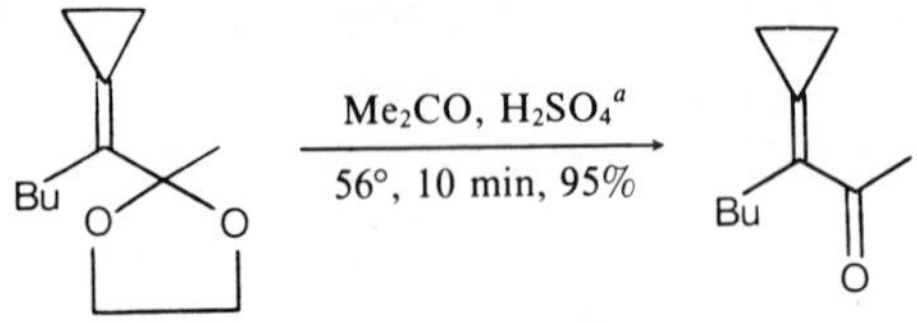
$$\xrightarrow[\text{56°, 10 min, 95\%}]{\text{Me}_2\text{CO, H}_2\text{SO}_4{}^a}$$

(2) $RR'C\underset{O}{\overset{O}{\big<}}(CH_2)_2 \xrightarrow[20°, 12\ h]{\text{Me}_2\text{CO, TsOH}^b} RR'CO$

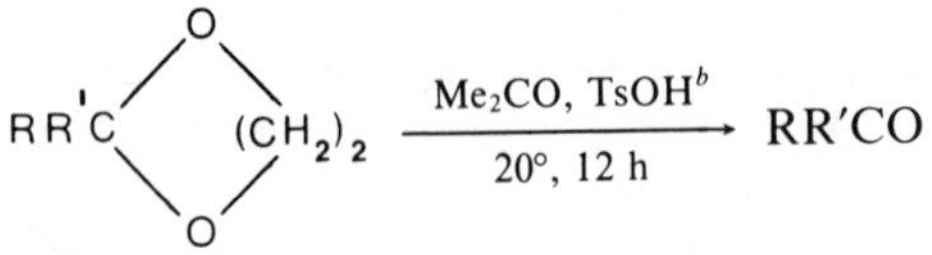

The reactant is a 3,6,17-tris(ethylenedioxy) steroid; the product has carbonyl groups at C_6 and C_{17}.[b]

(3)   **8** $\xrightarrow[\text{reflux, 1–3 h, 90–95\%}]{\text{Me}_2\text{CO–H}_2\text{O, Py}\cdot\text{TsOH}^c} RR'CO$

Acid-Catalyzed Hydrolysis

(1) 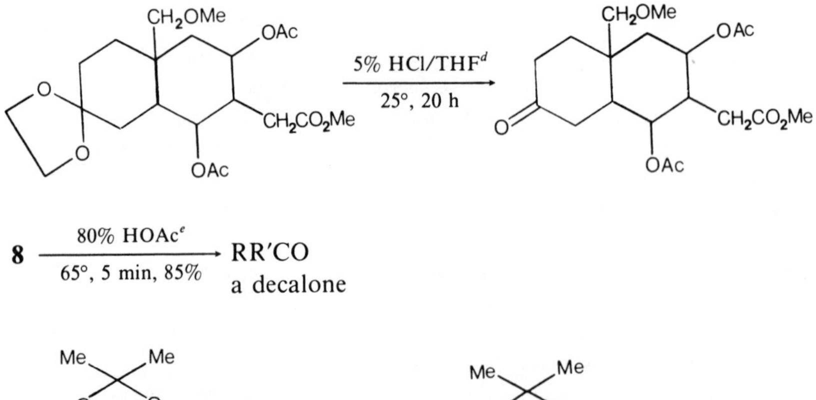
$$\xrightarrow[\text{25°, 20 h}]{\text{5\% HCl/THF}^d}$$

(2)   **8** $\xrightarrow[\text{65°, 5 min, 85\%}]{\text{80\% HOAc}^e} RR'CO$ a decalone

(3) 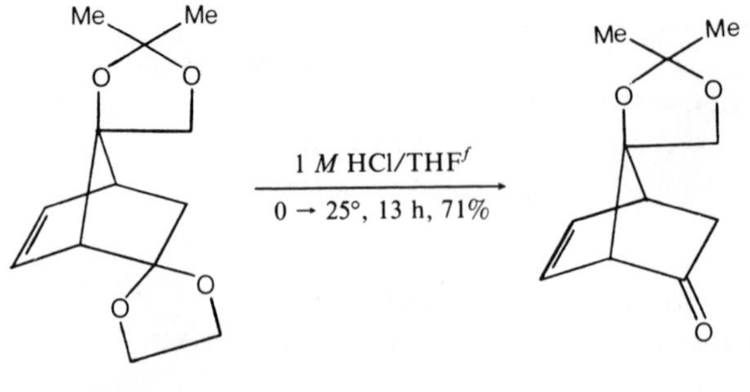
$$\xrightarrow[\text{0} \rightarrow \text{25°, 13 h, 71\%}]{\text{1 }M\text{ HCl/THF}^f}$$

Note that an acetonide was stable to 1 M HCl in this case.[f] Some variations have been reported in this system (including the use of 30% HOAc, 90°, high yield).[g]

(4) **8** $\xrightarrow[\;90\text{–}95\%\;]{\text{SiO}_2\text{–}\text{H}_2\text{O}/\text{CH}_2\text{Cl}_2 \text{ or } (\text{HOOC})_2 \text{ or } \text{H}_2\text{SO}_4{}^h}$ RR'CO

In the absence of SiO_2 compound **8** is cleaved in lower yields (65–75%).

(5) Wet magnesium sulfate (C_6H_6, 20°, 1 h) effects selective, quantitative cleavage of a 1,3-dioxolane in the presence of an α,β-unsaturated 1,3-dioxolane.[i]

(6) Aqueous tartaric acid (25°, 24 h) cleaves a 1,3-dioxolane (69% yield) in the presence of an acetate.[j]

(7) Perchloric acid (70% $HClO_4$/CH_2Cl_2, 0°, 1 h → 25°, 3 h, 87% yield)[k] and periodic acid (aq dioxane, 3 h, quant yield)[l] cleave 1,3-dioxolanes; the latter drives the reaction to completion by oxidation of the ethylene glycol that forms. Yields are substantially higher from cleavage with perchloric acid (3 N $HClO_4$/THF, 25°, 3 h, 80% yield) than with hydrochloric acid (HCl/HOAc, 65% yield).[m]

Oxidative Cleavage

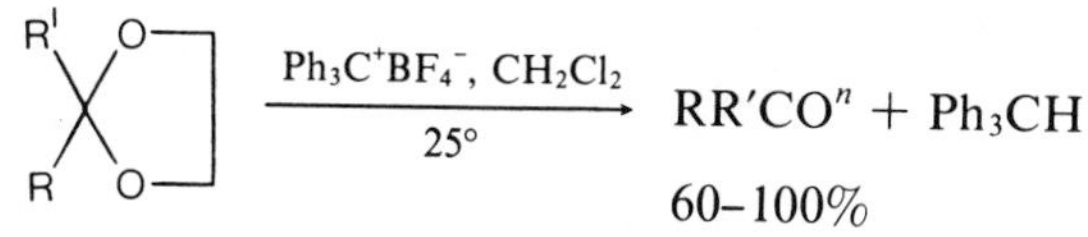

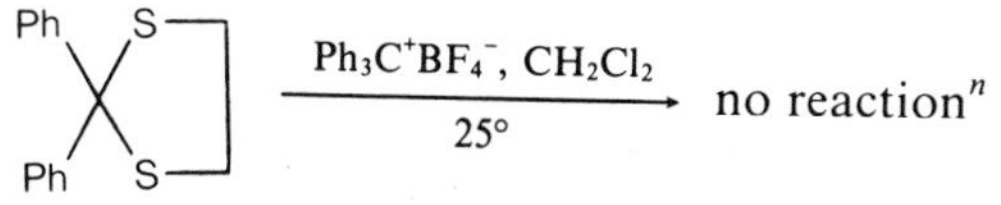

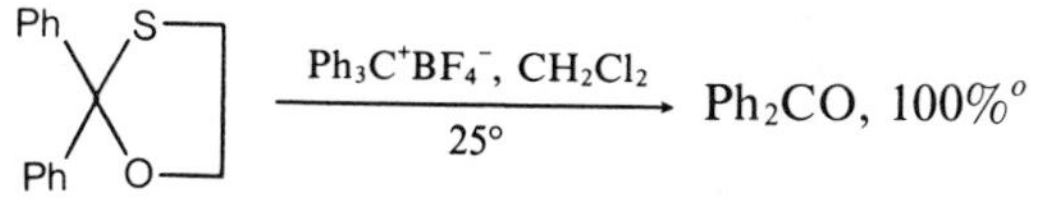

$\xrightarrow[25°]{Ph_3C^+BF_4^-,\ CH_2Cl_2}$ $RR'CO$[n] + Ph_3CH

60–100%

$\xrightarrow[25°]{Ph_3C^+BF_4^-,\ CH_2Cl_2}$ no reaction[n]

$\xrightarrow[25°]{Ph_3C^+BF_4^-,\ CH_2Cl_2}$ Ph_2CO, 100%[o]

[a] F. Huet, A. Lechevallier, and J. M. Conia, *Tetrahedron Lett.*, 2521 (1977).

[b] G. Bauduin, D. Bondon, Y. Pietrasanta, and B. Pucci, *Tetrahedron*, **34**, 3269 (1978).

[c] R. Sterzycki, *Synthesis*, 724 (1979).

[d] P. A. Grieco, M. Nishizawa, T. Oguri, S. D. Burke, and N. Marinovic, *J. Am. Chem. Soc.*, **99**, 5773 (1977).

[e] J. H. Babler, N. C. Malek, and M. J. Coghlan, *J. Org. Chem.*, **43**, 1821 (1978).

[f] P. A. Grieco, Y. Yokoyama, G. P. Withers, F. J. Okuniewicz, and C.-L. J. Wang, *J. Org. Chem.*, **43**, 4178 (1978).

[g] P. A. Grieco, Y. Ohfune, and G. Majetich, *J. Am. Chem. Soc.*, **99**, 7393 (1977).

[h] F. Huet, A. Lechevallier, M. Pellet, and J. M. Conia, *Synthesis*, 63 (1978).

[i] J. J. Brown, R. H. Lenhard, and S. Bernstein, *J. Am. Chem. Soc.*, **86**, 2183 (1964).

[j] B. Willhalm, U. Steiner, and H. Schinz, *Helv. Chim. Acta*, **41**, 1359 (1958).

[k] P. A. Grieco, T. Oguri, S. Gilman, and G. R. DeTitta, *J. Am. Chem. Soc.*, **100**, 1616 (1978).

[l] H. M. Walborsky, R. H. Davis, and D. R. Howton, *J. Am. Chem. Soc.*, **73**, 2590 (1951).

[m] J. A. Zderic and D. C. Limon, *J. Am. Chem. Soc.*, **81**, 4570 (1959).

[n] D. H. R. Barton, P. D. Magnus, G. Smith, and D. Zurr, *J. Chem. Soc., Chem. Commun.*, 861 (1971).

[o] D. H. R. Barton, P. D. Magnus, G. Smith, G. Streckert, and D. Zurr, *J. Chem. Soc., Perkin Trans. 1*, 542 (1972).

9.   4-Bromomethyl-1,3-dioxolane: **, 9**

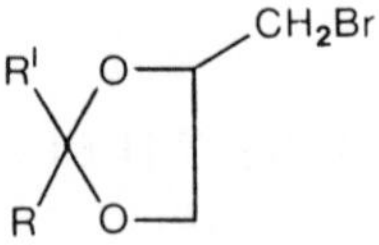

Formation (→) / Cleavage (←)[a]

$$\dfrac{HOCH_2CH(OH)CH_2Br,\ TsOH,\ C_6H_6}{\text{reflux, 5 h, 93–98\%}}$$

RR′CO   **9**

$$\dfrac{\text{activated Zn, MeOH}}{\text{reflux, 12 h, 89–96\%}}$$

Compound **9** is stable to several reagents that attack carbonyl groups (e.g., *m*-ClC₆H₄CO₃H, NH₃, NaBH₄, MeLi, and CrO₃/HOAc). It is cleaved under neutral conditions.[a]

[a] E. J. Corey and R. A. Ruden, *J. Org. Chem.*, **38**, 834 (1973).

10.   4-*o*-Nitrophenyl-1,3-dioxolane: **, 10**

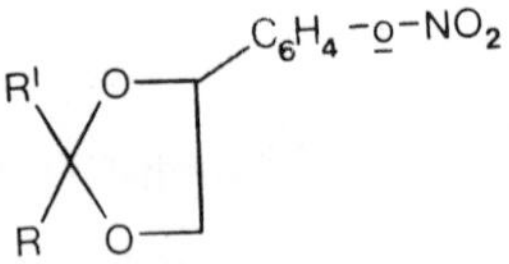

Compound **10,** readily formed from the glycol (TsOH, C₆H₆, reflux, 70–95% yield), is cleaved by irradiation (350 nm, C₆H₆, 25°, 6 h, 75–90% yield).[a]

[a] J. Hébert and D. Gravel, *Can. J. Chem.*, **52**, 187 (1974).

O,O′-**Phenylenedioxy Ketal:**   **, B**

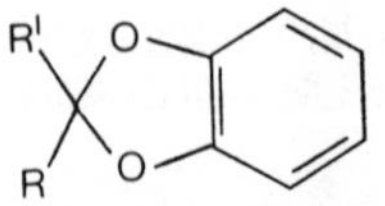

Formation $(\rightarrow)^a$ */ Cleavage* $(\leftarrow)^b$

$$\xrightarrow[\text{90°, 30 h, 85\%}]{\text{catechol, TsOH}}$$

RR'CO O,O'-phenylenedioxy ketal

$$\xleftarrow[\text{reflux, 6 h}]{\text{5 }N\text{ HCl, dioxane}}$$

In some steroid syntheses an ethylene ketal provided unsatisfactory protection for a carbonyl group since the ketal was hydrolyzed during Jones oxidation.[b]

[a] M. Rosenberger, D. Andrews, F. DiMaria, A. J. Duggan, and G. Saucy, *Helv. Chim. Acta,* **55,** 249 (1972).

[b] M. Rosenberger, A. J. Duggan, and G. Saucy, *Helv. Chim. Acta,* **55,** 1333 (1972).

Dithio Acetals and Ketals

A carbonyl group can be protected as a sulfur derivative, for example, a dithio acetal or ketal, 1,3-dithiane, or 1,3-dithiolane, by reaction of the carbonyl compound in the presence of an acid catalyst with a thiol or dithiol. The derivatives are in general cleaved by reaction with Hg(II) salts or oxidation; acidic hydrolysis is unsatisfactory. Representative examples of formation and cleavage are shown below.

ACYCLIC DITHIO ACETALS AND KETALS

11. *S,S'*-Dimethyl Acetals and Ketals: $RR'C(SCH_3)_2$, 11

12. *S,S'*-Diethyl Acetals and Ketals: $RR'C(SC_2H_5)_2$, 12

13. *S,S'*-Dipropyl Acetals and Ketals: $RR'C(SC_3H_7)_2$, 13

14. *S,S'*-Dibutyl Acetals and Ketals: $RR'C(SC_4H_9)_2$, 14

15. *S,S'*-Dipentyl Acetals and Ketals: $RR'C(SC_5H_{11})_2$, 15

16. *S,S'*-Diphenyl Acetals and Ketals: $RR'C(SC_6H_5)_2$, 16

17. *S,S'*-Dibenzyl Acetals and Ketals: $RR'C(SCH_2C_6H_5)_2$, 17

General Methods of Formation

$$(1)\quad RR'CO + R''SH \xrightarrow[20°,\ 30\ min]{concd\ HCl} RR'C(SR'')_2$$

11, 12

$RR'CO$ = an aldose; R'' = Me or Et[a]

$$(2)\quad RR'CO + R''S{-}SiMe_3 \xrightarrow[0\text{–}25°,\ 70\text{–}95\%]{ZnI_2,\ Et_2O} RR'C(SR'')_2$$ [b]

11, 12, 19

R,R′ = H, alkyl, cycloalkyl, aryl, steroid

R″ = Me, Et, or $(-CH_2-)_3$

$$(3)\quad RR'CO + R''SH \xrightarrow[20°,\ 1\ h,\ >80\%]{Me_3SiCl,\ CHCl_3} RR'C(SR'')_2$$ [c]

12, 13, 16

R, R′ = H, alkyl, cycloalkyl, aryl

R″ = Et, Pr, or Ph

$$(4)\quad RR'CO + B(SR'')_3 \xrightarrow[25°,\ 18\ h,\ 75\text{–}85\%]{reflux,\ 2\ h\ or} RR'C(SR'')_2$$ [d]

12, 14, 15

R, R′ = H, alkyl, aryl

R″ = Et, C_4H_9, or C_5H_{11}

$$(5)\quad ArCHO + PhSH \xrightarrow[0°,\ 10\ min,\ 86\%]{BF_3 \cdot Et_2O,\ CHCl_3} ArCH(SPh)_2$$ [e]

16

General Methods of Cleavage

(1)
OH
BuS
BuS
O
C5H11
OTHP
AgNO3/Ag2O, CH3CN-H2O[f]
0°, 2 h, 85%
OH
O
O

14a

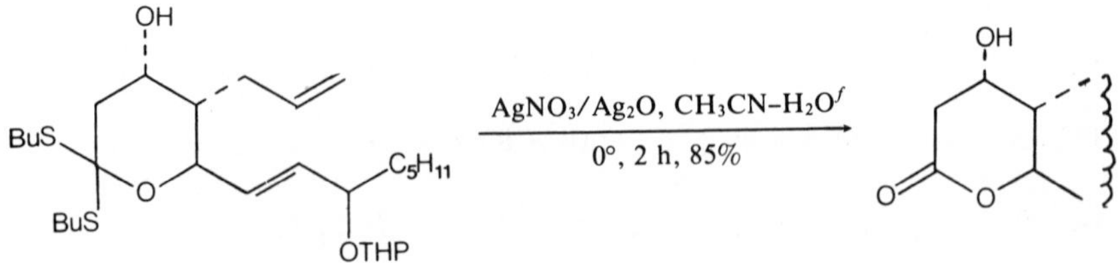

The S,S'-dibutyl group is stable to acids (e.g., HOAc/H_2O–THF, 45°, 3 h; TsOH/CH_2Cl_2, 0°, 0.5 h).[f]

(2) $RR'C(SR'')_2 \xrightarrow[\text{25°, 4 h, 80–100\%}]{\text{AgClO}_4/\text{H}_2\text{O–C}_6\text{H}_6{}^{g}} RR'CO$

12, 16

$R'' = $ Et or Ph

(3) $RR'C(SEt)_2 \xrightarrow[\text{HgCl}_2/\text{CaCO}_3,\ \text{CH}_3\text{CN–H}_2\text{O}^{i}]{\text{HgCl}_2/\text{CdCO}_3,\ \text{aq acetone}^{h}\ \text{or}} RR'CO$

12   high yield

(4) $RR'C(SR'')_2 \xrightarrow[\text{25°, 15 min}]{\text{Me}_2\text{CH(CH}_2)_2\text{ONO/CH}_2\text{Cl}_2} \xrightarrow[\text{63–93\%}]{\text{H}_2\text{O}^{j}} RR'CO$

12, 17, 19

$R'' = $ Et, PhCH$_2$, or (—CH$_2$—)$_3$

Isoamyl nitrite cleaves aromatic dithio acetals in preference to aliphatic dithio acetals, and dithio acetals in preference to dithio ketals. It also cleaves 1,3-oxa-thiolanes (1 h, 65–90% yield).[j]

(5) $RR'C(SR'')_2 \xrightarrow[\text{25°, 5 min, 73–98\%}]{\text{Tl(NO}_3)_3/\text{CH}_3\text{OH–H}_2\text{O}^{e}} RR'CO$

16, 19, 20

$R'' = $ Ph, (—CH$_2$—)$_3$, or (—CH$_2$—)$_2$

(6) $RR'C(SR'')_2 \xrightarrow[\text{25°, 2–3 h, 90–100\%}]{\text{SO}_2\text{Cl}_2/\text{SiO}_2\text{–H}_2\text{O, CH}_2\text{Cl}_2{}^{k}} RR'CO$

11, 12, 16, 19, 20

$R'' = $ Me, Et, Ph, (-CH$_2$-)$_3$, or (-CH$_2$-)$_2$

(7) $RCH(SMe)_2 \xrightarrow[\text{25°, 4.5 h, 80–95\%}]{\text{I}_2/\text{NaHCO}_3,\ \text{dioxane–H}_2\text{O}^{l}} RCHO$

11

$RR'C(SPh)_2 \xrightarrow[\substack{\text{reflux, 2 h}\\79\%}]{\text{I}_2/\text{MeOH}} [RR'C(OMe)_2] \xrightarrow[\text{25°, 16 h, 87\%}]{\text{HClO}_4/\text{H}_2\text{O}^{m}} RR'CO$

16

(8) $RR'C(SMe)_2$ $\xrightarrow[\text{or NaIO}_4/\text{H}_2\text{O, 25°}]{\text{H}_2\text{O}_2/\text{aq acetone}}$ $[RR'C(SMe)SMe]$ $\xrightarrow[\text{0°, 50–70\%}]{\text{g HCl/CHCl}_3{}^n}$ $RR'CO$

$$\underset{\text{O}}{\overset{\parallel}{}}$$

11

(9) $RR'C(SR'')_2$ $\xrightarrow[\text{2–5 h, 60–80\%}]{\text{O}_2,\ h\nu/\text{hexane, Ph}_2\text{CO}^o}$ $RR'CO$

17, 20

$R'' = PhCH_2$, or $(—CH_2—)_2$

1,3-Oxathiolanes are also cleaved by irradiation.[o]

[a] H. Zinner, *Chem. Ber.,* **83**, 275 (1950).

[b] D. A. Evans, L. K. Truesdale, K. G. Grimm, and S. L. Nesbitt, *J. Am. Chem. Soc.,* **99**, 5009 (1977).

[c] B. S. Ong and T. H. Chan, *Synth. Commun.,* **7**, 283 (1977).

[d] F. Bessette, J. Brault, and J. M. Lalancette, *Can. J. Chem.,* **43**, 307 (1965).

[e] E. Fujita, Y. Nagao, and K. Kaneko, *Chem. Pharm. Bull.,* **26**, 3743 (1978).

[f] E. J. Corey, M. Shibasaki, J. Knolle, and T. Sugahara, *Tetrahedron Lett.,* 785 (1977).

[g] T. Mukaiyama, S. Kobayashi, K. Kamio, and H. Takei, *Chem. Lett.,* 237 (1972).

[h] J. English, Jr., and P. H. Griswold, Jr., *J. Am. Chem. Soc.,* **67**, 2039 (1945).

[i] A. I. Meyers, D. L. Comins, D. M. Roland, R. Henning, and K. Shimizu, *J. Am. Chem. Soc.,* **101**, 7104 (1979).

[j] K. Fuji, K. Ichikawa, and E. Fujita, *Tetrahedron Lett.,* 3561 (1978).

[k] M. Hojo and R. Masuda, *Synthesis,* 678 (1976).

[l] G. A. Russell and L. A. Ochrymowycz, *J. Org. Chem.,* **34**, 3618 (1969).

[m] B. M. Trost, T. N. Salzmann, and K. Hiroi, *J. Am. Chem. Soc.,* **98**, 4887 (1976).

[n] H. Nieuwenhuyse and R. Louw, *Tetrahedron Lett.,* 4141 (1971).

[o] T. T. Takahashi, C. Y. Nakamura, and J. Y. Satoh, *J. Chem. Soc., Chem. Commun.,* 680 (1977).

18. *S,S'*-Diacetyl Acetals and Ketals: $RR'C(SCOCH_3)_2$, 18

Formation[a]

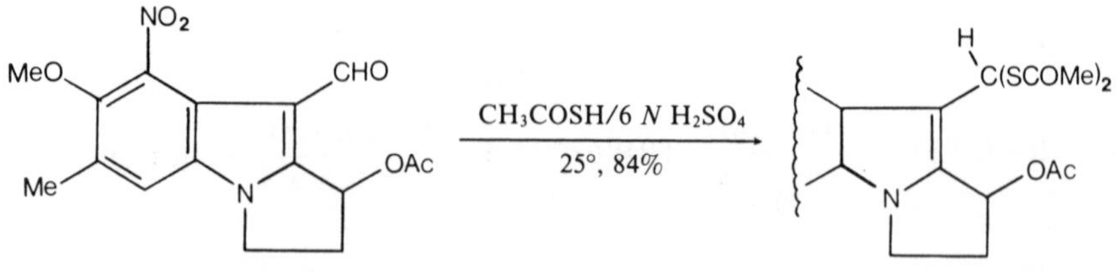

Cleavage[a]

$$18a \xrightarrow[\text{25°, 40 h, 96%}]{\text{NaOMe/MeOH}} \text{—CH(OMe)}_2 \xrightarrow[\text{or SiO}_2\text{, facile}]{\text{CHCl}_3\text{/cat. H}_2\text{O}}$$

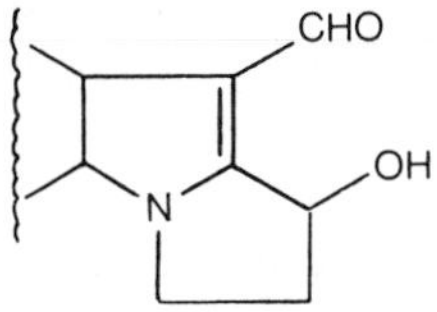

The formyl group was lost during attempted protection by ethylene glycol/ TsOH.[a]

[a] T. Kametani, Y. Kigawa, K. Takahashi, H. Nemoto, and K. Fukumoto, *Chem Pharm. Bull.,* **26,** 1918 (1978).

CYCLIC DITHIO ACETALS AND KETALS

19. **1,3-Dithiane Derivative:** 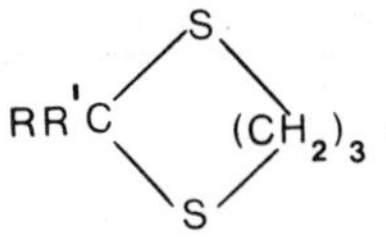 , **19**

20. **1,3-Dithiolane Derivative:** 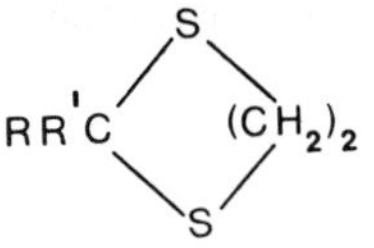 , **20**

General Methods of Formation

(1) $RR'CO + HS(CH_2)_nSH \xrightarrow[\text{25°, 12 h, high yield}]{\text{BF}_3\cdot\text{Et}_2\text{O, CH}_2\text{Cl}_2} \textbf{19, 20}$

   n = 2,[a] 3[b]

(2)

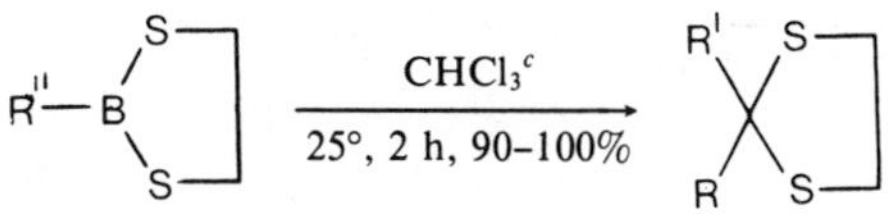

   R, R' = H, cycloalkyl, aryl   **20**

   R" = Cl or Ph

When R″ = Ph, the reaction is selective for unhindered ketones. Diaryl ketones, in general unreactive compounds, react rapidly when R″ = Cl.[c]

(3) $RR'CO + Me_3Si—S(CH_2)_2S—SiMe_3 \xrightarrow[\text{0–25°, 12–24 h, high yields}]{ZnI_2/Et_2O^d}$
RR'C
S
S
(CH2)2

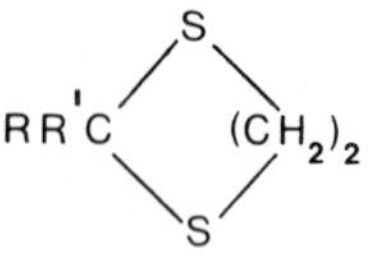

20

α,β-Unsaturated ketones are selectively protected (94 : 1, 94 : 4) in the presence of saturated ketones by this reagent.[d]

[a] R. P. Hatch, J. Shringarpure, and S. M. Weinreb, *J. Org. Chem.*, **43**, 4172 (1978).

[b] J. A. Marshall and J. L. Belletire, *Tetrahedron Lett.*, 871 (1971).

[c] D. R. Morton and S. J. Hobbs, *J. Org. Chem.*, **44**, 656 (1979).

[d] D. A. Evans, L. K. Truesdale, K. G. Grimm, and S. L. Nesbitt, *J. Am. Chem. Soc.*, **99**, 5009 (1977).

General Methods of Cleavage

(1)
ArCH
S
S
(CH2)3
Hg(ClO4)2/MeOH-CHCl3a
25°, 5 min, 93%
ArCHO

A 1,3-dithiane is stable to the conditions ($HgCl_2$, $CaCO_3$, $CH_3CN–H_2O$, 25°, 1–2 h) used to cleave a methylthiomethyl ether (e.g., a hemithio acetal).[b]

(2)
S
S
Ph
N
H
CuCl2/CuO, acetonec
reflux, 90 min, 85%
COPh
N
H

(3)
S
S
AgNO3/EtOH-H2Od
50°, 20 min, 55%
O

Attempted cleavage using Hg(II) salts gave material that could not be distilled.[d]

1,3-Dithiolanes can be cleaved with Ag_2O ($MeOH–H_2O$, reflux, 16 h to 4 days, 75–85% yield).[e]

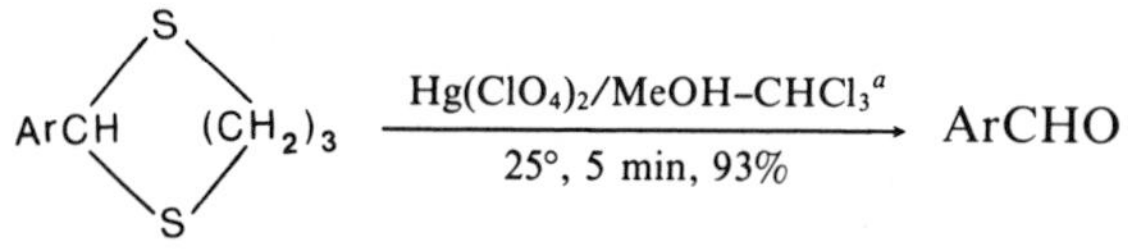

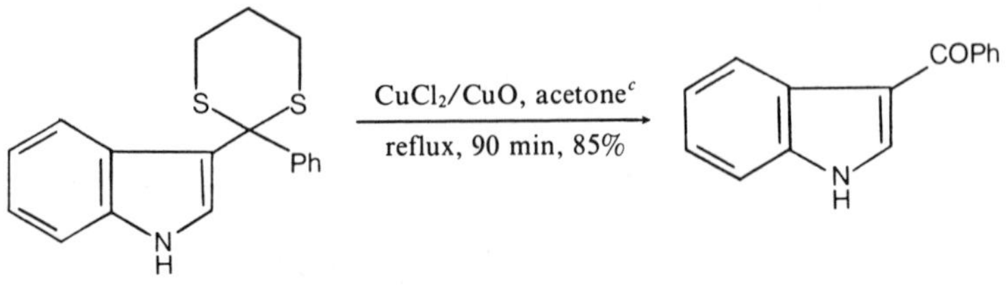

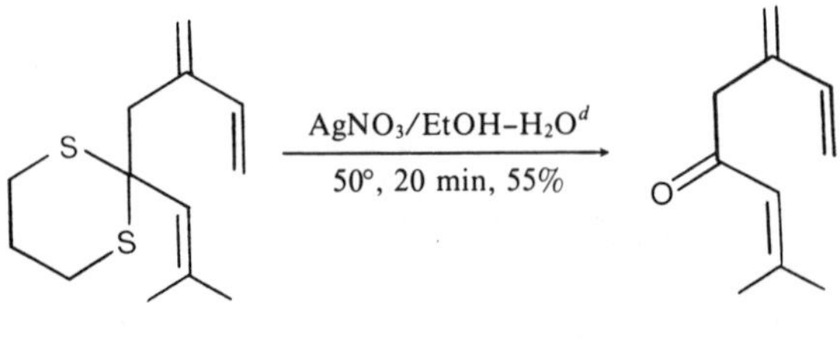

(4)

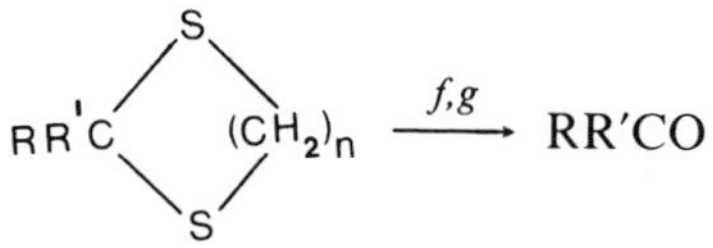

RR'C— (CH₂)ₙ dithiolane $\xrightarrow{f,g}$ RR'CO

19, 20

n = 2: NBS, aq acetone, 0°, 20 min, 80% yield[f]

n = 3: NCS, AgNO₃, CH₃CN–H₂O, 25°, 5–10 min, 70–100% yield[g]

NCS = *N*-chlorosuccinimide

(5)

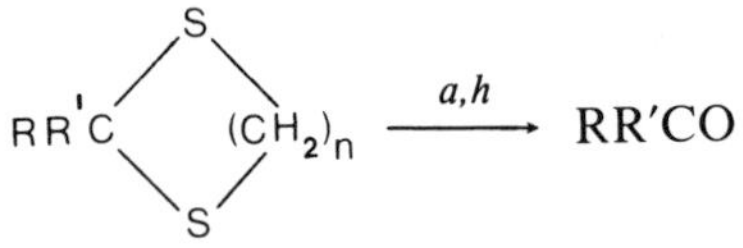

$\xrightarrow{a,h}$ RR'CO

19, 20

n = 2, 3: Tl(NO₃)₃, CH₃OH, 25°, 5 min, 73–99% yield[a]

n = 2, 3: Tl(OCOCF₃)₃, THF, 25°, 1 min, 83–95% yield[h]

α,β-Unsaturated 1,3-dithiolanes are selectively cleaved in the presence of saturated 1,3-dithiolanes [Tl(NO₃)₃, 5 min, 97% yield].[i]

(6)

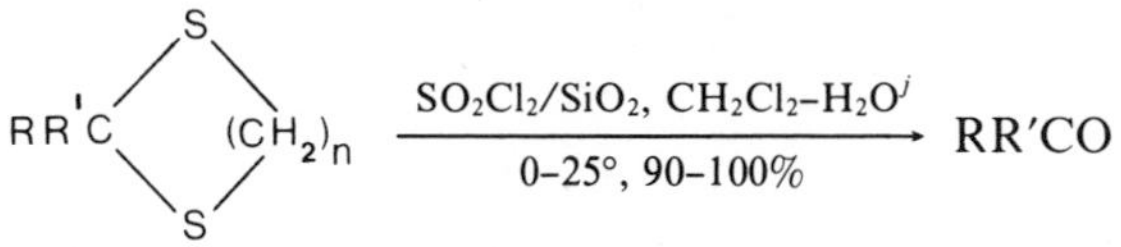

$\xrightarrow[\text{0–25°, 90–100%}]{\text{SO}_2\text{Cl}_2/\text{SiO}_2,\ \text{CH}_2\text{Cl}_2\text{–H}_2\text{O}^j}$ RR'CO

19, 20

n = 2, 3

(7)

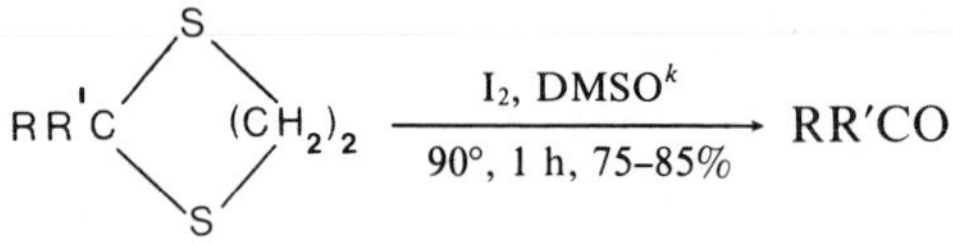

$\xrightarrow[\text{90°, 1 h, 75–85\%}]{\text{I}_2,\ \text{DMSO}^k}$ RR'CO

20

(8)

$\xrightarrow[\text{75–100\%}]{p\text{-MeC}_6\text{H}_4\text{SO}_2\text{NClNa, aq MeOH}}$ RR'CO

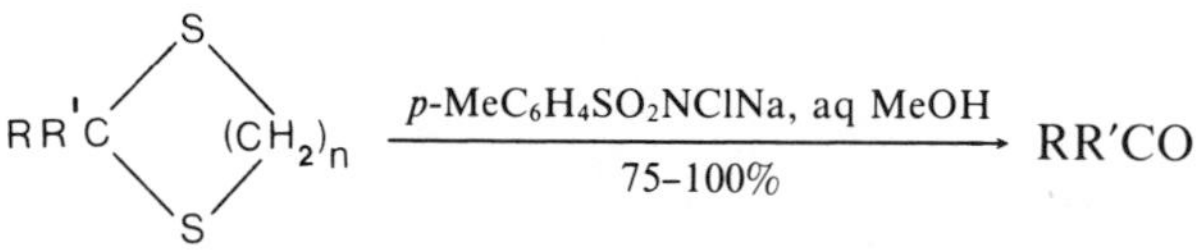

19, 20

n = 2,[l] 3[m]

1,3-Oxathiolanes are also cleaved by Chloramine-T.[m]

(9) 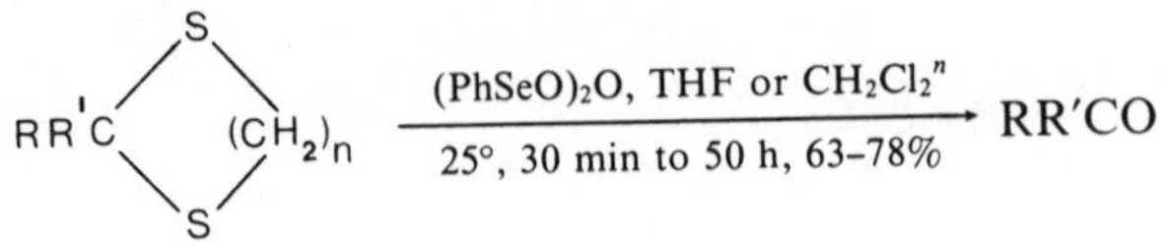 $RR'C\underset{S}{\overset{S}{<}}(CH_2)_n$ $\xrightarrow[\text{25°, 30 min to 50 h, 63–78\%}]{(PhSeO)_2O,\ THF\ or\ CH_2Cl_2{}^n}$ $RR'CO$

19, 20

n = 2, 3

(10) 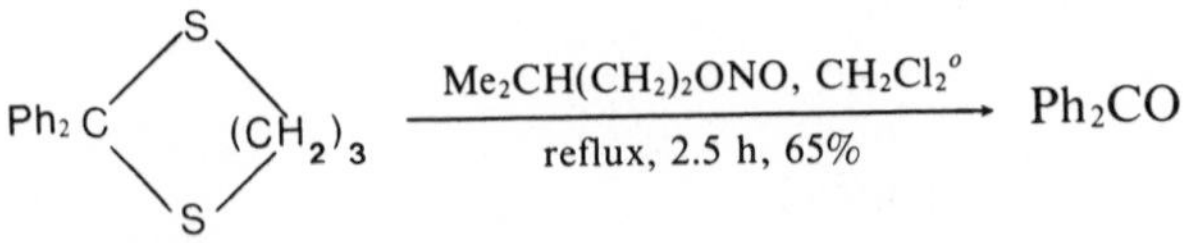 $Ph_2C\underset{S}{\overset{S}{<}}(CH_2)_3$ $\xrightarrow[\text{reflux, 2.5 h, 65\%}]{Me_2CH(CH_2)_2ONO,\ CH_2Cl_2{}^o}$ Ph_2CO

1,3-Oxathiolanes are also cleaved by isoamyl nitrite.[o]

(11) 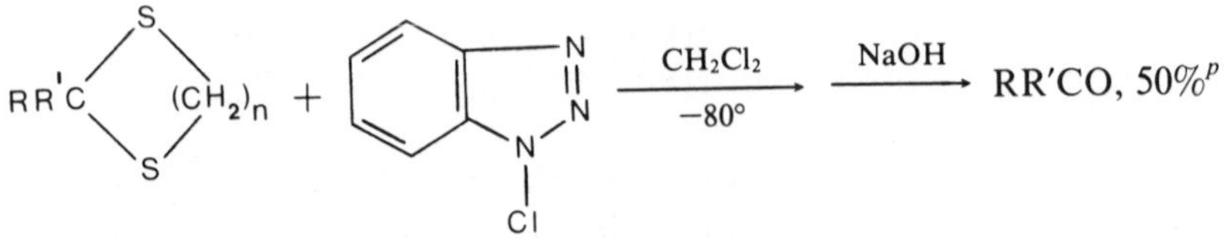 $RR'C\underset{S}{\overset{S}{<}}(CH_2)_n$ + [benzotriazole-N-Cl structure] $\xrightarrow[-80°]{CH_2Cl_2}$ $\xrightarrow{NaOH}$ $RR'CO,\ 50\%{}^p$

19, 20

n = 2, 3

1,3-Dithianes and 1,3-dithiolanes, used in this example to protect C_3-keto steroids, were not cleaved by $HgCl_2$–$CdCO_3$.[p]

(12) 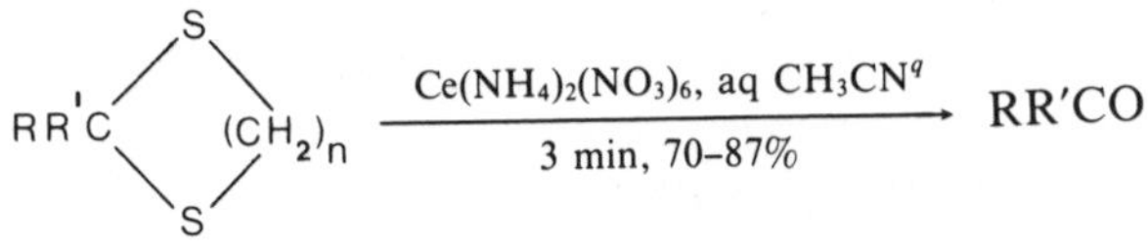 $RR'C\underset{S}{\overset{S}{<}}(CH_2)_n$ $\xrightarrow[\text{3 min, 70–87\%}]{Ce(NH_4)_2(NO_3)_6,\ aq\ CH_3CN{}^q}$ $RR'CO$

19, 20

n = 2, 3

(13) 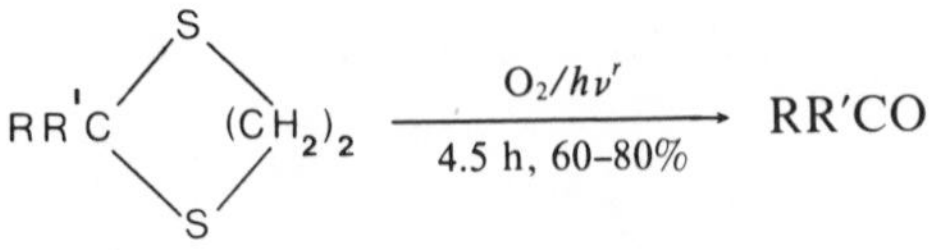 $RR'C\underset{S}{\overset{S}{<}}(CH_2)_2$ $\xrightarrow[\text{4.5 h, 60–80\%}]{O_2/h\nu{}^r}$ $RR'CO$

20

1,3-Oxathiolanes are also cleaved by $O_2/h\nu$.[r]

(14) $RR'C\overset{S}{\underset{S}{\diamond}}(CH_2)_3$ $\xrightarrow[\text{LiClO}_4 \text{ or Bu}_4\text{N}^+\text{ClO}_4^-]{\text{1.5 V, CH}_3\text{CN–H}_2\text{O}^s}$ RR'CO, 50–75%

19

1,3-Dithiolanes were not cleaved efficiently by electrolytic oxidation.[s]

(15) $RR'C\overset{S}{\underset{S}{\diamond}}(CH_2)_n$ $\xrightarrow{t-x}$ RR'CO

19, 20

n = 2, 3: MeOSO$_2$F (C$_6$H$_6$, 25°, 1 h, 62–88% yield[t] or liq SO$_2$, 70–85% yield[u])

n = 2:   MeI, aq MeOH, reflux, 2–20 h, 60–80% yield[u]

n = 3:   MeI, aq CH$_3$CN, 25°[v]

n = 2:   Et$_3$O$^+$BF$_4^-$, followed by 3% aq CuSO$_4$, 81% yield[w]

n = 2:   Me$_2$S$^+$Br Br$^-$, CH$_2$Cl$_2$, 25°, 1 h → reflux, 8 h, followed by H$_2$O, 55–91% yield[x]

(16) $\underset{R}{\overset{R^1}{\diagdown}}\text{C}\underset{\text{S}}{\overset{\text{S}}{\diagup}}$ $\xrightarrow[\text{25°, 15 min to 20 h}]{\text{OHCCOOH, HOAc}^y}$ RR'CO + $\underset{HO_2C}{\overset{H}{\diagdown}}\text{C}\underset{\text{S}}{\overset{\text{S}}{\diagup}}$

20   60–90%

(17) $RR'C\overset{S}{\underset{S}{\diamond}}(CH_2)_2$ $\xrightarrow[\text{25°, 45 min}]{\text{NO}^+\text{HSO}_4^-, \text{CH}_2\text{Cl}_2}$ $\xrightarrow{\text{H}_2\text{O}}$ RR'CO, 56–82%[z]

Cleavage by nitrosyl sulfate avoids the use of toxic methyl fluorosulfonate.[z]

[a] E. Fujita, Y. Nagao, and K. Kaneko, *Chem. Pharm. Bull.*, **26**, 3743 (1978).

[b] E. J. Corey and M. G. Bock, *Tetrahedron Lett.*, 2643 (1975).

[c] P. Stütz and P. A. Stadler, *Org. Synth.*, **56**, 8 (1977).

[d] C. A. Reece, J. O. Rodin, R. G. Brownlee, W. G. Duncan, and R. M. Silverstein, *Tetrahedron*, **24**, 4249 (1968).

[e] D. Gravel, C. Vaziri, and S. Rahal, *J. Chem. Soc., Chem. Commun.*, 1323 (1972).

[f] E. N. Cain and L. L. Welling, *Tetrahedron Lett.*, 1353 (1975).

[g] E. J. Corey and B. W. Erickson, *J. Org. Chem.*, **36**, 3553 (1971).

[h] T.-L. Ho and C. M. Wong, *Can. J. Chem.*, **50**, 3740 (1972).

[i] R. A. J. Smith and D. J. Hannah, *Synth. Commun.*, **9**, 301 (1979).

[j] M. Hojo and R. Masuda, *Synthesis*, 678 (1976).

[k] J. B. Chattopadhyaya and A. V. Rama Rao, *Tetrahedron Lett.*, 3735 (1973).

[l] W. F. J. Huurdeman, H. Wynberg, and D. W. Emerson, *Tetrahedron Lett.*, 3449 (1971).

[m] D. W. Emerson and H. Wynberg, *Tetrahedron Lett.*, 3445 (1971).

[n] D. H. R. Barton, N. J. Cussans, and S. V. Ley, *J. Chem. Soc., Chem. Commun.*, 751 (1977).

[o] K. Fuji, K. Ichikawa, and E. Fujita, *Tetrahedron Lett.*, 3561 (1978).

[p] P. R. Heaton, J. M. Midgley, and W. B. Whalley, *J. Chem. Soc., Chem. Commun.*, 750 (1971).

[q] T.-L. Ho, H. C. Ho, and C. M. Wong, *J. Chem. Soc., Chem. Commun.*, 791 (1972).

[r] T. T. Takahashi, C. Y. Nakamura, and J. Y. Satoh, *J. Chem. Soc., Chem. Commun.*, 680 (1977).

[s] Q. N. Porter and J. H. P. Utley, *J. Chem. Soc., Chem. Commun.*, 255 (1978).

[t] T.-L. Ho and C. M. Wong, *Synthesis*, 561 (1972).

[u] M. Fetizon and M. Jurion, *J. Chem. Soc., Chem. Commun.*, 382 (1972).

[v] S. Takano, S. Hatakeyama, and K. Ogasawara, *J. Chem. Soc., Chem. Commun.*, 68 (1977).

[w] T. Oishi, K. Kamemoto, and Y. Ban, *Tetrahedron Lett.*, 1085 (1972).

[x] G. A. Olah, Y. D. Vankar, M. Arvanaghi, and G. K. S. Prakash, *Synthesis*, 720 (1979).

[y] H. Muxfeldt, W.-D. Unterweger, and G. Helmchen, *Synthesis*, 694 (1976).

[z] G. A. Olah, S. C. Narang, G. F. Salem, and B. G. B. Gupta, *Synthesis*, 273 (1979).

21.   4*H*,7*H*-1,3-Dithiepin Derivative: 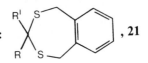 **, 21**

Dithiepin derivatives, prepared in high yield from 1,2-bis(mercaptomethyl)benzenes, are cleaved by $HgCl_2$ (80% yield). Neither reagents nor products have unpleasant odors.[a]

[a] I. Shahak and E. D. Bergmann, *J. Chem. Soc., C,* 1005 (1966).

Hemithio Acetals and Ketals

ACYCLIC HEMITHIO ACETALS AND KETALS

Acyclic hemithio acetals and ketals can be prepared directly from a carbonyl compound (e.g., formation of **22**) or by transketalization, a reaction that does not involve a free carbonyl group, from a 1,3-dithiane or 1,3-dithiolane (e.g., formation of **23**). They are cleaved by acidic hydrolysis (**22**) or Hg(II) salts (**23**).

22.   *O*-Trimethylsilyl-*S*-alkyl Acetals and Ketals: RR′C(SR″)OSi(CH₃)₃, 22

Formation (→) / Cleavage (←)

(1) $RR'CO + R''SSiMe_3 \xrightarrow[\text{25°, 30 min, 80-90\%}]{ZnI_2{}^{a}}$ **22**

    R″ = Me, Et, or Ph

When a carbonyl compound is treated for 24 h with 2 eq of the silyl reagent, the dithio acetal or ketal is obtained in quantitative yield.[a]

(2)

$$RR'CO \xrightarrow[25°,\ 3\ h,\ 75\text{–}90\%]{Me_3SiCl\ +\ R''SH,\ Py^{b}} \mathbf{22}$$

$$\mathbf{22} \xrightarrow{\text{dil HCl}^{b}} RR'CO$$

R″ = Et, Ph

In ether or tetrahydrofuran organolithium reagents cleave the silicon–oxygen bond; in hexamethylphosphoramide, they react at the carbon atom.[b]

[a] D. A. Evans, L. K. Truesdale, K. G. Grimm, and S. L. Nesbitt, *J. Am. Chem. Soc.*, **99**, 5009 (1977).

[b] T. H. Chan and B. S. Ong, *Tetrahedron Lett.*, 319 (1976).

23. *O*-Methyl-*S*-2-(methylthio)ethyl Acetals and Ketals: $RR'C(OCH_3)SCH_2CH_2SCH_3$, 23

Formation[a]

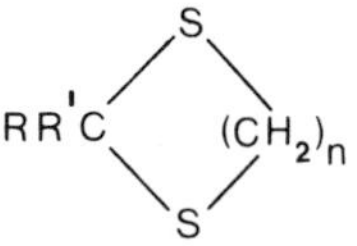

$$\xrightarrow[0°,\ 10\ min\ \to\ 23°,\ 2\ h]{MeOSO_2F/CH_2Cl_2} \xrightarrow[23°,\ 2\ h]{MeOH,\ CH_2Cl_2} \mathbf{23},\ \sim 72\%$$

n = 2, 3

Cleavage[a]

$$\mathbf{23} \xrightarrow[0°,\ \text{rapid}]{HgCl_2/CaCO_3,\ THF\text{–}H_2O} RR'CO$$

1,3-Dithianes and 1,3-dithiolanes, compounds that can be oxidized or hydrogenolyzed, are replaced by compounds that are stable to these conditions.[a]

[a] E. J. Corey and T. Hase, *Tetrahedron Lett.*, 3267 (1975).

CYCLIC HEMITHIO ACETALS AND KETALS

24. 1,3-Oxathiolanes: 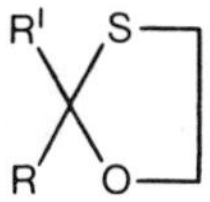, 24

Formation[a]

$$RR'CO + HS(CH_2)_2OH \xrightarrow[25°,\ 20\ h,\ 60\text{–}90\%]{ZnCl_2,\ NaOAc,\ dioxane} \mathbf{24}$$

Cleavage

Some reagents (including Chloramine-T, $O_2/h\nu$, isoamyl nitrite, and triphenyl-methyl fluoroborate) have been used to cleave both 1,3-oxathiolanes, and 1,3-dithianes and 1,3-dithiolanes (cf. pp. 135, 136, 136, and 127, respectively).

(1) $\mathbf{24}$ $\xrightarrow[\text{100°, 1 h, 83\%}]{\text{HgCl}_2/\text{HOAc-KOAc}}$ or $\xrightarrow[\text{25°, 30 min, 91\%}]{\text{HgCl}_2/\text{NaOH, EtOH-H}_2\text{O}}$ or

$\xrightarrow[\text{100°, 90 min, 92\%}]{\text{Raney Ni/HOAc-KOAc}}$ RR'CO[b]

C_3-keto steroid

(2) $\mathbf{24}$ $\xrightarrow[\text{reflux, 22 h, 60\%}]{\text{HCl/HOAc}^c}$ RR'CO

$C_{3,7,12}$-triketo steroid

[a] J. Romo, G. Rosenkranz, and C. Djerassi, *J. Am. Chem. Soc.*, **73**, 4961 (1951).
[b] C. Djerassi, M. Shamma, and T. Y. Kan, *J. Am. Chem. Soc.*, **80**, 4723 (1958).
[c] R. H. Mazur and E. A. Brown, *J. Am. Chem. Soc.*, **77**, 6670 (1955).

Diseleno Acetals and Ketals: RR'C(SeR″)$_2$, C

Formation

(1) RR'CO + R″SeH $\xrightarrow[\text{20°, 3 h, 70-95\%}]{\text{ZnCl}_2/\text{N}_2, \text{CCl}_4{}^a}$ C

R' = H, alkyl; R″ = Me or Ph

(2) RR'C(OR″) + (PhSe)$_3$B $\xrightarrow[\text{20°, 20 min to 24 h}]{\text{CF}_3\text{CO}_2\text{H, CHCl}_3{}^b}$ RR'C(SePh)$_2$

R″ = Me or Ph

Cleavage[a]

C + HgCl$_2$/CaCO$_3$, CH$_3$CN–H$_2$O $\xrightarrow{\text{20°, 2-4 h}}$ RR'CO, 65–80%

C + CuCl$_2$/CuO, Me$_2$CO–H$_2$O $\xrightarrow{\text{20°, 5 min to 2 h}}$ RR'CO, 73–99%

C + H$_2$O$_2$/THF $\xrightarrow{\text{0°, 15 min} \rightarrow \text{20°, 3 h}}$ RR'CO, 60–65%

C + (PhSeO)$_2$O/THF $\xrightarrow{\text{20° or 60°, 5 min} \rightarrow \text{6 h}}$ RR'CO, 60–90%

Diseleno acetals and ketals are cleaved more rapidly than the dithio acetals and ketals; a methyl derivative is cleaved more rapidly than a phenyl derivative. Methyl iodide or ozone converts diseleno acetals and ketals to vinyl selenides.[a]

[a] A. Burton, L. Hevesi, W. Dumont, A. Cravador, and A. Krief, *Synthesis*, 877 (1979).
[b] D. L. J. Clive and S. M. Menchen, *J. Org. Chem.*, **44**, 4279 (1979).

MISCELLANEOUS DERIVATIVES

O-Substituted Cyanohydrins

25. *O*-Acetyl Cyanohydrin: RR′C(CN)OCOCH$_3$, 25

***Formation* (→) / *Cleavage* (←)**[a]

$$\text{RR'CO} \xrightarrow[25°,\ 2\ \text{h},\ 82\%]{Me_2C(CN)OH,\ Et_3N} \xrightarrow[25°,\ 40\ \text{h},\ 82\%]{Ac_2O,\ Py} \mathbf{25}$$

C$_{17}$-keto steroid

$$\xleftarrow[25°,\ 5\ \text{min},\ 84\%]{KOH/CH_3OH\text{-}H_2O} \xleftarrow{Li(O\text{-}t\text{-}Bu)_3AlH/THF}$$

Compound **25** is stable to peracids, CrO$_3$, and bromine.

[a] P. D. Klimstra and F. B. Colton, *Steroids*, **10**, 411 (1967).

26. *O*-Trimethylsilyl Cyanohydrin: RR′C(CN)OSi(CH$_3$)$_3$, 26

***Formation* (→) / *Cleavage* (←)**

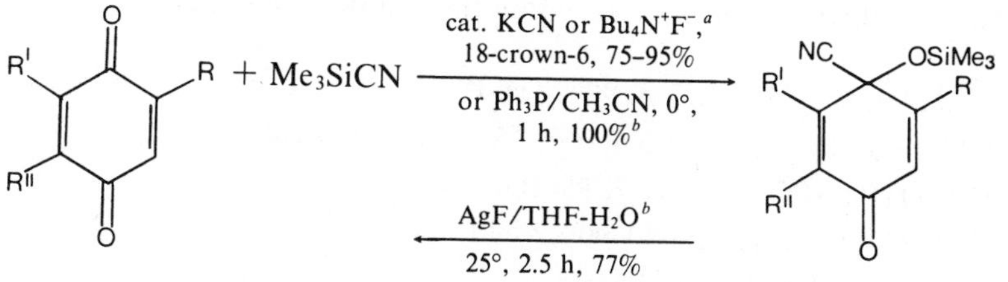

R'
R
RII
+ Me3SiCN
cat. KCN or Bu4N+F−,a
18-crown-6, 75–95%
or Ph3P/CH3CN, 0°,
1 h, 100%b
AgF/THF-H2Ob
25°, 2.5 h, 77%
NC
OSiMe3
R
R'
RII

R, R′ = H, Me, OMe, or *t*-Bu; R″ = H or Me[a]

R, R′ = Br; R″ = H[b]

O-Trimethylsilylcyanohydrins are also cleaved by dilute acid or dilute base.[c]

[a] D. A. Evans, J. M. Hoffman, and L. K. Truesdale, *J. Am. Chem. Soc.*, **95**, 5822 (1973).

[b] D. A. Evans and R. Y. Wong, *J. Org. Chem.*, **42**, 350 (1977).

[c] D. A. Evans, L. K. Truesdale, and G. L. Carroll, *J. Chem. Soc., Chem. Commun.*, 55 (1973).

27. *O*-1-Ethoxyethyl Cyanohydrin: RR'C(CN)OCH(OC$_2$H$_5$)CH$_3$, 27

Compound **27** was prepared (NaCN/HCl–THF, 0°, 75% yield, followed by EtOCH=CH$_2$/HCl, 50% yield) to convert an aldehyde (R' = H, R"X/LDA) to a protected ketone. It was cleaved by hydrolysis (0.01 *N* HCl/MeOH, 25°, followed by NaOH, 0°, 85% yield).[a]

[a] G. Stork and L. Maldonado, *J. Am. Chem. Soc.*, **93**, 5286 (1971).

28. *O*-Tetrahydropyranyl Cyanohydrin: RR'C(CN)O-tetrahydropyranyl, 28

Compound **28** was prepared from a steroid cyanohydrin (dihydropyran, TsOH, reflux, 1.5 h) and cleaved by hydrolysis (cat. concd HCl/acetone, reflux, 15 min, followed by aq Py, reflux, 1 h).[a]

[a] P. deRuggieri and C. Ferrari, *J. Am. Chem. Soc.*, **81**, 5725 (1959).

Substituted Hydrazones

29. *N,N*-Dimethylhydrazone: RR'C=NN(CH$_3$)$_2$, 29

Formation[a]

$$RR'CO \xrightarrow[\text{reflux, 24 h, 90–94\%}]{\text{H}_2\text{NNMe}_2/\text{EtOH–HOAc}} \textbf{29}$$

Cleavage

N,N-Dimethylhydrazones are cleaved by the following conditions:

NaIO$_4$/MeOH, pH 7, 2–3 h, 90% yield[b]

Cu(OAc)$_2$/H$_2$O–THF, pH 5.4, 25°, 15 min, 97% yield[c]

CuCl$_2$/THF, HPO$_4^-$, → pH 7, 85–100% yield[c]

CH$_3$I/95% EtOH, reflux, 80–90% yield[d]

O$_3$/CH$_2$Cl$_2$, −78°, 60–100% yield[e]

O$_2$/*hν*, rose bengal, MeOH, −78° → −20°, followed by Ph$_3$P or Me$_2$S, 48–88% yield[f]

CoF$_3$ (CHCl$_3$, reflux, 67–93% yield);[g] MoOCl$_3$ or MoF$_6$ (H$_2$O–THF, 25°, 4 h, 80–90% yield)[h]; WF$_6$ (CHCl$_3$, 0 → 25°, 1 h, 84–95% yield)[i]; UF$_6$ (50–95% yield)[j]

N,N-Dimethylhydrazones are stable to CrO_3/H_2SO_4 (0°, 3 min), to $NaBH_4$ (EtOH, 25°), to $LiAlH_4$ (THF, 25°), and to B_2H_6 followed by H_2O_2/OH^-. They are cleaved by CrO_3/Py and by p-$NO_2C_6H_4CO_3H/CHCl_3$, 25°.[d]

[a] G. R. Newkome and D. L. Fishel, *Org. Synth.*, **50**, 102 (1970).

[b] E. J. Corey and D. Enders, *Tetrahedron Lett.*, 3 (1976).

[c] E. J. Corey and S. Knapp, *Tetrahedron Lett.*, 3667 (1976).

[d] M. Avaro, J. Levisalles, and H. Rudler, *J. Chem. Soc., Chem. Commun.*, 445 (1969).

[e] R. E. Erickson, P. J. Andrulis, J. C. Collins, M. L. Lungle, and G. D. Mercer, *J. Org. Chem.*, **34**, 2961 (1969).

[f] E. Friedrich, W. Lutz, H. Eichenauer, and D. Enders, *Synthesis*, 893 (1977).

[g] G. A. Olah, J. Welch, and M. Henninger, *Synthesis*, 308 (1977).

[h] G. A. Olah, J. Welch, G. K. S. Prakash, and T.-L. Ho, *Synthesis*, 808 (1976).

[i] G. A. Olah and J. Welch, *Synthesis*, 809 (1976).

[j] G. A. Olah, J. Welch, and T.-L. Ho, *J. Am. Chem. Soc.*, **98**, 6717 (1976).

30.   2,4-Dinitrophenylhydrazone (2,4-DNP Group): $RR'C{=}NNHC_6H_3$-2,4-$(NO_2)_2$, 30

Formation[a]

$$\frac{H_2SO_4 \cdot H_2NNHC_6H_3\text{-}2,4\text{-}(NO_2)_2,\ EtOH\text{–}H_2O}{25°,\ 10\ min,\ 80\%}$$

RR'CO   **30**

$$\frac{O_3/EtOAc}{-78°,\ 70\%}$$

In a synthesis of sativene a carbonyl group was protected as a 2,4-DNP while a double bond was reduced with BH_3/H_2O_2–OH^-. Attempted protection of the carbonyl group as a ketal caused migration of the double bond; protection as an oxime or oxime acetate was unsatisfactory since they would be reduced by BH_3.[a]

Cleavage

2,4-Dinitrophenylhydrazones are cleaved by many oxidizing and reducing agents. Methods that are useful in synthetic schemes are described below.

(1)   **30** $\xrightarrow[\text{reflux, 80–95\%}]{\text{TiCl}_3/\text{DME–H}_2\text{O, N}_2{}^b}$ RR'CO

(2)   **30** $\xrightarrow[\text{75°, 20 h, 80–85\%}]{\text{Me}_2\text{CO, sealed tube}^c}$ RR'CO

a J. E. McMurry, *J. Am. Chem. Soc.*, **90**, 6821 (1968).

b J. E. McMurry and M. Silvestri, *J. Org. Chem.*, **40**, 1502 (1975).

c S. R. Maynez, L. Pelavin, and G. Erker, *J. Org. Chem.*, **40**, 3302 (1975).

31.   Oxime Derivatives: RR'C=NOH, 31

Formation (→) / Cleavage (←)[a]

$$\underset{60°}{\overset{HCl \cdot H_2NOH/Py}{\longrightarrow}}$$

RR'CO   **31**

$$\underset{0°, 3 \text{ h}, 76\%}{\overset{NaNO_2/1 \text{ } N \text{ HCl, } CH_3OH-H_2O}{\longleftarrow}}$$

In the last step of a synthesis of erythronolide A, acid-catalyzed hydrolysis of an acetonide failed because the carbonyl-containing precursor was unstable to acidic hydrolysis (3% MeOH-HCl, 0°, 30 min, conditions developed for the synthesis of erythronolide B). Consequently the carbonyl group was protected as an oxime, the acetonide was cleaved, and the carbonyl group was regenerated.[a]

Cleavage

Oximes are cleaved by oxidation, reduction, or hydrolysis in the presence of another carbonyl compound. Some synthetically useful methods are shown below.

(1)   RR'C=NOH $\underset{25°, 3 \text{ h}, 94\%}{\overset{CH_3CO(CH_2)_2COOH/1 \text{ } N \text{ HCl}^b}{\longrightarrow}}$ RR'CO

Pyruvic acid (HOAc, reflux, 1–3 h, 77% yield)[c] and acetone (80°, 100 h, 72% yield)[d] effect cleavage in a similar manner.

(2)   RR'C=NOH $\underset{50°, 1-3 \text{ h}, 80-95\%}{\overset{(PhSeO)_2O/THF^e}{\longrightarrow}}$ RR'CO

An *O*-methyloxime is stable to phenylselenic anhydride.[e]

(3)   RR'C=NOH $\underset{25°, 12 \text{ h or } 40°, \text{ few hours } \sim95\%}{\overset{Na_2S_2O_4/H_2O^f}{\longrightarrow}}$ RR'CO

(4)   RR'C=NOH $\underset{\text{reflux, 2-16 h}}{\overset{NaHSO_3/EtOH-H_2O}{\longrightarrow}}$ $\underset{30 \text{ min}}{\overset{\text{dil HCl}}{\longrightarrow}}$ RR'CO, $\sim85\%^g$

(5) $RR'C{=}NOH \xrightarrow[20°]{Ac_2O} RR'C{=}NOAc \xrightarrow[25-65°]{Cr(OAc)_2/THF-H_2O} RR'CO, 75-95\%^h$

Chromous acetate also cleaves unsubstituted oximes, but the reaction is slow and requires high temperatures.[h]

[a] E. J. Corey, P. B. Hopkins, S. Kim, S. Yoo, K. P. Nambiar, and J. R. Falck, *J. Am. Chem. Soc.,* **101,** 7131 (1979).

[b] C. H. Depuy and B. W. Ponder, *J. Am. Chem. Soc.,* **81,** 4629 (1959).

[c] E. B. Hershberg, *J. Org. Chem.,* **13,** 542 (1948).

[d] S. R. Maynez, L. Pelavin, and G. Erker, *J. Org. Chem.,* **40,** 3302 (1975).

[e] D. H. R. Barton, D. J. Lester, and S. V. Ley, *J. Chem. Soc., Chem. Commun.,* 445 (1977).

[f] P. M. Pojer, *Aust. J. Chem.,* **32,** 201 (1979).

[g] S. H. Pines, J. M. Chemerda, and M. A. Kozlowski, *J. Org. Chem.,* **31,** 3446 (1966).

[h] E. J. Corey and J. E. Richman, *J. Am. Chem. Soc.,* **92,** 5276 (1970).

32.   *O*-Benzyl Oxime: RR'C=NOCH₂C₆H₅, 32

The reactions shown below were used in a synthesis of perhydrohistrionicotoxin; the carbonyl groups were protected as an oxime and an *O*-benzyl oxime.[a]

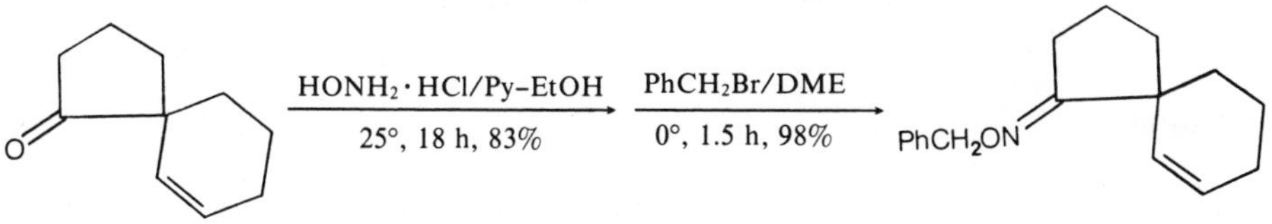

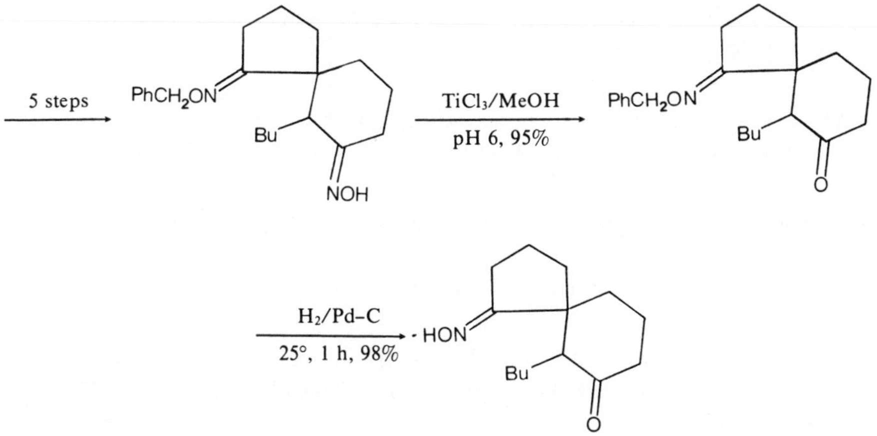

[a] E. J. Corey, M. Petrzilka, and Y. Ueda, *Helv. Chim. Acta,* **60,** 2294 (1977).

33. *O*-Phenylthiomethyl Oxime: $RR'C=NOCH_2SC_6H_5$, 33

In a prostaglandin synthesis a carbonyl group was protected as an oxime that had its hydroxyl group protected against Collins oxidation by the phenylthiomethyl group. The phenylthiomethyl group is readily removed to give an oxime that is then cleaved to the carbonyl compound.[a]

Formation[a]

$$RR'CO + H_2NOCH_2SPh \xrightarrow[\text{25°, 24 h, 100\%}]{\text{Py}} 33$$

Cleavage[a]

$$33 \xrightarrow[\text{25–50°, 0.5–48 h, 75\%}]{\text{HgCl}_2/\text{HgO, HOAc–KOAc}} \xrightarrow[\text{25°, 5 min, 100\%}]{\text{K}_2\text{CO}_3/\text{MeOH}} RR'C=NOH$$

$$\xrightarrow[\text{10°, 1 h}]{\text{NaNO}_2/\text{HOAc}} RR'CO$$

[a] I. Vlattas, L. Della Vecchia, and J. J. Fitt, *J. Org. Chem.*, **38**, 3749 (1973).

Imines

In general, imines are too reactive to be used to protect carbonyl groups. In a synthesis of juncusol,[a] however, a bromo- and an iodocyclohexylimine of two identical aromatic aldehydes were coupled by an Ullman coupling reaction modified by Ziegler.[b] The imines were cleaved by acidic hydrolysis (aq oxalic acid/THF, 20°, 1 h, 95% yield).

[a] A. S. Kende and D. P. Curran, *J. Am. Chem. Soc.*, **101**, 1857 (1979).
[b] F. E. Ziegler, K. W. Fowler, and S. Kanfer, *J. Am. Chem. Soc.*, **98**, 8282 (1976).

34. Substituted Methylene Derivatives: $RR'C=C(CN)R''$, 34

RR' = substituted pyrrole; R'' = —CN,[a] —CO$_2$Et[b]

Compound **34**, prepared from a 2-formylpyrrole and a malonic acid derivative, was used in a synthesis of chlorophyll.[a] It is cleaved under drastic conditions (concd alkali).[a,b]

[a] R. B. Woodward and 17 co-workers, *J. Am. Chem. Soc.*, **82**, 3800 (1960).
[b] J. B. Paine, R. B. Woodward, and D. Dolphin, *J. Org. Chem.*, **41**, 2826 (1976).

Cyclic Derivatives

35. Oxazolidines: 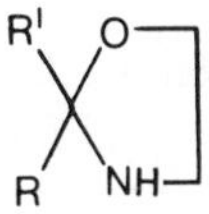 , **35**

Compound **35** was used to protect the carbonyl group in an α,β-unsaturated alde-hyde during reduction of the carbon–carbon double bond by H_2/Raney Ni. It was prepared in ~60% yield from 2-aminoethanol and cleaved by acidic hydrol-ysis (10% HCl, 25°, 12 h, 60% yield).[a]

[a] E. P. Goldberg and H. R. Nace, *J. Am. Chem. Soc.*, **77**, 359 (1955).

36. Imidazolidines: 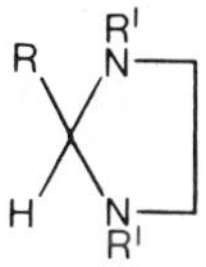 , **36**

N,N'-Disubstituted ethylenediamines react selectively with an aldehyde in the presence of a ketone to form compounds such as **36** (e.g., $R' =$ Ph, 65–90% yield). Compound **36** is cleaved by acidic hydrolysis (10% HCl, 30 min, 90% yield).[a]

[a] H.-W. Wanzlick and W. Löchel, *Chem. Ber.*, **86**, 1463 (1953).

37. Thiazolidines: 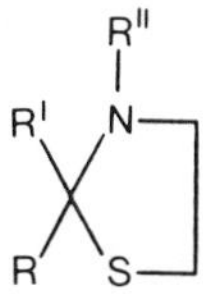 , **37**

Compounds such as **37** ($R'' =$ Me, c-C_6H_{11}, $PhCH_2$), prepared in satisfactory yields from an aminoethanethiol, have been used to protect the more reactive of two aldehyde groups during syntheses of pyridoxals. Compound **37** is cleaved by basic hydrolysis (NaOH, 25°, 95% yield).[a]

[a] K. Ueno, F. Ishikawa, and T. Naito, *Tetrahedron Lett.*, 1283 (1969).

MONOPROTECTION OF DICARBONYL COMPOUNDS

38. 5-(p-Ⓟ-Benzyloxymethyl)-1,3-dioxane:  , **38**

Compound **38** was prepared by reaction with polymer-bound 1,3-diol in the presence of an acid catalyst to protect one aldehyde group in an aromatic dialde-hyde. It is cleaved by acid (HCl/dioxane, 25°, 48 h, 70–85% yield).[a]

[a] C. C. Leznoff and S. Greenberg, *Can. J. Chem.*, **54**, 3824 (1976).

Selective Protection of α- and β-Diketones

A hindered β-diketone can be protected as an enol ether, **41a,** (eq. 1)[a] or an enamine, **39a,** (eq. 2).[b] Reduction followed by hydrolysis gives a less or more hindered monocarbonyl compound, respectively. Similarly an α-diketone can be selectively protected as an enol acetate, **40b,** (eq. 3)[c] or an enamine, **39b,** (eq. 4).[d]

39.   Enamines: $-\overset{\Vert}{\underset{O}{C}}-\overset{\vert}{\underset{N}{C}}=\overset{\vert}{C}-$, 39a, and $-\overset{\Vert}{\underset{O}{C}}-\overset{\vert}{\underset{N}{C}}=\overset{\vert}{C}-$, 39b

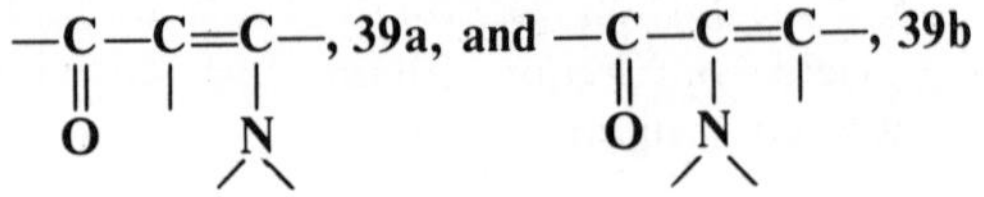

40.   Enol Acetates: $-\overset{\Vert}{\underset{O}{C}}-\overset{\vert}{\underset{OCOCH_3}{C}}=\overset{\vert}{C}-$, 40a, and $-\overset{\Vert}{\underset{O}{C}}-\overset{\vert}{\underset{OCOCH_3}{C}}=\overset{\vert}{C}-$, 40b

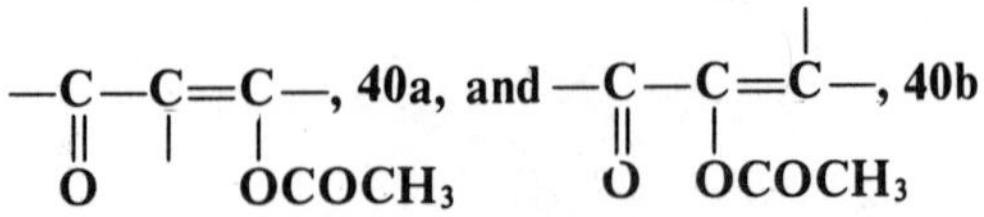

41.   Enol Ethers: $-\overset{\Vert}{\underset{O}{C}}-\overset{\vert}{\underset{OR''}{C}}=\overset{\vert}{C}-$, 41a, and $-\overset{\Vert}{\underset{O}{C}}-\overset{\vert}{\underset{OR''}{C}}=\overset{\vert}{C}-$, 41b

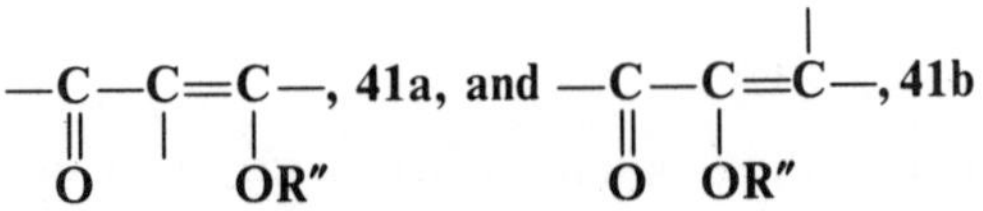

(eq. 1)[a]

R"OH/H⁺ or R"N₂ → LiAlH₄ →

dil H₂SO₄ → , 95%

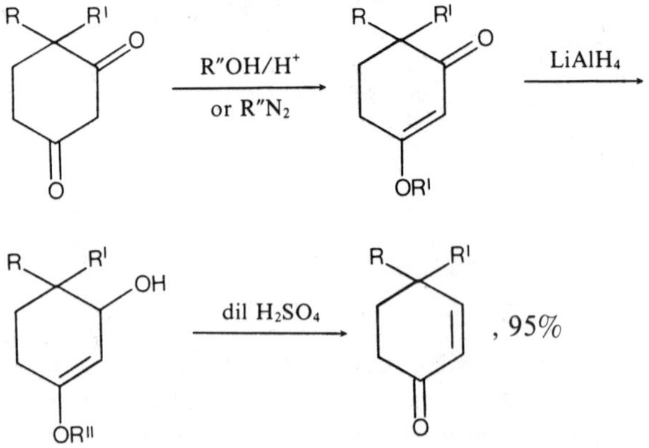

R"OH: R" = Me (HCl, 25°, 8 h, 83% yield)[a]

R" = Et (TsOH, C₆H₆, reflux, 6–8 h, 70–75% yield)[e]

R"N₂ = CH₂N₂, Et₂O/EtOH, 87% yield[a]

(eq. 2)[b]

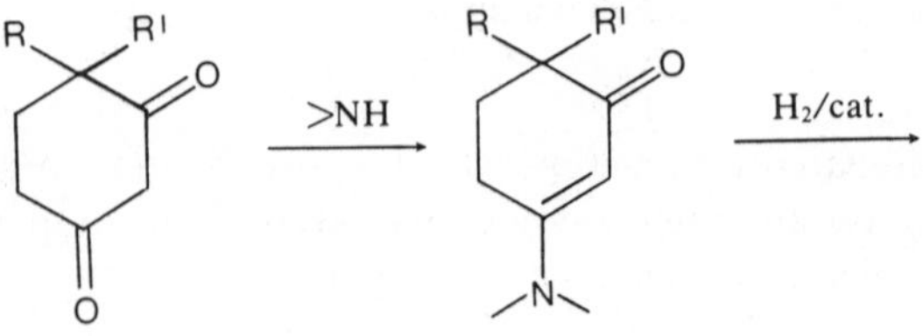

>NH →   H₂/cat. →

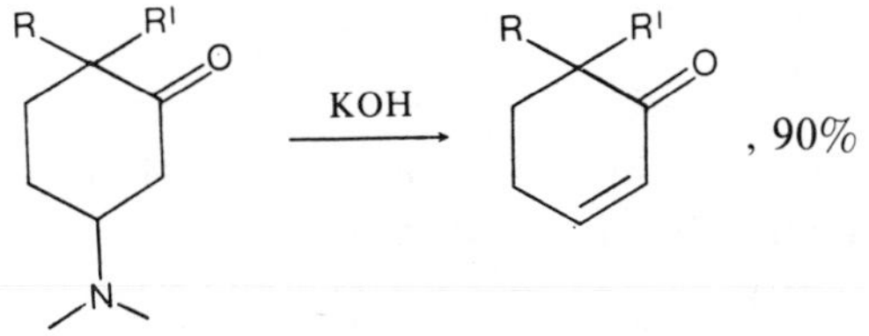

$>$NH = piperidine, TsOH, C$_6$H$_6$, reflux, 92% yield[b]

$>$NH = morpholine, TsOH, PhCH$_3$, reflux, 4–5 h, 72–80%[f]

(eq. 3)[c]

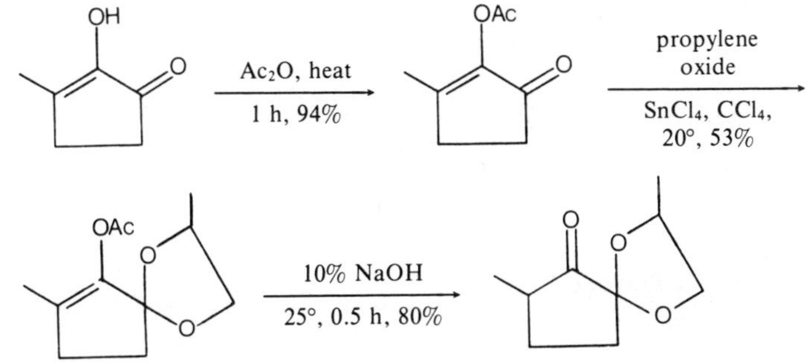

(eq. 4)[d]

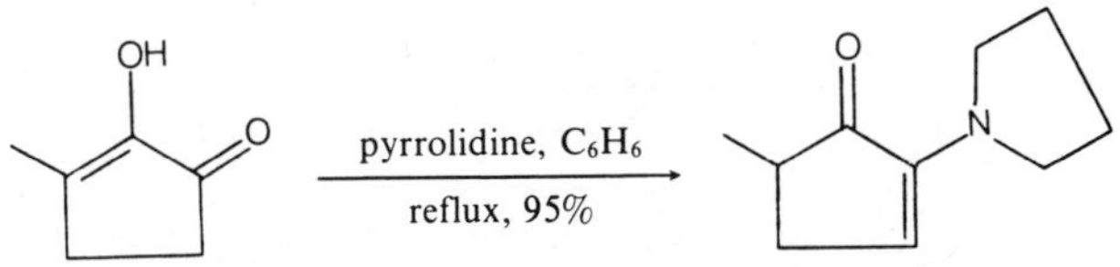

[a] H. O. House and G. H. Rasmusson, *J. Org. Chem.*, **28**, 27 (1963).

[b] P. Kloss, *Chem. Ber.*, **97**, 1723 (1964).

[c] J. L. E. Erickson and F. E. Collins, Jr., *J. Org. Chem.*, **30**, 1050 (1965).

[d] E. Gordon, F. Martens, and H. Gault, *C. R. Hebd. Seances Acad. Sci., Ser. C*, **261**, 4129 (1965).

[e] W. F. Gannon and H. O. House, *Org. Synth., Collect. Vol. V*, 539 (1973).

[f] S. Hünig, E. Lücke, and W. Brenninger, *Org. Synth., Collect. Vol. V*, 808 (1973).

42. S-Butylthio Enol Ether: RC(=CR¹R²)SC₄H₉, 42

Formation (→) / Cleavage (←)[a]

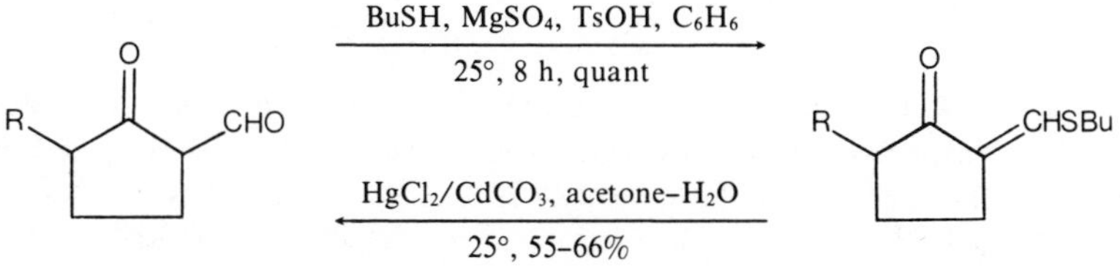

[a] P. R. Bernstein, *Tetrahedron Lett.*, 1015 (1979).

Cyclic Ketals, Hemithio and Dithio Ketals

Cyclohexane-1,2-dione reacts with ethylene glycol (TsOH, C_6H_6, reflux, 6 h) to form the diprotected compound. Monoprotected 1,3-oxathiolanes and 1,3-dithiolanes are isolated on reaction under similar conditions with 2-mercaptoethanol and ethanedithiol, respectively.[a]

[a] R. H. Jaeger and H. Smith, *J. Chem. Soc.,* 160, 646 (1955).

43.   Bismethylenedioxy Derivatives:

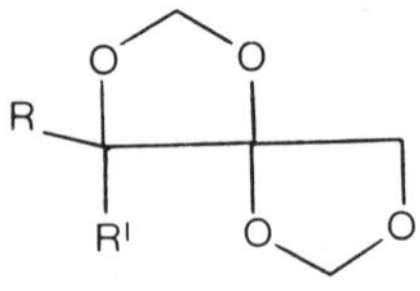

, **43**

Formation (→) / Cleavage (←)[a]

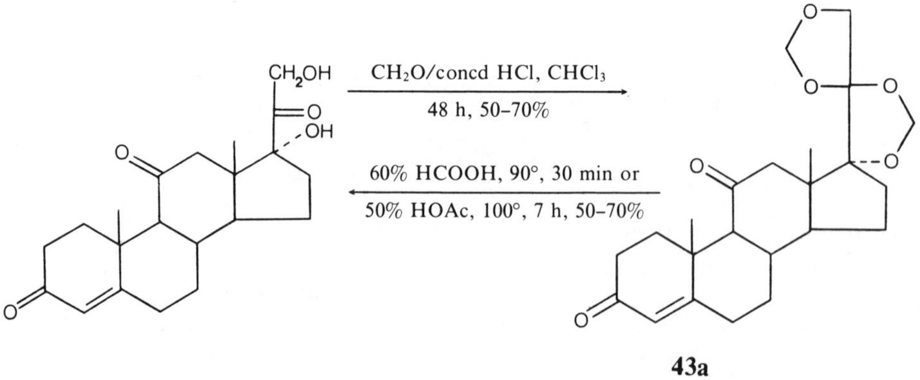

CH₂OH
=O
.OH

CH₂O/concd HCl, CHCl₃
48 h, 50–70%

60% HCOOH, 90°, 30 min or
50% HOAc, 100°, 7 h, 50–70%

43a

Compound **43** is stable to TsOH/C_6H_6 at reflux, and to CrO_3/H^+.[b] It is stable to NBS/hv.[c] In the formation of compound **43a** formaldehyde from formalin can react with a C_{11}-hydroxyl group to form a methoxymethyl ether. Paraformaldehyde can be used to avoid formation of the ethers.[d]

[a] R. E. Beyler, F. Hoffman, R. M. Moriarty, and L. H. Sarett, *J. Org. Chem.,* **26**, 2421 (1961).
[b] J. F. W. Keana, in *Steroid Reactions,* C. Djerassi, Ed., Holden-Day, San Francisco, 1963, pp. 56–61.
[c] D. Duval, R. Condom, and R. Emiliozzi, *C. R. Hebd. Seances Acad. Sci., Ser. C,* **285**, 281 (1977).
[d] J. A. Edwards, M. C. Calzada, and A. Bowers, *J. Med. Chem.,* **7**, 528 (1964).

44. **Tetramethylbismethylenedioxy Derivatives:** 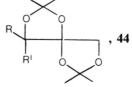 , **44**

A bismethylenedioxy group in a 4-chloro or 11-keto steroid is stable to cleavage by formic acid or glacial acetic acid (100°, 6 h). The tetramethyl compound (e.g., compound **44**) is readily hydrolyzed (50% HOAc, 90°, 3–4 h, 80–90% yield).[a]

[a] A. Roy, W. D. Slaunwhite, and S. Roy, *J. Org. Chem.,* **34,** 1455 (1969).

5

Protection for
The Carboxyl Group

*Included in Reactivity Chart 6.

*Included in Reactivity Chart 6.

Esters

The carboxyl group is present in a number of compounds of biological and synthetic interest (e.g., amino acids, penicillins, macrolide precursors). In peptide syntheses a terminal carboxyl group in an amino acid is either protected (often as a polymer-supported ester) so that coupling can occur at the α-amino group, or converted to an activated ester so that coupling will occur at the carboxyl group. To effect selective protection when a side-chain carboxyl group is present, an amino acid can be treated with formaldehyde to form a 5-oxo-1,3-oxazolidine. Macrolide precursors are often converted to activated esters (e.g., *S-t*-butyl, *S*-2-pyridyl) to facilitate lactonization.

Two examples of polymer-supported esters, **44a, b,** are included in this chapter; extensive information is reported annually[a] on new polymer-supported esters that are used in peptide syntheses.[b] Some activated esters (**50–52**) that have been used as macrolide precursors and some (**29, 53–56**) that have been used in peptide syntheses are also described in this chapter; the many activated esters that are used in peptide syntheses are discussed elsewhere.[c] A useful list, with references, reports all known protected amino acids (e.g., —NH₂, —COOH, and side chain protected compounds).[d]

Some newer, synthetically useful methods to form and cleave esters are described at the beginning of the chapter[e]; conditions that are unique to a protective group are described with that group. Esters that have been used extensively as protective groups are included in Reactivity Chart 6.[f]

[a] *Specialist Periodical Reports,* "Amino–Acids, Peptides, and Proteins," The Chemical Society, London, Vols. 1–10, (1969–1979).

[b] See also: P. Hodge, "Polymer-Supported Protecting Groups," *Chem. Ind. (London),* 624 (1979); R. B. Merrifield, G. Barany, W. L. Cosand, M. Engelhard, and S. Mojsov, "Some Recent Developments in Solid Phase Peptide Synthesis," in *Peptides: Proceedings of the Fifth American Peptide Symposium,* M. Goodman and J. Meienhofer, Eds., Wiley, New York, 1977, pp. 488–502.

[c] *Specialist Periodical Reports,* "Amino–Acids, Peptides, and Proteins," The Chemical Society, London, Vols. 1–7, (1969–1976).

[d] G. A. Fletcher and J. H. Jones, *Int. J. Pept. Protein Res.,* **4,** 347–371 (1972).

[e] For classical methods see C. A. Buehler and D. E. Pearson, *Survey of Organic Syntheses,* Wiley-Interscience, New York, 1970, Vol. 1, pp. 801–830; 1977, Vol. 2, pp. 711–726.

[f] See also: E. Haslam, "Recent Developments in Methods for the Esterification and Protection of the Carboxyl Group," *Tetrahedron,* **36,** 2409–2433 (1980). E. Haslam, "Activation and Protection of the Carboxyl Group," *Chem. Ind.* (*London*), 610–617 (1979); E. Haslam, "Protection of Carboxyl Groups," in *Protective Groups in Organic Chemistry,* J. F. W. McOmie, Ed., Plenum, New York and London, 1973, pp. 183–215.

Newer Methods of Formation

(1) $RCOOH + R'X \xrightarrow[\text{25–80}°,\ \text{1–10 h}]{\text{DBU, } C_6H_6{}^a} RCOOR'$, 70–95%

$RCOOH =$ alkyl, aryl, hindered acids
$R' =$ Et, *n-* and *s-*Bu, CH_3SCH_2 . . .
$X =$ Cl, Br, I

This reaction also proceeds well in acetonitrile, allowing lower temperatures (25°) and shorter times.[b]

(2) $RCOOH + R'OH \xrightarrow[\text{25°, 1–24 h, 70–95\%}]{\text{DCC/DMAP, } Et_2O{}^c} RCOOR'$

$DMAP =$ 4-dimethylaminopyridine
$RCOOH =$ alkyl, aryl, α-amino, hindered acids
$R' =$ Et, *t-*Bu, Ph, $PhCH_2$, aryl

(3) $\underset{\underset{NHPG}{|}}{RCHCOOH} \xrightarrow[\text{pH 7}]{Cs_2CO_3} \xrightarrow[\text{6 h}]{R'X,\ DMF{}^d} \underset{\underset{NHPG}{|}}{RCHCOOR'}$

$R' =$ Me, 80%; $PhCH_2$, 70–90%; o-$NO_2C_6H_4CH_2$, 90%; p-$MeOC_6H_4CH_2$, 70%; Ph_3C, 40–60%; *t-*Bu, 14%; $PhCOCH(Me)$, 80%; *N*-phthalimidomethyl, 80% yield[d]

A study of relative rates of this reaction indicates that $Cs^+ > K^+ > Na^+ > Li^+$; $I^- \gg Br^- \gg Cl^-$; HMPA > DMSO > DMF.[e]

(4) $\underset{\underset{NHPG}{|}}{RCHCOOH} + R'X \xrightarrow[\text{25°, 24 h, 90–95\%}]{\text{NaHCO}_3\text{/DMF}{}^f} \underset{\underset{NHPG}{|}}{RCHCOOR'}$

$R' =$ Et, *n-*Bu, *s-*Bu
$X =$ Br, I

(5) $\text{RCHCOOH} + \text{R'X}$ $\xrightarrow[\substack{25°, \ 3-24 \ h, \ 70-95\% \\ \text{phase transfer reaction}}]{(C_8H_{17})_3N^+MeCl^-, \ aq \ NaHCO_3/CH_2Cl_2{}^g}$ RCHCOOR'
      | |
      NHPG   NHPG

(6) $\text{RCOOH} + \text{R'}_3O^+BF_4^-$ $\xrightarrow[20°, \ 1-24 \ h]{EtN\text{-}i\text{-}Pr_2, \ CH_2Cl_2{}^h}$ $\text{RCOOR'}, \ 70-95\%$

RCOOH = hindered acids
R' = Me, Et

(7) $\text{RCOOH} + \text{Me}_2\text{NCH(OR')}_2$ $\xrightarrow[1-36 \ h]{25-80°{}^i}$ $\text{RCOOR'}, \ 80-95\%$

RCOOH = Ph, 2,4,6-Me$_3$C$_6$H$_2$-, N-protected amino acids
R' = Me, Et, PhCH$_2$, s-Bu

(8) $\text{RCOOH} + \text{R'OH}$ $\xrightarrow[0-20°, \ 24 \ h]{R''NC{}^j}$ $\text{RCOOR'}, \ 36-98\%$

RCOOH = amino, dicarboxylic acids; $\neq$ PhCOOH
R' = Me, Et, t-Bu
R" = t-Bu

(9) $\text{RCOOH} + \text{R'OH}$ $\xrightarrow[25°, \ 12 \ h, \ 75-80\%]{Ph_3P(OSO_2CF_3)_2, \ CH_2Cl_2{}^k}$ RCOOR'

R = aryl
R' = Et

Newer Methods of Cleavage

(1) $\text{RCOOR'} + \text{Nu}^-$ $\xrightarrow{\text{aprotic solvent}{}^l}$ RCOOH

Nu$^-$ = LiS-n-Pr: HMPA, 25°, 1 h, ca. quant yield^m
Nu$^-$ = NaSePh: HMPA–THF, reflux, 7 h, 90–100% yield^n
Nu$^-$ = LiCl: DMF or Py, reflux, 1–18 h, 60–90% yield^o
Nu$^-$ = KO-t-Bu: DMSO, 50–100°, 1–24 h, 65–95% yield^p
Nu$^-$ = NaCN (for decarboxylation of malonic esters): DMSO, 160°, 4 h, 70–80%^{q}

(2) $\text{RCOOR'} + \text{Me}_3\text{SiCl}$ $\xrightarrow[\text{reflux, 5-35 h}]{NaI, \ CH_3CN{}^{r-t}}$ $\text{RCOOH}, \ 70-90\%$

RCOOH = alkyl, aryl, hindered acids
R' = Me, Et, i-Pr, t-Bu, PhCH$_2$

(3) $\quad$ RCOOR$'$ + KO$_2$ $\xrightarrow[\text{25}°, \text{8-72 h}]{\text{18-crown-6, C}_6\text{H}_6{}^u}$ RCOOH, 80–95%

Potassium superoxide cleaves hindered esters of hindered acids; it does not cleave amides.[u]

(4) $\quad$ RCOOR$'$ + KO-t-Bu/H$_2$O $\xrightarrow[\text{80-100\%}]{\text{25}°, \text{2-48 h}^v}$ RCOOH

$\quad\quad$ 1 eq $\quad\quad$ 8 eq $\quad\quad$ 2 eq

RCOOH = Ph, aryl, hindered acids
R$'$ = Me, t-Bu, alkyl

"Anhydrous hydroxide" also cleaves tertiary amides.[v]

(5) $\quad$ RCHCOOR$'$ + BBr$_3$ $\xrightarrow[-10°, \text{1 h} \to 25°, \text{2 h}]{\text{CH}_2\text{Cl}_2{}^w}$ RCHCOOH, 60–85%
$\quad\quad\;\;|$ $\quad\quad\quad\quad\quad\quad\quad\quad\quad\quad\quad\quad\quad\quad\quad\;\;|$
$\quad\quad$ NHPG $\quad\quad\quad\quad\quad\quad\quad\quad\quad\quad\quad\quad\quad\quad\;$ NH$_2$

R$'$ = Me, Et, t-Bu, PhCH$_2$
PG = —COOCH$_2$Ph, —COO-t-Bu; OMe, OEt, O-t-Bu, OCH$_2$Ph side
$\quad\quad$ chain ethers

(6) $\quad$ RCOOR$'$ + AlX$_3$ + R$''$SH $\xrightarrow[\text{70-95\%}]{\text{25}°, \text{5-50 h}^x}$ RCOOH

R = Ph, steroid side chain, . . .
R$'$ = Me, Et, PhCH$_2$
R$''$ = Et, HO(CH$_2$)$_2$—
X = Cl, Br

[a] N. Ono, T. Yamada, T. Saito, K. Tanaka, and A. Kaji, *Bull, Chem. Soc. Jpn.,* **51,** 2401 (1978).

[b] C. G. Rao, *Org. Prep. Proc. Int.,* **12,** 225 (1980).

[c] A. Hassner and V. Alexanian, *Tetrahedron Lett.,* 4475 (1978).

[d] S.-S. Wang, B. F. Gisin, D. P. Winter, R. Makofske, I. D. Kulesha, C. Tzougraki, and J. Meienhofer, *J. Org. Chem.,* **42,** 1286 (1977).

[e] P. E. Pfeffer and L. S. Silbert, *J. Org. Chem.,* **41,** 1373 (1976).

[f] V. Bocchi, G. Casnati, A. Dossena, and R. Marchelli, *Synthesis,* 961 (1979).

[g] V. Bocchi, G. Casnati, A. Dossena, and R. Marchelli, *Synthesis,* 957 (1979).

[h] D. J. Raber, P. Gariano, A. O. Brod, A. Gariano, W. C. Guida, A. R. Guida, and M. D. Herbst, *J. Org. Chem.,* **44,** 1149 (1979).

[i] H. Brechbühler, H. Büchi, E. Hatz, J. Schreiber, and A. Eschenmoser, *Helv. Chim. Acta,* **48,** 1746 (1965).

[j] D. Rehn and I. Ugi, *J. Chem. Res., Synop.,* 119 (1977).

[k] J. B. Hendrickson and S. M. Schwartzman, *Tetrahedron Lett.,* 277 (1975).

[l] J. McMurry, "Ester Cleavages via S$_N$2-Type Dealkylation," *Org. Reactions,* **24,** 187–224 (1976).

[m] P. A. Bartlett and W. J. Johnson, *Tetrahedron Lett.,* 4459 (1970).

[n] D. Liotta, W. Markiewicz, and H. Santiesteban, *Tetrahedron Lett.*, 4365 (1977).

[o] F. Elsinger, J. Schreiber, and A. Eschenmoser, *Helv. Chim. Acta*, **43**, 113 (1960).

[p] F. C. Chang and N. F. Wood, *Tetrahedron Lett.*, 2969 (1964).

[q] A. P. Krapcho, G. A. Glynn, and B. J. Grenon, *Tetrahedron Lett.*, 215 (1967).

[r] M. E. Jung and M. A. Lyster, *J. Am. Chem. Soc.*, **99**, 968 (1977).

[s] T. Morita, Y. Okamoto, and H. Sakurai, *J. Chem. Soc., Chem. Commun.*, 874 (1978).

[t] G. A. Olah, S. C. Narang, B. G. B. Gupta, and R. Malhotra, *J. Org. Chem.*, **44**, 1247 (1979).

[u] J. San Filippo, L. J. Romano, C.-I. Chern, and J. S. Valentine, *J. Org. Chem.*, **41**, 586 (1976).

[v] P. G. Gassman and W. N. Schenk, *J. Org. Chem.*, **42**, 918 (1977).

[w] A. M. Felix, *J. Org. Chem.*, **39**, 1427 (1974).

[x] M. Node, K. Nishide, M. Sai, and E. Fujita, *Tetrahedron Lett.*, 5211 (1978).

1.   Methyl Ester: $RCOOCH_3$, 1

See also p. 155, ref. *e*, and pp. 155–158.

Formation

(1) $ArCOOH + H_2NCON(NO)Me \xrightarrow[\text{0°, 75\%}]{\text{KOH, DME-H}_2\text{O}^a} ArCOOMe$

This method generates diazomethane *in situ.*[a]

(2) $\underset{\underset{NH_2}{|}}{RCHCOOH} + Me_2C(OMe)_2 \xrightarrow[\text{25°, 18 h}]{\text{cat. HCl}^b} \underset{\underset{NH_2 \cdot HCl}{|}}{RCHCOOMe}, 80\text{–}95\%$

Cleavage

(1)

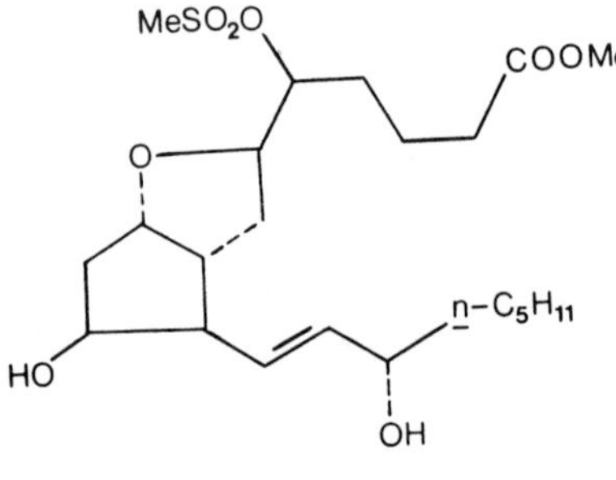

$$\xrightarrow[\text{5°, 15 h}]{\text{LiOH, CH}_3\text{OH–H}_2\text{O (3:1)}^c} RCOOH$$

1a

Lithium hydroxide cleaves only the ester in **1a**.

(2) $RCOOMe + RCOOEt \xrightarrow[\text{75°, 24 h}]{\text{NaCN, HMPA}^d} \underset{75\text{–}92\%}{RCOOH} + \underset{92\%}{RCOOEt}$

(3) $RCOOMe$ $\xrightarrow[\text{reflux, 6 h, 94\%}]{\text{DBN, xylene}^{e}}$ or $\xrightarrow[165°, 48 \text{ h, } 90\%]{\text{DBU, xylene}^{f}}$ $RCOOH$

$RCOOH$ = hindered acids

[a] For example, see S. M. Hecht and J. W. Kozarich, *Tetrahedron Lett.,* 1397 (1973).
[b] J. R. Rachelle, *J. Org. Chem.,* **28,** 2898 (1963).
[c] E. J. Corey, I. Székely, and C. S. Shiner, *Tetrahedron Lett.,* 3529 (1977).
[d] P. Müller and B. Siegfried, *Helv. Chim. Acta,* **57,** 987 (1974).
[e] D. H. Miles and E. J. Parish, *Tetrahedron Lett.,* 3987 (1972).
[f] E. J. Parish and D. H. Miles, *J. Org. Chem.,* **38,** 1223 (1973).

Substituted Methyl Esters

2.   Methoxymethyl Ester: $RCOOCH_2OCH_3$, 2

Formation

(1) $RCOO^{-} H\overset{+}{N}Et_3 + ClCH_2OMe$ $\xrightarrow[25°, 1 \text{ h}]{\text{DMF}^{a}}$ 2

$RCOOH$ = a penicillin

(2) $RCOOH + [MeOCH_2OMe, Zn/BrCH_2COOEt]$ $\xrightarrow[0 \rightarrow 20°, 2 \text{ h}]{0° \quad CH_3COCl}$ **2,** 75–85%[b]

A number of methoxymethyl esters were prepared by this method, which avoids the use of the carcinogen chloromethyl methyl ether.[b]

Cleavage

(1) $RCOOCH_2OCH_3$ $\xrightarrow{R'_3SiBr, \text{ trace MeOH}^{c}}$ $RCOOH$

R' = Et or *n*-Bu

Methoxymethyl ethers are stable to these cleavage conditions.[c]
Methoxymethyl esters are unstable to silica gel chromatography, but are stable to mild acid (0.01 N HCl/EtOAc, MeOH, 25°, 16 h).[d]

[a] A. B. A. Jansen and T. J. Russell, *J. Chem. Soc.,* 2127 (1965).
[b] F. Dardoize, M. Gaudemar, and N. Goasdoue, *Synthesis,* 567 (1977).
[c] S. Masamune, *Aldrichimica Acta,* **11,** 23–30 (1978), see p. 30.
[d] L. M. Weinstock, S. Karady, F. E. Roberts, A. M. Hoinowski, G. S. Brenner, T. B. K. Lee, W. C. Lumma, and M. Sletzinger, *Tetrahedron Lett.,* 3979 (1975).

3. Methylthiomethyl Ester: $RCOOCH_2SCH_3$, 3

Formation

(1) $$RCOOK + CH_3SCH_2Cl \xrightarrow[\text{reflux, 6 h, 85–97\%}]{\text{NaI, 18-crown-6, } C_6H_6{}^{a}} 3$$

(2) $$RCOO^- \, H\overset{+}{N}Et_3 + Me_2\overset{+}{S}Cl \; X^- \xrightarrow[-70 \to 25^\circ\text{, 80–85\%}]{Et_3N\text{, 0.5 h}^{b}} 3$$

Cleavage

(1) $$3 \xrightarrow[\text{reflux, 6 h}]{HgCl_2/CH_3CN\text{–}H_2O} \xrightarrow[20^\circ\text{, 30 min}]{H_2S} RCOOH\text{, 82–98\%}^{a}$$

(2) $$3 \xrightarrow[\text{reflux, 24 h}]{\text{MeI, acetone}} \xrightarrow{1\,N\text{ NaOH}} RCOO^-\text{, 87–97\%}^{c}$$

(3) $$3 \xrightarrow[25^\circ\text{, 15 min}]{CF_3COOH} RCOOH\text{, 80–90\%}^{d}$$

(4) $$3 \xrightarrow{H_2O_2\text{, }(NH_4)_6Mo_7O_{24}} \xrightarrow{\text{NaOH, pH 11}} RCOO^-\text{, 97\%}^{c}$$

[a] L. G. Wade, J. M. Gerdes, and R. P. Wirth, *Tetrahedron Lett.*, 731 (1978).
[b] T.-L. Ho, *Synth. Commun.*, **9**, 267 (1979).
[c] J. M. Gerdes and L. G. Wade, *Tetrahedron Lett.*, 689 (1979).
[d] T.-L. Ho and C. M. Wong, *J. Chem. Soc., Chem. Commun.*, 224 (1973).

4. Tetrahydropyranyl Ester (THP Ester): RCOO-2-tetrahydropyranyl, 4

Formation (→) / *Cleavage* (←)[a]

dihydropyran, TsOH, CH₂Cl₂
20°, 1.5 h, quant
(CH₂)₆COOH
HO
(CH₂)₆CO₂THP
THPO
HOAc, THF–H₂O (4:2:1)
45°, 3.5 h

[a] K. F. Bernady, M. B. Floyd, J. F. Poletto, and M. J. Weiss, *J. Org. Chem.*, **44**, 1438 (1979).

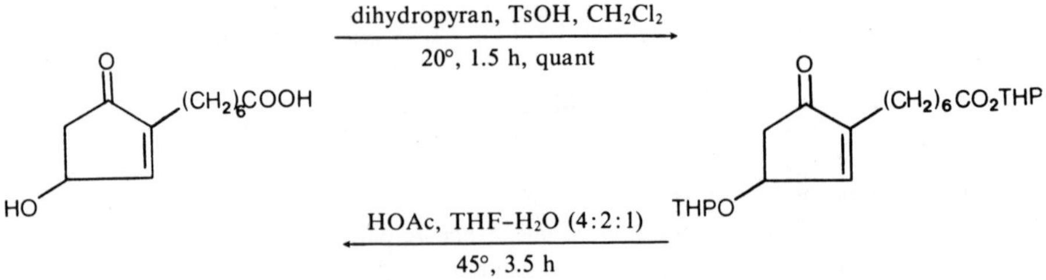

5. Tetrahydrofuranyl Ester: RCOO-2-tetrahydrofuranyl, 5

Formation ($\rightarrow$) / Cleavage ($\leftarrow$)[a]

$$\text{Et}_3\text{N, THF}$$
$$20\text{--}50°, 85\text{--}95\%$$

$$\text{RCOOH} + \qquad\qquad\qquad\qquad\qquad\qquad \textbf{5}$$

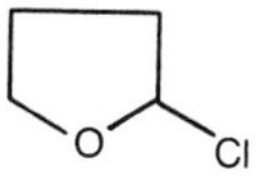

$$\text{HOAc, H}_2\text{O, THF (3:1:1)}$$
$$25°$$

[a] C. G. Kruse, N. L. J. M. Broekhof, and A. van der Gen, *Tetrahedron Lett.*, 1725 (1976).

6. Methoxyethoxymethyl Ester (MEM Ester): RCOOCH$_2$OCH$_2$CH$_2$OCH$_3$, 6

Formation ($\rightarrow$) / Cleavage ($\leftarrow$)[a]

$$\text{MeOCH}_2\text{CH}_2\text{OCH}_2\text{Cl, }i\text{-Pr}_2\text{NEt, CH}_2\text{Cl}_2$$
$$0°, 2\text{ h, high yield}$$

$$\text{RCOOH} \qquad\qquad\qquad\qquad\qquad\qquad \textbf{6}$$

$$3\ N\text{ HCl, THF}$$
$$40°, 12\text{ h}$$

In an attempt to synthesize the macrolide antibiotic chlorothricolide, an un-hindered —COOH group was selectively protected, in the presence of a hindered —COOH group, as a MEM ester that was then reduced to an alcohol group.[b]

[a] A. I. Meyers and P. J. Reider, *J. Am. Chem. Soc.*, **101**, 2501 (1979).
[b] R. E. Ireland and W. J. Thompson, *Tetrahedron Lett.*, 4705 (1979).

7. Benzyloxymethyl Ester: RCOOCH$_2$OCH$_2$C$_6$H$_5$, 7

Formation[a]

$$\text{RCOONa} + \text{PhCH}_2\text{OCH}_2\text{Cl} \xrightarrow[25°, 70\%]{\text{HMPA}} \textbf{7}$$

Cleavage[a]

(1) $7 \xrightarrow[25°, 70\text{--}100\%]{\text{H}_2/\text{Pd--C, EtOH}} \text{RCOOH}$

(2) $7 \xrightarrow[\text{25°, 2 h, 75–95\%}]{\text{aq HCl–THF}}$ RCOOH

[a] P. A. Zoretic, P. Soja, and W. E. Conrad, *J. Org. Chem.*, **40**, 2962 (1975).

8.   Phenacyl Ester: $RCOOCH_2COC_6H_5$, 8

Formation

(1) $\underset{\overset{|}{NHCOOCH_2Ph}}{RCHCOOH} \xrightarrow[\text{20°, 12 h, 83\%}]{\text{PhCOCH}_2\text{Br, Et}_3\text{N, EtOAc}^{[a]}} \underset{\overset{|}{NHCOOCH_2Ph}}{RCHCOOCH_2COPh}$

$$\textbf{8a}$$

(2) $RCOOH \xrightarrow[\text{25°, 10 min, 90–99\%}]{\text{PhCOCH}_2\text{Br, KF/DMF}^{[b]}} \textbf{8}$

R = alkyl, aryl, or hindered acids (at 100°)

Cleavage

(1) $\textbf{8} \xrightarrow[\text{25°, 1 h, 90\%}]{\text{Zn/HOAc}^{[c]}}$ RCOOH

(2) $\textbf{8} \xrightarrow[\text{20°, 1 h, 72\%}]{\text{H}_2\text{/Pd–C, aq MeOH}^{[a]}}$ RCOOH

(3) $\textbf{8a} \xrightarrow[\text{20°, 30 min, 72\%}]{\text{PhSNa, DMF}^{[a]}} \underset{\overset{|}{NHCOOCH_2Ph}}{RCHCOOH}$

As illustrated in eq. 3, a phenacyl ester is much more readily cleaved by nucleophiles than are other esters. Phenacyl esters are stable to acidic hydrolysis (e.g., concd $HCl^{[a]}$; $HBr/HOAc^{[a]}$; 50% $CF_3COOH/CH_2Cl_2^{[d]}$; HF, 0°, 1 $h^{[d]}$).

Under basic coupling conditions an aspartyl peptide that has a β-phenacyl ester *is* converted to a succinimide.[e]

[a] G. C. Stelakatos, A. Paganou, and L. Zervas, *J. Chem. Soc. C*, 1191 (1966).
[b] J. H. Clark and J. M. Miller, *Tetrahedron Lett.*, 599 (1977).
[c] J. B. Hendrickson and C. Kandall, *Tetrahedron Lett.*, 343 (1970).
[d] C. C. Yang and R. B. Merrifield, *J. Org. Chem.*, **41**, 1032 (1976).
[e] M. Bodanszky and J. Martinez, *J. Org. Chem.*, **43**, 3071 (1978).

9. *p*-Bromophenacyl Ester: $RCOOCH_2COC_6H_4\text{-}p\text{-}Br$, 9

In a penicillin synthesis the carboxyl group was protected as a *p*-bromophenacyl ester that was cleaved by nucleophilic displacement (PhSK, DMF, 20°, 30 min, 64% yield). Hydrogenolysis of a benzyl ester was difficult (perhaps because of catalyst poisoning by sulfur); basic hydrolysis of methyl or ethyl esters led to attack at the β-lactam ring.[a]

[a] P. Bamberg, B. Eckström, and B. Sjöberg, *Acta Chem. Scand.*, **21**, 2210 (1967).

10. α-Methylphenacyl Ester: $RCOOCH(CH_3)COC_6H_5$, 10

11. *p*-Methoxyphenacyl Ester: $RCOOCH_2COC_6H_4\text{-}p\text{-}OCH_3$, 11

Compounds **10** and **11** can be cleaved by irradiation (e.g., **10**: 313 nm, dioxane or EtOH, 20°, 6 h, R = amino acids, 80–95% yield[a]; **11**: 313 nm, dioxane or EtOH, 20°, 5–17 h, R = amino acids, 80–95% yield[a]; **11**: >300 nm, 30°, 8 h, R = a gibberellic acid, 36–62% yield[b]). Another phenacyl derivative, $RCOOCH(COC_6H_5)\text{-}C_6H_3\text{-}3,5\text{-}(OCH_3)_2$, cleaved by irradiation, has also been reported.[c]

[a] J. C. Sheehan and K. Umezawa, *J. Org. Chem.*, **38**, 3771 (1973).
[b] E. P. Serebryakov, L. M. Suslova, and V. K. Kucherov, *Tetrahedron*, **34**, 345 (1978).
[c] J. C. Sheehan, R. M. Wilson, and A. W. Oxford, *J. Am. Chem. Soc.*, **93**, 7222 (1971).

12. Diacylmethyl Ester: $RCOOCH(COR')COR''$, 12

Formation[a]

$$RCOOK \xrightarrow[\text{25°, 12 h, satisfactory yield}]{\text{ClCH(COR')COR'', DMF}} 12$$

Cleavage[a]

$$12 \xrightarrow[0 \to 25°, 3\ h]{NaNO_2/H_2O\text{–acetone or } n\text{-PrONO}} \xrightarrow[pH\ 2]{\text{dil HCl}} RCOOH, 90\%$$

R = penicillins, aryl acids
R', R'' = Me, Ph, OMe, OEt

These β-keto esters are moderately stable in acidic and basic media; they can be cleaved (by nitrites) under mild conditions that do not attack a β-lactam ring.[a]

[a] T. Ishimaru, H. Ikeda, M. Hatamura, H. Nitta, and M. Hatanaka, *Chem. Lett.*, 1313 (1977).

13. _N_-Phthalimidomethyl Ester: RCOOCH$_2$N , **13**

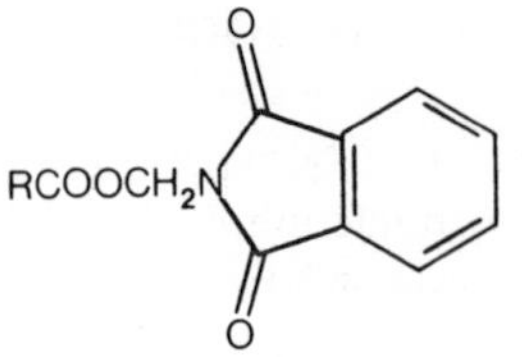

Formation

RCOOH + XCH$_2$-_N_-phthalimido → **13**

X = OH: Et$_2$NH, EtOAc, 37°, 12 h, 70–80% yield[a]
    = Cl: (c-C$_6$H$_{11}$)$_2$NH, DMF or DMSO, 60°, few minutes, 70–80% yield[a]
    = Cl, Br: KF, DMF, 80°, 2 h, 65–75% yield[b]

Cleavage

(1) **13** $\xrightarrow[\text{20°, 3 h, 90\%}]{\text{H}_2\text{NNH}_2/\text{MeOH}^a}$ or $\xrightarrow[\text{25°, 24 h or reflux, 2 h, 82\%}]{\text{Et}_2\text{NH/MeOH, H}_2\text{O}^a}$ RCOOH

(2) **13** $\xrightarrow[\text{20°, 45 min, 77\%}]{\text{NaOH/MeOH, H}_2\text{O}^a}$ RCOOH

(3) **13** $\xrightarrow[\text{25°, 12 h, 80\%}]{\text{Zn/HOAc}^c}$ RCOOH

(4) **13** $\xrightarrow[\text{20°, 16 h, 83\%}]{\text{g HCl/EtOAc}^a}$ or $\xrightarrow[\text{20°, 10–15 min, 80\%}]{\text{HBr/HOAc}^a}$ RCOOH

[a] G. H. L. Nefkens, G. I. Tesser, and R. J. F. Nivard, _Recl. Trav. Chim. Pays-Bas_, **82**, 941 (1963).
[b] K. Horiki, _Synth. Commun._, **8**, 515 (1978).
[c] D. L. Turner and E. Baczynski, _Chem. Ind._ (_London_), 1204 (1970).

14. Ethyl Ester: RCOOC$_2$H$_5$, 14

See also p. 155, ref. _e_, and pp. 155–158.

Formation[a]

$$\text{RCHCOO}^- + \text{EtOTs} \xrightarrow[\text{24–30 h, 80–100\%}]{\text{EtOH, reflux}} \text{RCHCOOEt}$$
$$\overset{|}{\underset{\overset{+}{\text{N}}\text{H}_3}{}} \qquad\qquad \overset{|}{\underset{\overset{+}{\text{N}}\text{H}_3\text{OTs}^-}{}}$$

[a] K. Ueda, _Bull. Chem. Soc. Jpn._, **52**, 1879 (1979).

2-Substituted Ethyl Esters

A number of 2-substituted ethyl esters have been prepared; cleavage of these esters generally occurs by a fragmentation reaction to generate ethylene or an ethylene derivative (eq. 1).

(eq. 1) 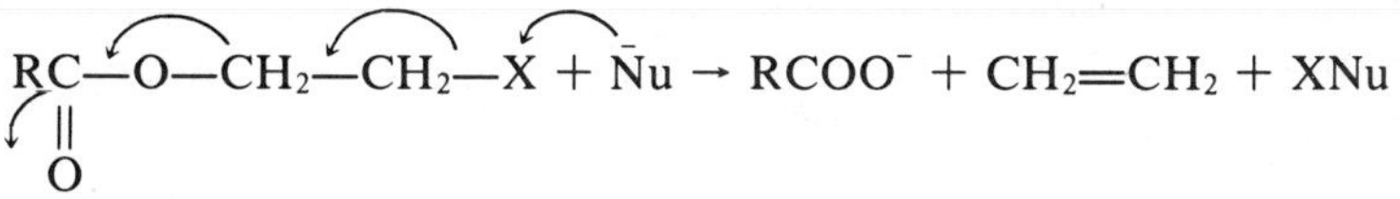 $RC-O-CH_2-CH_2-X + \bar{N}u \rightarrow RCOO^- + CH_2{=}CH_2 + XNu$

15.   2,2,2-Trichloroethyl Ester: RCOOCH₂CCl₃, 15

Formation

$$RCOOH + HOCH_2CCl_3 \xrightarrow{DCC/Py^a} \text{ or } \xrightarrow[\text{reflux}]{TsOH/C_6H_5CH_3^{a,b}} 15$$

Cleavage

(1) $15 \xrightarrow[0°, 2.5\ h]{Zn/HOAc^a} RCOOH$

Trichloroethyl esters are cleaved with zinc/THF buffer at pH 4.2–7.2 (20°, 10 min, 75–95% yield).[c]

(2) $15 \xrightarrow[LiClO_4,\ MeOH]{-1.65\ V^d} RCOOH,\ 87–91\%$

A tribromoethyl ester is cleaved by electrolytic reduction at -0.70 V (85% yield)[d]; a dichloroethyl ester is cleaved at -1.85 V (78% yield).[d]

[a] R. B. Woodward, K. Heusler, J. Gosteli, P. Naegeli, W. Oppolzer, R. Ramage, S. Ranganathan, and H. Vorbrüggen, *J. Am. Chem. Soc.*, **88**, 852 (1966).
[b] J. F. Carson, *Synthesis*, 24 (1979).
[c] G. Just and K. Grozinger, *Synthesis*, 457 (1976).
[d] M. F. Semmelhack and G. E. Heinsohn, *J. Am. Chem. Soc.*, **94**, 5139 (1972).

16.   2-Haloethyl Ester: RCOOCH₂CH₂Cl, 16

2-Haloethyl esters have been cleaved by a variety of nucleophiles:

Li⁺ or Na⁺ Co(I)phthalocyanine/MeOH, 0–20°, 40 min to 60 h, 60–98% yield[a]
NaS(CH₂)₂SNa/CH₃CN, reflux, 2 h, 80–85% yield[b]
NaSeH/EtOH, 25°, 1 h → reflux, 6 min, 92–99% yield[c]

(NaS)$_2$CS/CH$_3$CN, reflux, 1.5 h, 75–86% yield[d]

Me$_3$SnLi/THF, 3 h → n-Bu$_4$N$^+$F$^-$, reflux, 15 min, 78–86% yield[e]

[a] H. Eckert and I. Ugi, *Angew Chem., Inter. Ed. Engl.*, **15**, 681 (1976).

[b] T.-L. Ho, *Synthesis*, 510 (1975).

[c] T.-L. Ho, *Synth. Commun.*, **8**, 301 (1978).

[d] T.-L. Ho, *Synthesis*, 715 (1974).

[e] T.-L. Ho, *Synth. Commun.*, **8**, 359 (1978).

17.   ω-Chloroalkyl Ester: RCOO(CH$_2$)$_n$Cl, 17

ω-Chloroalkyl esters, such as compound **17**, n = 4, 5, have been cleaved by sodium sulfide (reflux, 4 h, 58–85% yield).[a] The reaction proceeds by sulfide displacement of the chloride ion followed by intramolecular displacement of the carboxylate group by the (now) sulfhydryl group.

[a] T.-L. Ho and C. M. Wong, *Synth. Commun.*, **4**, 307 (1974).

18.   2-(Trimethylsilyl)ethyl Ester: RCOOCH$_2$CH$_2$Si(CH$_3$)$_3$, 18

Compound **18** can be prepared from an acid[a] or acid chloride[b] by reaction with the silyl alcohol (DCC/Py–CH$_3$CN, 0°, 5–15 h, 66–97% yield[a]; Py, 25°, 3 h[b]); it is cleaved by fluoride ion[a] (Et$_4$N$^+$F$^-$ or n-Bu$_4$N$^+$F$^-$, DMF or DMSO, 20–30°, 5–60 min, quant yield).[c]

[a] P. Sieber, *Helv. Chim. Acta*, **60**, 2711 (1977).

[b] H. Gerlach, *Helv. Chim. Acta*, **60**, 3039 (1977).

[c] P. Sieber, R. H. Andreatta, K. Eisler, B. Kamber, B. Riniker, and H. Rink, *Peptides: Proceedings of the Fifth American Peptide Symposium*, M. Goodman and J. Meienhofer, Eds., Halsted Press, New York, 1977, pp. 543–545.

19.   2-Methylthioethyl Ester: RCOOCH$_2$CH$_2$SCH$_3$, 19

Compound **19** is prepared from a carboxylic acid and methylthioethyl alcohol or methylthioethyl chloride (MeSCH$_2$CH$_2$OH/TsOH, C$_6$H$_6$, reflux, 55 h, 55% yield; MeSCH$_2$CH$_2$Cl/Et$_3$N, 65°, 12 h, 50–70% yield).[a] It is cleaved by oxidation [H$_2$O$_2$/(NH$_4$)$_6$Mo$_7$O$_{24}$, acetone, 25°, 2 h, 80–95% yield → pH 10–11, 25°, 12–24 h, 85–90% yield][b] and by alkylation followed by hydrolysis (MeI, 70–95% yield → pH 10, 5–10 min, 70–95% yield).[a]

[a] M. J. S. A. Amaral, G. C. Barrett, H. N. Rydon, and J. E. Willett, *J. Chem. Soc. C*, 807 (1966).

[b] P. M. Hardy, H. N. Rydon, and R. C. Thompson, *Tetrahedron Lett.*, 2525 (1968).

20.   2-(p-Nitrophenylsulfenyl)ethyl Ester: $RCOOCH_2CH_2SC_6H_4$-p-NO_2, 20

Compound **20** is similar to compound **19** in that it is prepared from a thioethyl alcohol and cleaved by oxidation [H_2O_2/$(NH_4)_6Mo_7O_{24}$].[a]

[a] M. J. S. A. Amaral, *J. Chem. Soc. C*, 2495 (1969).

21.   2-(p-Toluenesulfonyl)ethyl Ester: $RCOOCH_2CH_2SO_2C_6H_4$-p-CH_3, 21

Formation[a]

$$RCOOH + HOCH_2CH_2SO_2C_6H_4\text{-}p\text{-Me} \xrightarrow[0°,\ 1\ h\ \rightarrow\ 20°,\ 16\ h,\ 70\text{-}90\%]{DCC/Py} 21$$

Cleavage

(1) **21** $\xrightarrow[20°,\ 2\ h,\ 95\%]{Na_2CO_3/dioxane,\ H_2O^a}$ or $\xrightarrow[20°,\ 3\ min,\ 60\text{-}95\%]{1\ N\ NaOH,\ dioxane,\ H_2O^a}$ RCOOH

(2) **21** $\xrightarrow[20°,\ 2.5\ h,\ 60\text{-}85\%]{KCN,\ dioxane,\ H_2O^a}$ RCOOH

(3)

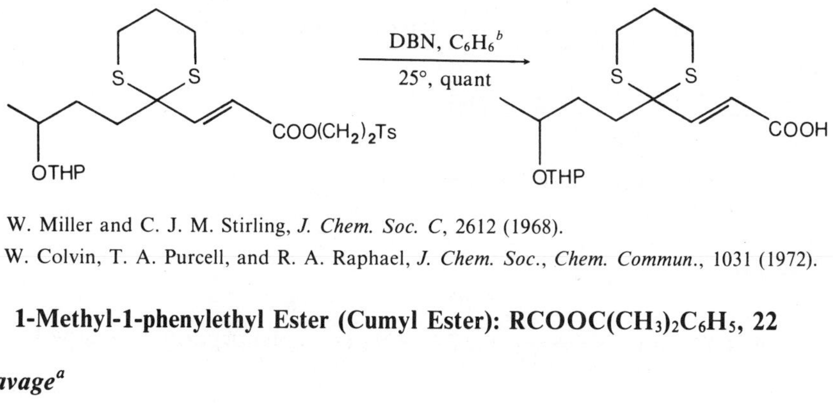

[a] A. W. Miller and C. J. M. Stirling, *J. Chem. Soc. C*, 2612 (1968).
[b] E. W. Colvin, T. A. Purcell, and R. A. Raphael, *J. Chem. Soc., Chem. Commun.*, 1031 (1972).

22.   1-Methyl-1-phenylethyl Ester (Cumyl Ester): $RCOOC(CH_3)_2C_6H_5$, 22

Cleavage[a]

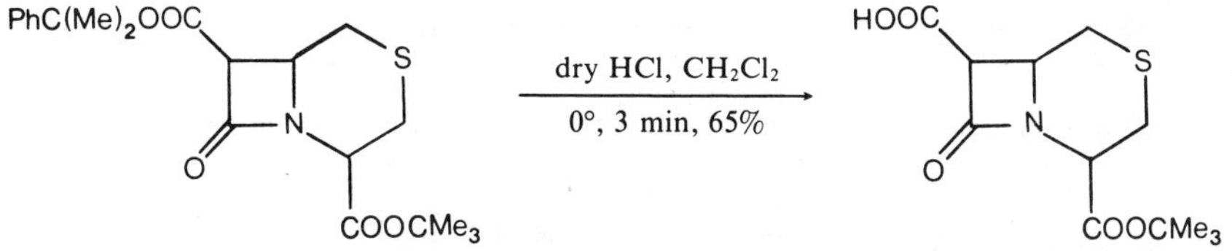

Note that a cumyl ester can be selectively cleaved in the presence of a *t*-butyl ester and a β-lactam.

[a] D. M. Brunwin and G. Lowe, *J. Chem. Soc., Perkin Trans. 1*, 1321 (1973).

23.   *t*-Butyl Ester: RCOOC(CH₃)₃, 23

See also p. 155, ref. *e*, and pp. 155–158.

t-Butyl esters are stable to mild basic hydrolysis, to hydrazine, and to ammonia; they are cleaved by moderately acidic hydrolysis. *t*-Butyl and tetrahydropyranyl esters have similar reactivities toward acids and bases.

Formation

(1)   RCOOH + CH₂=CMe₂ $\xrightarrow[\text{25°, 2–24 h, 50–60\%}]{\text{concd H}_2\text{SO}_4/\text{Et}_2\text{O}^a}$ **23**

   RCOOH = alkyl,[a] amino[b] acids; penicillins[c]

(2)   RCOOH $\xrightarrow[\text{7–10°, 45 min}]{\text{(COCl)}_2/\text{C}_6\text{H}_6\text{, DMF}}$ $\xrightarrow[\text{0°, 3 h}]{t\text{-BuOH, Et}_3\text{N, CH}_2\text{Cl}_2}$ **23**, 75%[d]

   RCOOH = a penicillin

(3)   ArCOCl + LiO-*t*-Bu $\xrightarrow[\text{79–82\%}]{\text{25°, 15 h}^e}$ ArCOO-*t*-Bu

(4)   RCOOH $\xrightarrow[\text{Et}_3\text{N, THF}]{\text{2,4,6-Cl}_3\text{C}_6\text{H}_2\text{COCl}}$ [RCOOCOC₆H₂-2,4,6-Cl₃]

   $\xrightarrow[\text{25°, 20 min, 90\%}]{t\text{-BuOH, DMAP, C}_6\text{H}_6}$ **23**[f]

   DMAP = 4-dimethylaminopyridine

Cleavage

(1)   RCOO-*t*-Bu $\xrightarrow[\text{20°, 3 h}]{\text{HCOOH}^g}$ RCOOH (= cephalosporin derivative)

(2)   RCOO-*t*-Bu $\xrightarrow[\text{25°, 1 h}]{\text{CF}_3\text{COOH, CH}_2\text{Cl}_2{}^h}$ RCOOH

   RCOOH = a cephalosporin derivative

(3)   RCHCOO-*t*-Bu $\xrightarrow[\text{10°, 10 min, 70\%}]{\text{HBr/HOAc}^b}$ RCHCOOH
   | |
   NHPG   NHPG

   PG = phthaloyl or trifluoroacetyl: stable
      = benzyloxycarbonyl or *t*-butoxycarbonyl: cleaved

(4) $\underset{\text{NHPG}}{\text{RCHCOO-}t\text{-Bu}} \xrightarrow[\text{reflux, 30 min, 76\%}]{\text{TsOH/C}_6\text{H}_6{}^b} \underset{\text{NHPG}}{\text{RCHCOOH}}$

A t-butyl ester is stable to the conditions needed to convert an α,β-unsaturated ketone to a dioxolane [$HO(CH_2)_2OH$, TsOH, C_6H_6, reflux].[i]

(5) $\text{ArCOO-}t\text{-Bu} \xrightarrow[\text{100°, 5 h, 94\%}]{\text{KOH/18-crown-6, PhMe}^j} \text{ArCOOH}$

(6) $\text{ArCOO-}t\text{-Bu} \xrightarrow[\text{15 min, 100\%}]{\text{190-200°}^k} \text{ArCOOH}$

[a] A. L. McCloskey, G. S. Fonken, R. W. Kluiber, and W. S. Johnson, *Org. Synth., Collect. Vol. IV*, 261 (1963).

[b] G. W. Anderson and F. M. Callahan, *J. Am. Chem. Soc.*, **82**, 3359 (1960).

[c] R. J. Stedman, *J. Med. Chem.*, **9**, 444 (1966).

[d] C. F. Murphy and R. E. Koehler, *J. Org. Chem.*, **35**, 2429 (1970).

[e] G. P. Crowther, E. M. Kaiser, R. A. Woodruff, and C. R. Hauser, *Org. Synth.*, **51**, 96 (1971).

[f] J. Inanaga, K. Hirata, H. Saeki, T. Katsuki, and M. Yamaguchi, *Bull. Chem. Soc. Jpn.*, **52**, 1989 (1979).

[g] S. Chandrasekaran, A. F. Kluge, and J. A. Edwards, *J. Org. Chem.*, **42**, 3972 (1977).

[h] D. B. Bryan, R. F. Hall, K. G. Holden, W. F. Huffman, and J. G. Gleason, *J. Am. Chem. Soc.*, **99**, 2353 (1977).

[i] A. Martel, T. W. Doyle, and B.-Y. Luh, *Can. J. Chem.*, **57**, 614 (1979).

[j] C. J. Pedersen, *J. Am. Chem. Soc.*, **89**, 7017 (1967).

[k] L. H. Klemm, E. P. Antoniades, and C. D. Lind, *J. Org. Chem.*, **27**, 519 (1962).

24. Cyclopentyl Ester: RCOO-c-C$_5$H$_9$, 24

25. Cyclohexyl Ester: RCOO-c-C$_6$H$_{11}$, 25

Compounds **24**[a] and **25**[b] have been used to protect the β-COOH group in aspartyl peptides to minimize aspartimide formation during acidic or basic reactions. They are cleaved in quantitative yield by HF, -20 to $0°$; they are stable to CF$_3$COOH.

[a] J. Blake, *Int. J. Pept. Protein Res.*, **13**, 418 (1979).

[b] J. P. Tam, T.-W. Wong, M. W. Riemen, F.-S. Tjoeng, and R. B. Merrifield, *Tetrahedron Lett.*, 4033 (1979).

26. Allyl Ester: RCOOCH$_2$CH=CH$_2$, 26

Formation[a]

$$\text{RCOOMe} + \text{HOCH}_2\text{CH=CR'R''} \xrightarrow[\text{1-3 days, 80-95\%}]{\text{NaH, THF}} \textbf{26}$$

Cleavage[b]

$$26 \xrightarrow[0°,\ 1\ h]{Me_2CuLi,\ Et_2O} \xrightarrow{H_3O^+} \begin{array}{c} RCOOH + MeCu + CH_2{=}CHCH_2CH_3 \\ 75\text{–}85\% \end{array}$$

[a] N. Engel, B. Kübel, and W. Steglich, *Angew. Chem., Inter. Ed. Engl.*, **16**, 394 (1977).
[b] T.-L. Ho, *Synth. Commun.*, **8**, 15 (1978).

27.   Cinnamyl Ester: RCOOCH₂CH=CHC₆H₅, 27

Compound **27** can be prepared from an activated carboxylic acid derivative and cinnamyl alcohol; it is cleaved under nearly neutral conditions.[a]

Cleavage[a]

$$27 \xrightarrow[23°,\ 2\text{–}4\ h]{Hg(OAc)_2,\ MeOH} [RCOOCH_2CH(HgOAc)CH(OMe)Ph]$$

$$\xrightarrow[23°,\ 12\text{–}16\ h]{KSCN,\ H_2O} RCOOH,\ 90\%$$

[a] E. J. Corey and M. A. Tius, *Tetrahedron Lett.*, 2081 (1977).

28.   Phenyl Ester: RCOOC₆H₅, 28

Formation (→)[a] / Cleavage (←)[b]

$$\begin{array}{c} \xrightarrow[-20\ \to\ 20°,\ 12\ h,\ 86\%]{PhOH,\ DCC,\ CH_2Cl_2} \\[4pt] RCHCOOH \qquad\qquad\qquad\qquad \mathbf{28} \\ | \\ NHPG \\[4pt] \xleftarrow[20°,\ 15\ min]{H_2O_2/H_2O,\ DMF,\ pH\ 10.5} \end{array}$$

Phenyl esters of amino acids have also been prepared by reaction with phenol and a catalytic amount of BOP (PhOH, BOP, Et₃N, CH₂Cl₂, 25°, 2 h, 73–97% yield).[c]

BOP =

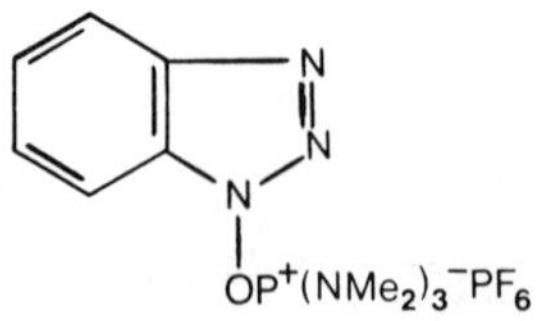

$$\overset{|}{OP^+(NMe_2)_3}{}^-PF_6$$

[a] I. J. Galpin, P. M. Hardy, G. W. Kenner, J. R. McDermott, R. Ramage, J. H. Seely, and R. G. Tyson, *Tetrahedron*, **35**, 2577 (1979).

[b] G. W. Kenner and J. H. Seely, *J. Am. Chem. Soc.*, **94**, 3259 (1972).

[c] B. Castro, G. Evin, C. Selve, and R. Seyer, *Synthesis*, 413 (1977).

29.   *p*-Methylthiophenyl Ester: $RCOOC_6H_4$-*p*-SCH_3, 29

Many phenyl esters substituted with electron-withdrawing groups, (e.g., Cl, $-NO_2$) serve as activated esters in peptide syntheses. Compound **29** is an unactivated, protected compound that is activated by oxidation.[a]

Formation[a]

$$RCHCOOH + HOC_6H_4\text{-}p\text{-}SMe \xrightarrow[0°, 1\text{ h} \to 20°, 12\text{ h}]{DCC/CH_2Cl_2} \textbf{29},\ 60\text{–}70\%$$
$$|$$
$$NHPG$$

Activation[a]

$$\textbf{29} \xrightarrow[20°, 12\text{ h}]{H_2O_2/HOAc} RCHCOOC_6H_4\text{-}p\text{-}SO_2Me,\ 60\text{–}80\%$$
$$|$$
$$NHPG$$

[a] B. J. Johnson and T. A. Ruettinger, *J. Org. Chem.*, **35**, 255 (1970).

30.   Benzyl Ester: $RCOOCH_2C_6H_5$, 30

Benzyl esters can be prepared by classical methods, see p. 155, ref. *e*, and by some of the newer methods described on pp. 155–158. They are useful as protective groups since they can be cleaved by hydrogenolysis.

Cleavage

(1) $RCOOCH_2Ph \xrightarrow[25°, 45\text{ min to 24 h}]{H_2/Pd\text{-}C^{a}} RCOOH + PhMe$

high yields

(2)   Catalytic transfer hydrogenation (eqs. 1 and 2 below) can be used to cleave benzyl esters in some compounds that contain sulfur, a poison for hydrogenolysis catalysts.

(eq. 1)   Pd–C/cyclohexene[b] or /1,4-cyclohexadiene,[c] 25°, 1.5–6 h, good yields

(eq. 2)   Pd–C/4.4% HCOOH, MeOH, 25°, 5–10 min in a column, 100% yield[d]

(3)

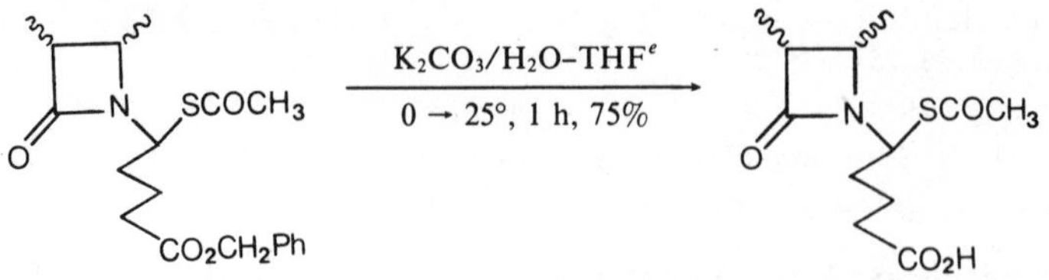

$$\xrightarrow[0 \to 25°, \text{1 h, 75\%}]{K_2CO_3/H_2O\text{–}THF^e}$$

(4) $RCOOCH_2Ph \xrightarrow[0 \to 25°, \text{5 h, 80–95\%}]{AlCl_3/PhOMe, CH_2Cl_2, CH_3NO_2{}^f} RCOOH$

RCOOH = penicillin derivatives

(5) $\underset{\underset{NHPG}{|}}{RCHCOOCH_2Ph} \xrightarrow[50\%]{Na/NH_3{}^g} \underset{\underset{NH_2}{|}}{RCHCOOH}$

PG = —COOCH$_2$Ph; side chain PG = —SCH$_2$Ph

(6) $\underset{\underset{NH_2}{|}}{PhCH_2OCO(CH_2)_2CHCOOCH_2Ph} \xrightarrow[32°, \text{60 min}]{\text{aq CuSO}_4, \text{ EtOH, pH 8}} \xrightarrow{\text{pH 3}} \xrightarrow{\text{EDTA}}$

$$\underset{\underset{NH_2}{|}}{PhCH_2OCO(CH_2)_2CHCOOH, \text{ }75\%^h}$$

EDTA = ethylenediaminetetraacetic acid

Dibenzyl esters of aspartic or glutamic acid are converted into the mono β- or γ-esters, respectively, by this reaction.[h]

(7)   Benzyl esters can be cleaved by electrolytic reduction at -2.7 V.[i]

[a] W. H. Hartung and R. Simonoff, *Org. Reactions*, **VII**, 263–326 (1953).

[b] G. M. Anantharamaiah and K. M. Sivanandaiah, *J. Chem. Soc., Perkin Trans. 1*, 490 (1977).

[c] A. M. Felix, E. P. Heimer, T. J. Lambros, C. Tzougraki, and J. Meienhofer, *J. Org. Chem.*, **43**, 4194 (1978).

[d] B. ElAmin, G. M. Anantharamaiah, G. P. Royer, and G. E. Means, *J. Org. Chem.*, **44**, 3442 (1979).

[e] W. F. Huffman, R. F. Hall, J. A. Grant, and K. G. Holden, *J. Med. Chem.*, **21**, 413 (1978).

[f] T. Tsuji, T. Kataoka, M. Yoshioka, Y. Sendo, Y. Nishitani, S. Hirai, T. Maeda, and W. Nagata, *Tetrahedron Lett.*, 2793 (1979).

[g] C. W. Roberts, *J. Am. Chem. Soc.*, **76**, 6203 (1954).

[h] R. L. Prestidge, D. R. K. Harding, J. E. Battersby, and W. S. Hancock, *J. Org. Chem.*, **40**, 3287 (1975).

[i] V. G. Mairanovsky, *Angew. Chem., Inter. Ed. Engl.*, **15**, 281 (1976).

Substituted Benzyl Esters

31.   Triphenylmethyl Ester: $RCOOC(C_6H_5)_3$, 31

Triphenylmethyl esters are unstable in aqueous solution, but are stable to oxymercuration.[a]

Formation[b]

$$RCOO^-M^+ + Ph_3CBr \xrightarrow[\text{3-5 h, 85-95\%}]{C_6H_6,\ reflux} RCOOCPh_3$$

$$M^+ = Ag^+,\ K^+,\ Na^+$$

Cleavage

(1) $HCl \cdot H_2NCH_2COOCPh_3 \xrightarrow{MeOH\ or\ H_2O/dioxane} HCl \cdot H_2NCH_2COOH$

$\qquad$ 18°, 5 h, 72%; 18°, 24 h, 98%; 100°, 1 min, 98%[c]

(2)   Triphenylmethyl esters have been cleaved by electrolytic reduction at −2.6 V.[d]

[a] W. A. Slusarchyk, H. E. Applegate, C. M. Cimarusti, J. E. Dolfini, P. Funke, and M. Puar, *J. Am. Chem. Soc.*, **100**, 1886 (1978).

[b] K. D. Berlin, L. H. Gower, J. W. White, D. E. Gibbs, and G. P. Sturm, *J. Org. Chem.*, **27**, 3595 (1962).

[c] G. C. Stelakatos, A. Paganou, and L. Zervas, *J. Chem. Soc. C*, 1191 (1966).

[d] V. G. Mairanovsky, *Angew. Chem., Inter. Ed. Engl.*, **15**, 281 (1976).

32.   Diphenylmethyl Ester: $RCOOCH(C_6H_5)_2$, 32

See also p. 155, ref. *e*, and pp. 155–158.
$\quad$ Diphenylmethyl esters are similar in acid lability to *t*-butyl esters and can be cleaved by acidic hydrolysis from *S*-containing peptides that poison hydrogenolysis catalysts.

Formation

(1) $RCOOH + N_2CPh_2 \xrightarrow[\text{0°, 30 min} \rightarrow \text{20°, 4 h}]{acetone[a]}$ **32**, 70%

(2) $RCOOH + Ph_2C{=}NNH_2 \xrightarrow[\text{>90\%}]{I_2,\ HOAc[b]} RCOOCHPh_2$

$\qquad$ RCOOH = a penicillin

(3) $\underset{\underset{OH}{|}}{RCHCOOH}$ + (Ph$_2$CHO)$_3$PO $\xrightarrow[\text{reflux, 1–5 h}]{\text{CF}_3\text{COOH, CH}_2\text{Cl}_2{}^c}$ $\underset{\underset{OCHPh_2}{|}}{RCHCOOCHPh_2}$

$$70–87\%$$

Cleavage

(1) $\underset{\underset{NHCOOCH_2Ph}{|}}{RCHCOOCHPh_2}$ $\xrightarrow[\text{3 h, 90\%}]{\text{H}_2/\text{Pd black, MeOH–THF}^a}$ $\underset{\underset{NH_2}{|}}{RCHCOOH}$

32a

(2)   **32a** $\xrightarrow[\text{20}°, \text{ 30 min, 82\%}]{\text{CF}_3\text{COOH, PhOH}^a}$ $\underset{\underset{NHCOOCH_2Ph}{|}}{RCHCOOH}$

(3)   RCOOCHPh$_2$ $\xrightarrow[\text{reflux, 6 h}]{\text{HOAc}^d}$ RCOOH

(4) $\underset{\underset{NHPG}{|}}{RCHCOOCHPh_2}$ $\xrightarrow[\substack{40°, 0.5 \text{ h} \rightarrow 10°, \\ \text{several hours}}]{\text{BF}_3 \cdot \text{Et}_2\text{O, HOAc}^e}$ $\underset{\underset{NHPG}{|}}{RCHCOOH, 65\%}$

The sulfur–sulfur bond in cystine is stable to these conditions.^{e}

(5)   PhCONHCH$_2$COOCHPh$_2$ $\xrightarrow[\text{reflux, 60 min, 100\%}]{\text{H}_2\text{NNH}_2, \text{MeOH}^f}$ PhCONHCH$_2$CONHNH$_2$

(6)   Diphenylmethyl esters are cleaved by electrolytic reduction at −2.6 V.^{g}

^{a} G. C. Stelakatos, A. Paganou, and L. Zervas, *J. Chem. Soc. C*, 1191 (1966).

^{b} R. Bywood, G. Gallagher, G. K. Sharma, and D. Walker, *J. Chem. Soc., Perkin Trans. 1*, 2019 (1975).

^{c} L. Lapatsanis, *Tetrahedron Lett.*, 4697 (1978).

^{d} E. Haslam, R. D. Haworth, and G. K. Makinson, *J. Chem. Soc.*, 5153 (1961).

^{e} R. G. Hiskey and E. L. Smithwick, *J. Am. Chem. Soc.*, **89**, 437 (1967).

^{f} R. G. Hiskey and J. B. Adams, *J. Am. Chem. Soc.*, **87**, 3969 (1965).

^{g} V. G. Mairanovsky, *Angew. Chem., Inter. Ed. Engl.*, **15**, 281 (1976).

33.   Bis(*o*-nitrophenyl)methyl Ester: RCOOCH(C$_6$H$_4$-*o*-NO$_2$)$_2$, 33

Bis(*o*-nitrophenyl)methyl esters are formed and cleaved by the same methods used for diphenylmethyl esters. They can also be cleaved by irradiation ($h\nu$ = 320 nm, dioxane, THF, . . . , 1–24 h, quant yield).^{a}

^{a} A. Patchornik, B. Amit, and R. B. Woodward, *J. Am. Chem. Soc.*, **92**, 6333 (1970).

34. 9-Anthrylmethyl Ester: RCOOCH$_2$-9-anthryl, 34

Formation

(1) RCOOH + 9-anthrylmethyl chloride $\xrightarrow[\text{reflux, 4–6 h, 70–90\%}]{\text{Et}_3\text{N, MeCN}^a}$ **34**

(2) RCOOH + N$_2$CH-9-anthryl $\xrightarrow[\text{25°, 10 min, 80\%}]{\text{hexane}^b}$ **34**

Cleavage

9-Anthrylmethyl esters are cleaved by acidic (2 N HBr/HOAc, 25°, 10–30 min, 100% yield)^a and basic (0.1 N NaOH/dioxane, 25°, 15 min, 87% yield)^a hydrolysis, and by nucleophiles (MeSNa, THF–HMPA, −20°, 1 h, 90–100% yield).^{c}

^{a} F. H. C. Stewart, *Aust. J. Chem.*, **18**, 1699 (1965).
^{b} M. G. Krakovyak, T. D. Amanieva, and S. S. Skorokhodov, *Synth. Commun.*, **7**, 397 (1977).
^{c} N. Kornblum and A. Scott, *J. Am. Chem. Soc.*, **96**, 590 (1974).

35. 2-(9,10-Dioxo)anthrylmethyl Ester:

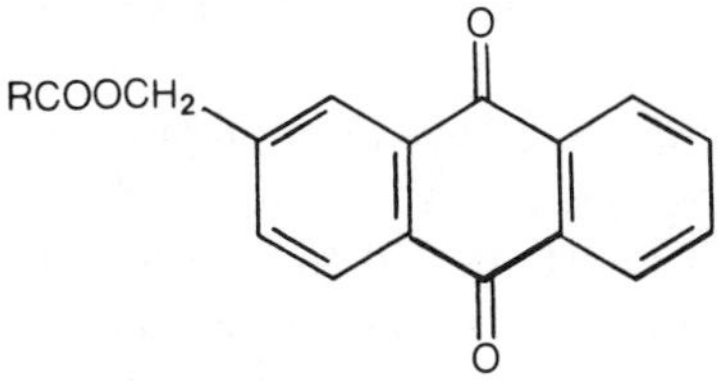

, **35**

Compound **35** is prepared from an *N*-protected amino acid and the anthrylmethyl alcohol in the presence of DCC/hydroxybenzotriazole. It is stable to moderately acidic conditions (e.g., CF$_3$COOH, 20°, 1 h; HBr/HOAc, $t_{1/2} = 65$ h; HCl/CH$_2$Cl$_2$, 20°, 1 h).^{a} Cleavage is effected by reduction of the quinone to the hydroquinone **a**; in the latter electron release from the —OH group of the hydroquinone results in facile cleavage of the methylene-to-carboxylate bond.

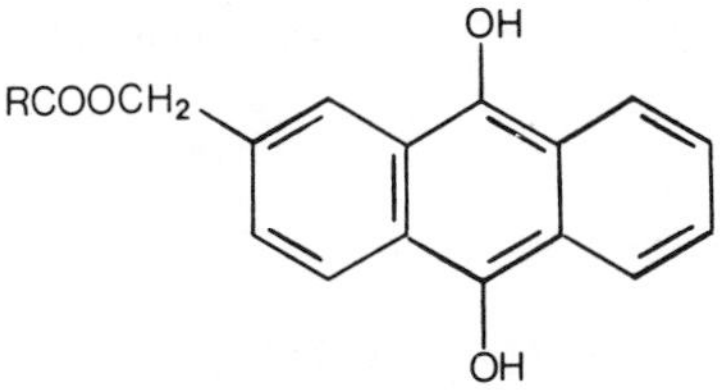

a

Cleavage^a

Compound **35** is cleaved by hydrogenolysis and by the following conditions:

Na$_2$S$_2$O$_4$, dioxane–H$_2$O, pH 7–8, 8 h, 100% yield
Irradiation, *i*-PrOH, 4 h, 99% yield

9-Hydroxyanthrone, Et_3N/DMF, 5 h, 99% yield

9,10-Dihydroxyanthracene/polystyrene resin, 1.5 h, 100% yield

[a] D. S. Kemp and J. Reczek, *Tetrahedron Lett.*, 1031 (1977).

36. 5-Dibenzosuberyl Ester: , 36

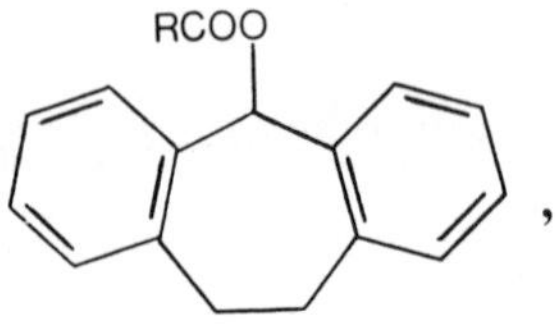

Compound **36** is prepared from dibenzosuberyl chloride (which is also used to protect —OH, —NH, and —SH groups)[a] and a carboxylic acid (Et_3N, reflux, 4 h, 45% yield). It can be cleaved by hydrogenolysis and, like *t*-butyl esters, by acidic hydrolysis (aq HCl/THF, 20°, 30 min, 98% yield).[a]

[a] J. Pless, *Helv. Chim. Acta*, **59**, 499 (1976).

37. 2,4,6-Trimethylbenzyl Ester: $RCOOCH_2C_6H_2$-2,4,6-$(CH_3)_3$, 37

Compound **37** has been prepared from an amino acid and the benzyl chloride (Et_3N, DMF, 25°, 12 h, 60–80% yield); it is cleaved by acidic hydrolysis (CF_3COOH, 25°, 60 min, 60–90% yield; 2 N HBr/HOAc, 25°, 60 min, 80–95% yield) and by hydrogenolysis. It is stable to methanolic hydrogen chloride used to remove N-*o*-nitrophenylsulfenyl groups or triphenylmethyl esters.[a]

[a] F. H. C. Stewart, *Aust. J. Chem.*, **21**, 2831 (1968).

38. *p*-Bromobenzyl Ester: $RCOOCH_2C_6H_4$-*p*-Br, 38

Compound **38** has been used to protect the β-COOH group in aspartic acid. It is cleaved by strong acidic hydrolysis (HF, 0°, 10 min, 100% yield), but is stable to 50% CF_3COOH/CH_2Cl_2 used to cleave *t*-butyl carbamates. It is 5–7 times more stable than a benzyl ester.[a]

[a] D. Yamashiro, *J. Org. Chem.*, **42**, 523 (1977).

39. *o*-Nitrobenzyl Ester: $RCOOCH_2C_6H_4$-*o*-NO_2, 39

40. *p*-Nitrobenzyl Ester: $RCOOCH_2C_6H_4$-*p*-NO_2, 40

Compound **39**, used in this example to protect penicillin precursors, can be cleaved by irradiation (H_2O/dioxane, pH 7). Reductive cleavage of benzyl or *p*-nitrobenzyl esters occurred in lower yields.[a]

p-Nitrobenzyl esters, **40**, have been prepared from the Hg(I) salt of penicillin precursors and the phenyldiazomethane.[b] They are much more stable to acidic hydrolysis (e.g., HBr) than *p*-chlorobenzyl esters and are recommended for terminal —COOH protection in solid-phase peptide synthesis.[c]

p-Nitrobenzyl esters of penicillin and cephalosporin precursors have been cleaved by alkaline hydrolysis with Na_2S (0°, aq acetone, 25–30 min, 75–85% yield).[d] They are also cleaved by electrolytic reduction at −1.2 V.[e]

[a] L. D. Cama and B. G. Christensen, *J. Am. Chem. Soc.*, **100**, 8006 (1978).
[b] W. Baker, C. M. Pant, and R. J. Stoodley, *J. Chem. Soc., Perkin Trans. 1*, 668 (1978).
[c] R. L. Prestidge, D. R. K. Harding, and W. S. Hancock, *J. Org. Chem.*, **41**, 2579 (1976).
[d] S. R. Lammert, A. I. Ellis, R. R. Chauvette, and S. Kukolja, *J. Org. Chem.*, **43**, 1243 (1978).
[e] V. G. Mairanovsky, *Angew Chem., Inter. Ed. Engl.*, **15**, 281 (1976).

41.   *p*-Methoxybenzyl Ester: $RCOOCH_2C_6H_4$-*p*-OCH_3, 41

p-Methoxybenzyl esters have been prepared from the Ag(I) salt of amino acids and the benzyl halide (Et_3N, $CHCl_3$, 25°, 24 h, 60% yield),[a] and from cephalosporin precursors and the benzyl alcohol [$Me_2NCH(OCH_2$-*t*-$Bu)_2$, CH_2Cl_2, 90% yield].[b] They are cleaved by acidic hydrolysis (CF_3COOH/PhOMe, 25°, 3 min, 98% yield[c]; HCOOH, 22°, 1 h, 81% yield).[a]

[a] G. C. Stelakatos and N. Argyropoulos, *J. Chem. Soc. C*, 964 (1970).
[b] J. A. Webber, E. M. Van Heyningen, and R. T. Vasileff, *J. Am. Chem. Soc.*, **91**, 5674 (1969).
[c] F. H. C. Stewart, *Aust. J. Chem.*, **21**, 2543 (1968).

42.   Piperonyl Ester:

RCOOCH₂— , **42**

Compound **42** can be prepared from an amino acid ester and the benzyl alcohol (imidazole/dioxane, 25°, 12 h, 85% yield)[a] or from an amino acid and the benzyl chloride (Et_3N, DMF, 25°, 57–95% yield).[a] It is cleaved, more readily than a *p*-methoxybenzyl ester, by acidic hydrolysis (CF_3COOH, 25°, 5 min, 91% yield).[a]

[a] F. H. C. Stewart, *Aust. J. Chem.*, **24**, 2193 (1971).

43.   4-Picolyl Ester: $RCOOCH_2$-4-pyridyl, 43

Compound **43** has been prepared from amino acids and picolyl alcohol (DCC/CH_2Cl_2, 20°, 16 h, 60% yield) or picolyl chloride (DMF, 90–100°, 2 h, 50% yield). It is cleaved by reduction (H_2/Pd–C, aq EtOH, 10 h, 98% yield; Na/NH_3, 1.5 h, 93% yield) and by basic hydrolysis (1 *N* NaOH, dioxane, 20°, 1 h, 93%

yield). The basic site in a picolyl ester allows its ready separation by extraction into an acidic medium.[a]

[a] R. Camble, R. Garner, and G. T. Young, *J. Chem. Soc. C*, 1911 (1969).

44.   *p*-Ⓟ-Benzyl Ester: RCOOCH$_2$C$_6$H$_4$-*p*-Ⓟ, 44

The first,[a] and still widely used, polymer-supported ester, **44a**, is formed from an amino acid and a chloromethylated copolymer of styrene-divinylbenzene.[a] Originally it was cleaved by basic hydrolysis (2 *N* NaOH, EtOH, 25°, 1 h).[a] Subsequently it has been cleaved by hydrogenolysis (H$_2$/Pd–C, DMF, 40°, 60 psi, 24 h, 71% yield)[b] and by HF, which concurrently removes many amine protective groups.[c]

Monoesterification of a symmetrical dicarboxylic acid chloride can be effected by reaction with a hydroxymethyl copolymer of styrene-divinylbenzene to give compound **44b**; the monosodium salt of a diacid is converted into a dibenzyl ester.[d]

[a] R. B. Merrifield, *J. Am. Chem. Soc.*, **85**, 2149 (1963).
[b] J. M. Schlatter and R. H. Mazur, *Tetrahedron Lett.*, 2851 (1977).
[c] J. Lenard and A. B. Robinson, *J. Am. Chem. Soc.*, **89**, 181 (1967).
[d] C. C. Leznoff and J. M. Goldwasser, *Tetrahedron Lett.*, 1875 (1977).

Silyl Esters

Silyl esters are stable to nonaqueous reaction conditions. A trimethylsilyl ester is cleaved by refluxing in alcohol; more substituted silyl esters are cleaved by mildly acidic or basic hydrolysis.

45.   Trimethylsilyl Ester: RCOOSi(CH$_3$)$_3$, 45

A carboxylic acid is converted to a trimethylsilyl ester by the reagents listed below. It is cleaved in aqueous solutions.

Me$_3$SiCl/Py, CH$_2$Cl$_2$, 30°, 2 h[a]
MeC(OSiMe)=NSiMe$_3$/Et$_2$O, 25°, 15 min, quant[b]
MeCH=C(OMe)OSiMe$_3$/CH$_2$Cl$_2$, 15–25°, 5–40 min, quant[c]
Me$_3$SiNHSO$_2$OSiMe$_3$/CH$_2$Cl$_2$, 30°, 0.5 h, 92–98% yield[d]

[a] B. Fechtig, H. Peter, H. Bickel, and E. Vischer, *Helv. Chim. Acta*, **51**, 1108 (1968).
[b] J. J. de Koning, H. J. Kooreman, H. S. Tan, and J. Verweij, *J. Org. Chem.*, **40**, 1346 (1975).
[c] Y. Kita, J. Haruta, J. Segawa, and Y. Tamura, *Tetrahedron Lett.*, 4311 (1979).
[d] B. E. Cooper and S. Westall, *J. Organomet. Chem.*, **118**, 135 (1976).

46.   Triethylsilyl Ester: $RCOOSi(C_2H_5)_3$, 46

Formation[a]

$$RCOOH \xrightarrow[60°,\ 0.5\ h,\ 95\%]{Et_3SiCl/Py}$$

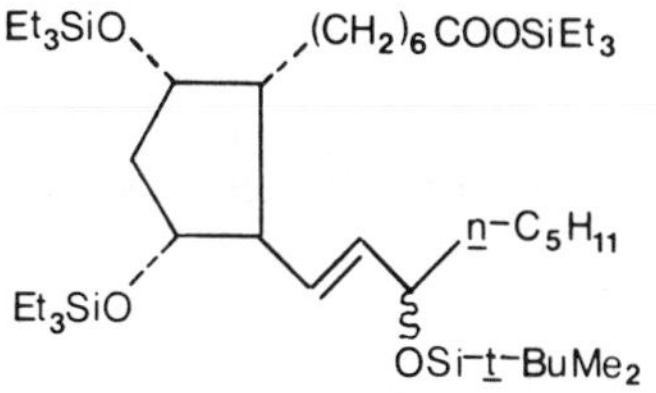

46a

$RCOOH = 15\text{-}O\text{-}Si\text{-}t\text{-}BuMe_2\text{-prostaglandin }F_{1\alpha}$

Cleavage[a]

$$\textbf{46a} \xrightarrow[20°,\ 4\ h,\ 76\%]{HOAc/THF/H_2O\ (8:8:1)}$$

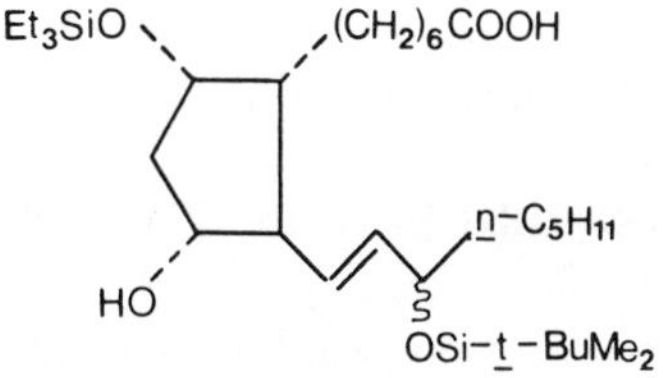

[a] T. W. Hart, D. A. Metcalfe, and F. Scheinmann, *J. Chem. Soc., Chem. Commun.*, 156 (1979).

47.   *t*-Butyldimethylsilyl Ester (TBDMS Group): $RCOOSi(CH_3)_2C(CH_3)_3$, 47

Formation[a]

$$RCOOH + t\text{-}BuMe_2SiCl \xrightarrow[25°,\ 48\ h,\ 88\%]{imidazole/DMF} \textbf{47}$$

Cleavage

(1) $\xrightarrow[25°,\ 20\ h]{HOAc/H_2O/THF^{a}\ (3:1:1)}$ RCOOH

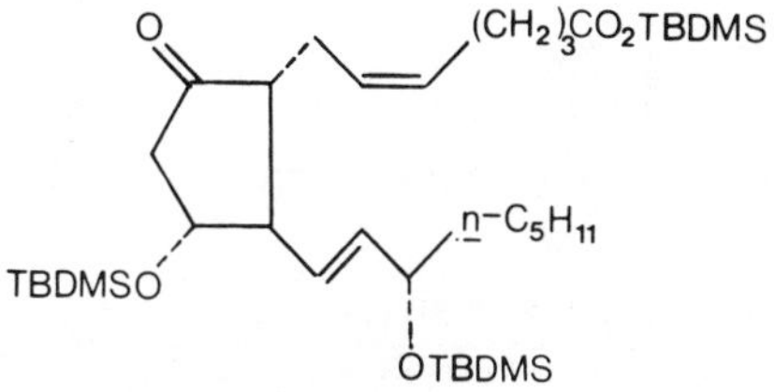

(2)   **47** $\xrightarrow[25°]{n\text{-}Bu_4N^+F^-/DMF^{a}}$ RCOOH

(3)

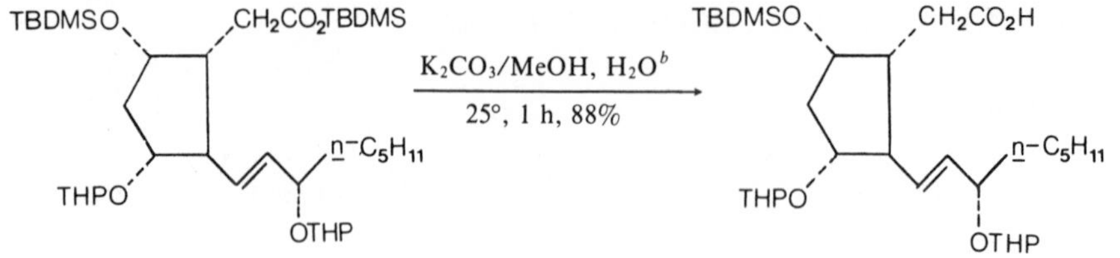

$$\text{K}_2\text{CO}_3/\text{MeOH, H}_2\text{O}^b$$
$$25°, 1 \text{ h, } 88\%$$

In eq. 1 the ester and both ether groups are removed.[a]

[a] E. J. Corey and A. Venkateswarlu, *J. Am. Chem. Soc.*, **94**, 6190 (1972).
[b] D. R. Morton and J. L. Thompson, *J. Org. Chem.*, **43**, 2102 (1978).

48. *i*-Propyldimethylsilyl Ester: RCOOSi(CH₃)₂CH(CH₃)₂, 48

Compound **48** is prepared from a carboxylic acid and the silyl chloride (Et_3N, 0°). It is cleaved at pH 4.5 by conditions that do not cleave a tetrahydropyranyl ether (HOAc–NaOAc, acetone–H_2O, 0°, 45 min → 25°, 30 min, 91% yield).[a]

[a] E. J. Corey and C. U. Kim, *J. Org. Chem.*, **38**, 1233 (1973).

49. Phenyldimethylsilyl Ester: RCOOSi(CH₃)₂C₆H₅, 49

Compound **49** has been prepared from an amino acid and phenyldimethylsilane (Ni/THF, reflux, 3–5 h, 62–92% yield).[a]

[a] M. Abe, K. Adachi, T. Takiguchi, Y. Iwakura, and K. Uno, *Tetrahedron Lett.*, 3207 (1975).

Activated Esters

Thiol esters, more reactive to nucleophiles than the corresponding oxygen esters, have been prepared to activate carboxyl groups, both for lactonization and peptide bond formation. For lactonization S-t-butyl[a] and S-2-pyridyl[b] esters are widely used. Some newer methods that are used to prepare thiol esters are shown below.

(1) RCOOH + R'SH $\xrightarrow[\text{0°, 5 min → 20°, 3 h}]{\text{DCC/DMAP, CH}_2\text{Cl}_2{}^c}$ RCOSR', 85–92%

R' = Et, t-Bu
DMAP = 4-dimethylaminopyridine (10^4 times more effective than pyridine)

(2) 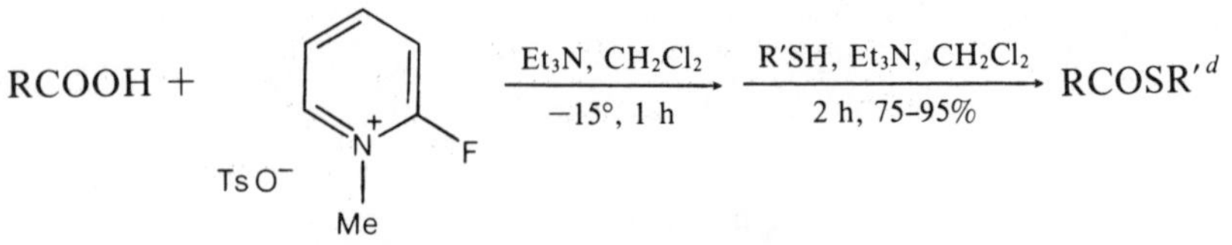 RCOOH +

$\xrightarrow[\text{−15°, 1 h}]{\text{Et}_3\text{N, CH}_2\text{Cl}_2}$ $\xrightarrow[\text{2 h, 75–95\%}]{\text{R'SH, Et}_3\text{N, CH}_2\text{Cl}_2}$ RCOSR'[d]

R' = n-Bu, s-Bu, t-Bu, Ph, 2-pyridyl

(3) $RCOOH + R'SH \xrightarrow[\text{25°, 1 h, 70–100%}]{\text{Me}_2\text{NPOCl}_2, \text{ Et}_3\text{N, DME}^e} RCOSR'$

$R' = Et, i\text{-Pr}, t\text{-Bu}, c\text{-C}_6\text{H}_{11}, Ph$

These neutral conditions can be used to prepare thiol esters of acid- or base-sensitive compounds including penicillins.[e]

(4) $\underset{\underset{NHPG}{|}}{RCHCOOH}$ +

$\xrightarrow{DCC} \xrightarrow{R'SH, Et_3N} \text{ or } \xrightarrow{R'STl}$

$\underset{\underset{NHPG}{|}}{RCHCOSR'}, 70\text{–}100\%^f$

$R' = t\text{-Bu}, Ph, PhCH_2$

(5) $\underset{\underset{NHPG}{|}}{RCHCOOH} + Ph_2POCl \xrightarrow[\text{0°, 30 min}]{\text{Et}_3\text{N, CH}_2\text{Cl}_2} \xrightarrow{R'SH, Et_3N} \text{ or } \xrightarrow[\text{25°, 1 h}]{R'STl}$

$\underset{\underset{NHPG}{|}}{RCHCOSR'}, 70\text{–}100\%^f$

$R' = t\text{-Bu}, Ph, PhCH_2$

(6) $RCOOH + R'SH \xrightarrow[\text{Et}_3\text{N/DMF, 25°, 3 h, 70–85%}]{\text{(EtO)}_2\text{POCN or (PhO)}_2\text{PON}_3} RCOSR'^g$

$R = $ alkyl, aryl, benzyl, amino acids; penicillins
$R' = Et, i\text{-Pr}, n\text{-Bu}, Ph, PhCH_2$

(7) $RCOCl + n\text{-Bu}_3\text{SnSR}' \xrightarrow{CHCl_3} RCOSR'^h$

$R' = t\text{-Bu}: 60°, 0.5 \text{ h}, 90\text{–}95\% \text{ yield}$
$\quad = Ph: 25°, 12 \text{ h}, 92\text{–}95\% \text{ yield}$
$\quad = PhCH_2: 25°, 0.5\text{–}1 \text{ h}, 87\text{–}96\% \text{ yield}$

[a] S. Masamune, S. Kamata, and W. Schilling, *J. Am. Chem. Soc.*, **97**, 3515 (1975).

[b] T. Mukaiyama, R. Matsueda, and M. Suzuki, *Tetrahedron Lett.*, 1901 (1970); E. J. Corey, P. Ulrich, and J. M. Fitzpatrick, *J. Am. Chem. Soc.*, **98**, 222 (1976).

[c] B. Neises and W. Steglich, *Angew. Chem., Inter. Ed. Engl.*, **17**, 522 (1978).

[d] Y. Watanabe, S.-i. Shoda, and T. Mukaiyama, *Chem. Lett.*, 741 (1976).

[e] H.-J. Liu, S. P. Lee, and W. H. Chan, *Synth. Commun.*, **9**, 91 (1979).

[f] K. Horiki, *Synth. Commun.*, **7**, 251 (1977).

[g] S. Yamada, Y. Yokoyama, and T. Shioiri, *J. Org. Chem.*, **39**, 3302 (1974).

[h] D. N. Harpp, T. Aida, and T. H. Chan, *Tetrahedron Lett.*, 2853 (1979).

50.   *S-t-*Butyl Ester: RCOSC(CH₃)₃, 50

See also pp. 180–181.

Formation[a]

$$RCOOR' + Me_2AlS\text{-}t\text{-}Bu \xrightarrow[25°, 75\text{-}100\%]{CH_2Cl_2} RCOS\text{-}t\text{-}Bu$$

R′ = Me, Et

This reaction avoids the use of toxic thallium compounds.

[a] R. P. Hatch and S. M. Weinreb, *J. Org. Chem.*, **42**, 3960 (1977).

51.   *S*-Phenyl Ester: RCOSC₆H₅, 51

See also pp. 180–181.

Formation[a]

$$RCOOH + PhSCN \xrightarrow[25°, 30 \text{ min}, 80\text{-}95\%]{Bu_3P, CH_2Cl_2} RCOSPh$$

[a] P. A. Grieco, Y. Yokoyama, and E. Williams, *J. Org. Chem.*, **43**, 1283 (1978).

52.   *S*-2-Pyridyl Ester: RCOS-2-pyridyl, 52

See also pp. 180–181.

Formation[a]

$$RCOOH + ClCOS\text{-}2\text{-}pyridyl \xrightarrow[0.5 \text{ h}, 95\text{-}100\%]{Et_3N, 0°} 52 + CO_2 + Et_3N \cdot HCl$$

[a] E. J. Corey and D. A. Clark, *Tetrahedron Lett.*, 2875 (1979).

53.   *N*-Hydroxypiperidine Ester: RCOO-*N*-piperidinyl, 53

Compound **53** is an example of an activated ester that is used in peptide syntheses.[a] No racemization was observed when an *N*-hydroxypiperidinyl amino acid ester was coupled by 16 different methods.[b]

Formation[a]

$$\underset{\underset{\text{NHPG}}{|}}{\text{RCHCOCl}} + \textit{N}\text{-hydroxypiperidine} \xrightarrow[\text{0°, 10 min, 75\%}]{\text{Et}_2\text{O}} \mathbf{53}$$

[a] B. O. Handford, J. H. Jones, G. T. Young, and T. F. N. Johnson, *J. Chem. Soc.*, 6814 (1965).
[b] F. Weygand, A. Prox, and W. König, *Chem. Ber.*, **99**, 1451 (1966).

54. *N*-Hydroxysuccinimide Ester: 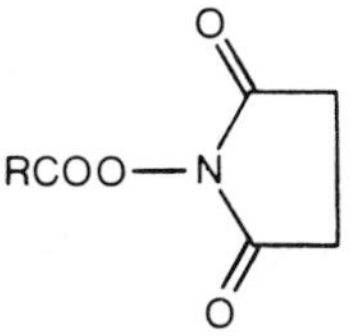 , **54**

Compounds such as **54** are prepared from an amino acid, *N*-hydroxysuccinimide, and DCC in 40–90% yield. *N*-Hydroxysuccinimide, regenerated during peptide formation from the activated ester, is a water-soluble by-product.[a]

[a] G. W. Anderson, J. E. Zimmerman, and F. C. Callahan, *J. Am. Chem. Soc.*, **86**, 1839 (1964).

55. *N*-Hydroxyphthalimide Ester: 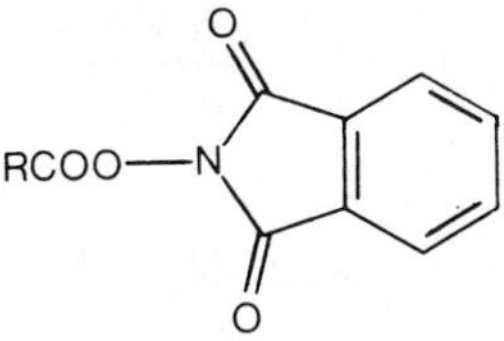 , **55**

Compounds such as **55** are prepared from an amino acid, *N*-hydroxyphthalimide, and DCC/DMF (0°, 12 h, 60–80% yield). *N*-Hydroxyphthalimide, regenerated during peptide formation from the activated ester, is removed as the sodium salt by shaking with aqueous sodium bicarbonate.[a]

[a] G. H. L. Nefkens, G. I. Tesser, and R. J. F. Nivard, *Recl. Trav. Chim. Pays-Bas*, **81**, 683 (1962).

56. *N*-Hydroxybenzotriazole Ester: 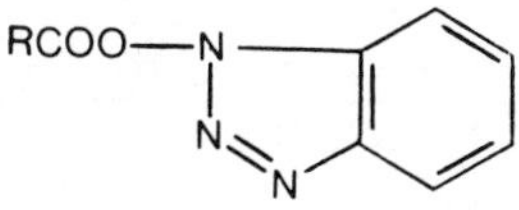 , **56**

Formation[a]

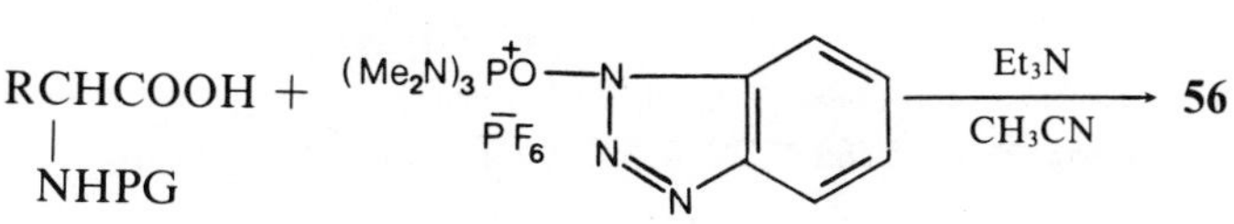

Dipeptides are formed in 80–90% yield with negligible racemization from compounds such as **56**.[a]

[a] B. Castro, J.-R. Dormoy, G. Evin, and C. Selve, *J. Chem. Res., Synop.*, 182 (1977).

Miscellaneous Esters

57. *O*-Acyl Oximes: RCOO—N=CHR′, 57

Formation[a]

$$RCOOK \xrightarrow[-5°, \ 30 \ min]{ClCOOEt/Py-acetone} \xrightarrow[20°, \ 2 \ h]{HON=CHR', \ acetone} 57$$

RCOOK = a penicillin derivative. (Conventional protective groups had disadvantages in this synthesis.)

R′ = Ph, 2-furanyl

Cleavage[a]

$$(1) \quad 57 \xrightarrow[20°, \ 1 \ h, \ 85\%]{NaI/acetone, \ Et_3N} or \xrightarrow[61\%]{NaSCN, \ same \ conditions} RCOOH$$

$$(2) \quad 57 \xrightarrow[20°, \ 2 \ h, \ 60\%]{PhSH/DMF, \ Et_3N} RCOOH$$

[a] G. R. Fosker, K. D. Hardy, J. H. C. Nayler, P. Seggery, and E. R. Stove, *J. Chem. Soc. C*, 1917 (1971).

58. 2,4-Dinitrophenylsulfenyl Ester: RCOOSC$_6$H$_3$-2,4-(NO$_2$)$_2$, 58

Formation (→) / *Cleavage* (←)

$$RCOONa \xrightarrow[25°, \ 22 \ h, \ 60\%]{2,4-(NO_2)_2-C_6H_3SCl, \ C_6H_6{}^{a}} 58$$

$$RCOOH \xleftarrow[1 \ h, \ 40-90\%]{h\nu, \ C_6H_6{}^{b}} 58$$

[a] A. J. Havlik and N. Kharasch, *J. Am. Chem. Soc.*, **78**, 1207 (1956).
[b] D. H. R. Barton, Y. L. Chow, A. Cox, and G. W. Kirby, *J. Chem. Soc.*, 3571 (1965).

59. 2-Alkyl-1,3-oxazoline: R— , **59**

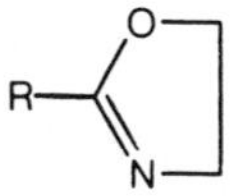

2-Alkyl-1,3-oxazolines, **59**, are prepared to protect both the carbonyl and hydroxyl groups of a carboxyl group. They are stable to Grignard reagents[a] and to lithium aluminum hydride (25°, 2 h).[b]

Formation[c]

$$RCOOH + HOCH_2C(Me)_2NH_2 \xrightarrow[\text{reflux, 70--80\%}]{\text{PhMe}} R—$$

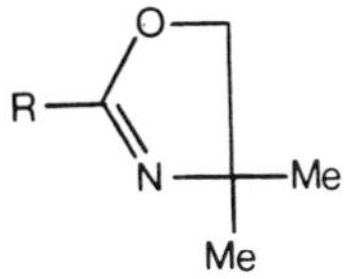

59a

Cleavage

(1) **59a** $\xrightarrow[90\%]{3\ N\ \text{HCl/EtOH}^a}$ RCOOH

(2)

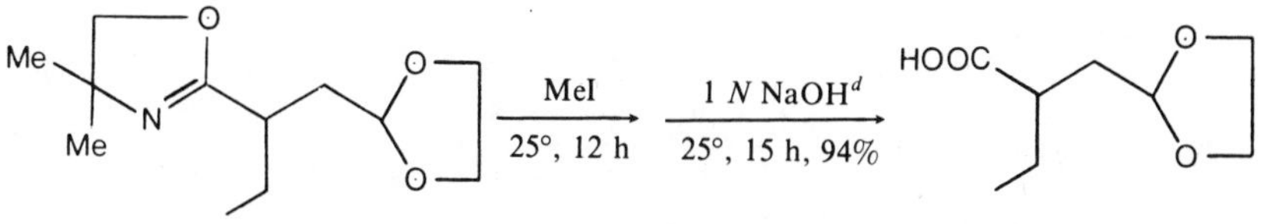

[a] A. I. Meyers and D. L. Temple, *J. Am. Chem. Soc.*, **92**, 6644 (1970).
[b] D. Haidukewych and A. I. Meyers, *Tetrahedron Lett.*, 3031 (1972).
[c] H. L. Wehrmeister, *J. Org. Chem.*, **26**, 3821 (1961).
[d] A. I. Meyers, D. L. Temple, R. L. Nolen, and E. D. Mihelich, *J. Org. Chem.*, **39**, 2778 (1974).

60. 4-Alkyl-5-oxo-1,3-oxazolidine: , **60**

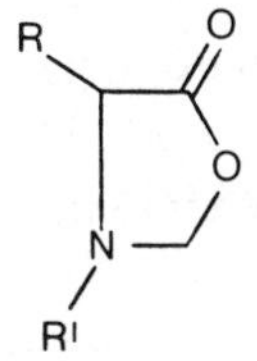

Compounds such as **60** are prepared to allow selective protection of the α- or ω-COOH groups in aspartic and glutamic acids.

Formation (→) / Cleavage (←)[a]

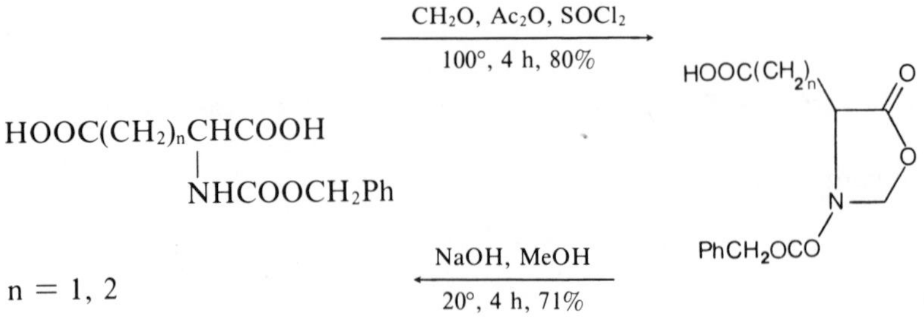

$$\text{CH}_2\text{O, Ac}_2\text{O, SOCl}_2$$
$$100°, 4\text{ h, }80\%$$

HOOC(CH₂)ₙCHCOOH
|
NHCOOCH₂Ph

$$\text{NaOH, MeOH}$$
$$20°, 4\text{ h, }71\%$$

n = 1, 2

[a] M. Itoh, *Chem. Pharm. Bull.*, **17**, 1679 (1969).

61.   5-Alkyl-4-oxo-1,3-dioxolane: , 61

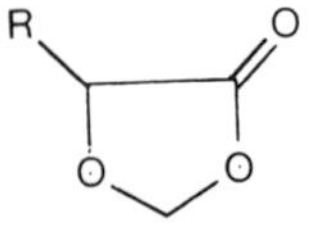

Compounds such as **61** are prepared to protect α-hydroxy carboxylic acids; they are cleaved by acidic hydrolysis of the acetal structure (HCl/DMF, 50°, 7 h, 71% yield), or basic hydrolysis of the lactone.[a]

Formation

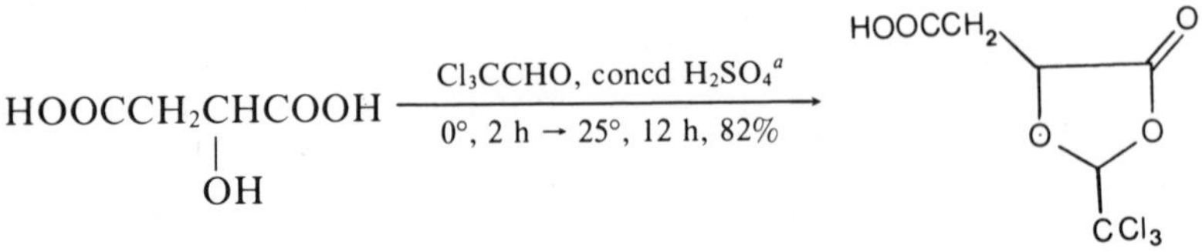

HOOCCH₂CHCOOH
|
OH

$$\text{Cl}_3\text{CCHO, concd H}_2\text{SO}_4{}^a$$
$$0°, 2\text{ h} \rightarrow 25°, 12\text{ h, }82\%$$

[a] H. Eggerer and C. Grünewälder, *Liebigs Ann. Chem.*, **677**, 200 (1964).

Stannyl Esters

62.   Triethylstannyl Ester: $RCOOSn(C_2H_5)_3$, 62

63.   Tri-*n*-butylstannyl Ester: $RCOOSn(n\text{-}C_4H_9)_3$, 63

Stannyl esters have been prepared to protect a —COOH group in the presence of an —NH₂ group.[a] Stannyl esters of N-acylamino acids are stable to reaction with anhydrous amines, and to water and alcohols[b]; aqueous amines convert them to ammonium salts.[b] Stannyl esters of amino acids are cleaved in quantitative yield by water or alcohols.[b]

Formation[a]

$$\xrightarrow[\text{C}_6\text{H}_6,\ \text{reflux, 88\%}]{(n\text{-Bu}_3\text{Sn})_2\text{O or }n\text{-Bu}_3\text{SnOH}}\ \textbf{63a}$$

a

Cleavage

(1) **63a** $\xrightarrow[\text{25°, 15 min, 63\%}]{\text{PhSK/DMF}^{a}}$ **a**

(2) $\underset{\text{NHPG}}{\text{RCHCOOSnEt}_3}$ $\xrightarrow[\text{25°, 30 min, 77\%}]{\text{HOAc/EtOH}^{b}}$ $\underset{\text{NHPG}}{\text{RCHCOOH}}$

[a] P. Bamberg, B. Ekström, and B. Sjöberg, *Acta Chem. Scand.*, **22**, 367 (1968).
[b] M. Frankel, D. Gertner, D. Wagner, and A. Zilkha, *J. Org. Chem.*, **30**, 1596 (1965).

AMIDES AND HYDRAZIDES

To a limited extent carboxyl groups have been protected as amides and hydrazides, derivatives that complement esters in the methods used for their cleavage. Amides and hydrazides are stable to the mild alkaline hydrolysis that cleaves esters. Esters are stable to nitrous acid, effective in cleaving amides, and to the oxidizing agents [including $Pb(OAc)_4$, MnO_2, SeO_2, CrO_3, and $NaIO_4{}^{a}$; $Ce(NH_4)_2(NO_3)_6{}^{b}$; Ag_2O^{c}; and $Hg(OAc)_2{}^{d}$] that have been used to cleave hydrazides.

Classically, amides and hydrazides have been prepared from an ester or an acid chloride and an amine or hydrazine, respectively; they can also be prepared directly from the acid as shown in eqs. 1 and 2.

(1a) $\underset{\text{NHPG}}{\text{RCHCOOH}} + \text{R'NH}_2$ $\xrightarrow[\text{20°, 4 h, 70–90\%}]{\text{DCC, THF or CH}_2\text{Cl}_2{}^{e}}$ $\underset{\text{NHPG}}{\text{RCHCONHR'}}$

(1b) $\underset{\text{NHPG}}{\text{RCHCOOH}}$ $\xrightarrow{\text{H}_2\text{NNH}_2,\ N\text{-hydroxybenzotriazole}^{f}}$ $\underset{\text{NHPG}}{\text{RCHCONHNH}_2}$

(2) $\text{RCOOH} + \text{R'R''NH}$ $\xrightarrow{\text{Ph}_3\text{P}^{g}\text{ or Bu}_3\text{P}/o\text{-NO}_2\text{—C}_6\text{H}_4\text{SCN}^{h}}$ RCONR'R''

Equations 3–5 illustrate three mild methods that can be used to cleave amides. Equations 3a and 3b indicate the conditions that were used by Woodward[i] and Eschenmoser,[j] respectively, in their synthesis of Vitamin B_{12}. Butyl nitrite,[k] nitrosyl chloride,[l] and nitrosonium tetrafluoroborate $(NO^+BF_4^-)$[m] have also been used to cleave amides. Since only tertiary amides are cleaved by potassium t-butoxide (eq. 4), this method can be used to effect selective cleavage of tertiary amides in the presence of primary or secondary amides.[n] (Esters, however, are cleaved by similar conditions.)[o] Photolytic cleavage of nitro amides (eq. 5) is discussed in a review.[p]

(3a) $RCONH_2 \xrightarrow{N_2O_4/CCl_4^{\,i}} RCOOH$

(3b) $RCONH_2 \xrightarrow{[ClCH_2CH=N(\rightarrow O)\text{-}c\text{-}C_6H_{11} + AgBF_4]^{\,j}} \xrightarrow{H_3O^+} RCOOH$

(4) $RCONR'R'' \xrightarrow[24°,\ 2\text{-}48\ h,\ 88\text{-}96\%]{KO\text{-}t\text{-}Bu/H_2O\ (6:2),\ Et_2O^{\,n}} RCOOH$

   $R', R'' \neq H$

(5)   **a, b,** or **c** $\xrightarrow[5\text{-}10\ h,\ 70\text{-}100\%]{350\ nm^{\,p}}$ RCOOH

   a = o-nitroanilides, for example, compound **68**,[q]
   b = N-acyl-7-nitroindoles, for example, compound **69**,[r]
   c = N-acyl-8-nitrotetrahydroquinolines, for example, compound **70**,[s]

Hydrazides have been used in penicillin[a] and peptide syntheses; in the latter syntheses they are converted by nitrous acid to azides to facilitate coupling.

   Some amides and hydrazides that have been prepared to protect carboxyl groups are included in Reactivity Chart 6.

[a] M. J. V. O. Baptista, A. G. M. Barrett, D. H. R. Barton, M. Girijavallabhan, R. C. Jennings, J. Kelly, V. J. Papadimitriou, J. V. Turner, and N. A. Usher, *J. Chem. Soc., Perkin Trans. 1,* 1477 (1977).

[b] T.-L. Ho, H. C. Ho, and C. M. Wong, *Synthesis,* 562 (1972).

[c] Y. Wolman, P. M. Gallop, A. Patchornik, and A. Berger, *J. Am. Chem. Soc.,* **84,** 1889 (1962).

[d] J. B. Aylward and R. O. C. Norman, *J. Chem. Soc. C,* 2399 (1968).

[e] J. C. Sheehan and G. P. Hess, *J. Am. Chem. Soc.,* **77,** 1067 (1955).

[f] For example, see S. S. Wang, I. D. Kulesha, D. P. Winter, R. Makofske, R. Kutny, and J. Meienhofer, *Int. J. Pept. Protein Res.,* **11,** 297 (1978).

[g] L. E. Barstow and V. J. Hruby, *J. Org. Chem.,* **36,** 1305 (1971).

[h] P. A. Grieco, D. S. Clark, and G. P. Withers, *J. Org. Chem.,* **44,** 2945 (1979).

[i] R. B. Woodward, *Pure Appl. Chem.,* **33,** 145 (1973).

[j] U. M. Kempe, T. K. Das Gupta, K. Blatt, P. Gygax, D. Felix, and A. Eschenmoser, *Helv. Chim. Acta,* **55,** 2187 (1972).

[k] N. Sperber, D. Papa, and E. Schwenk, *J. Am. Chem. Soc.*, **70**, 3091 (1948).

[l] M. E. Kuehne, *J. Am. Chem. Soc.*, **83**, 1492 (1961).

[m] G. A. Olah and J. A. Olah, *J. Org. Chem.*, **30**, 2386 (1965).

[n] P. G. Gassman, P. K. G. Hodgson, and R. J. Balchunis, *J. Am. Chem. Soc.*, **98**, 1275 (1976).

[o] P. G. Gassman and W. N. Schenk, *J. Org. Chem.*, **42**, 918 (1977).

[p] B. Amit, U. Zehavi, and A. Patchornik, *Isr. J. Chem.*, **12**, 103 (1974).

[q] B. Amit and A. Patchornik, *Tetrahedron Lett.*, 2205 (1973).

[r] B. Amit, D. A. Ben-Efraim, and A. Patchornik, *J. Am. Chem. Soc.*, **98**, 843 (1976).

[s] B. Amit, D. A. Ben-Efraim, and A. Patchornik, *J. Chem. Soc., Perkin Trans. 1*, 57 (1976).

AMIDES

64. *N,N*-Dimethylamide: $RCON(CH_3)_2$, 64

See also pp. 187–188.

Formation (→) / Cleavage (←)[a]

$$RCOOH \xrightarrow[\text{70°, 3 h}]{\text{SOCl}_2} \xrightarrow{\text{Me}_2\text{NH}} RCONMe_2$$

$$RCONMe_2 \xrightarrow[\text{170°, 6 h}]{\text{KOH, HO(CH}_2)_2\text{OH}} RCOOH$$

In these papers the carboxylic acid to be protected was a stable, unsubstituted compound. Harsh conditions were acceptable for both formation and cleavage of the amide.[a]

[a] D. E. Ames and P. J. Islip, *J. Chem. Soc.*, 351 (1961); 4363 (1963).

65. Pyrrolidinamide: $RCONR'R''$, $[R'R'' = (—CH_2—)_4]$, 65

Formation (→) / Cleavage (←)[a]

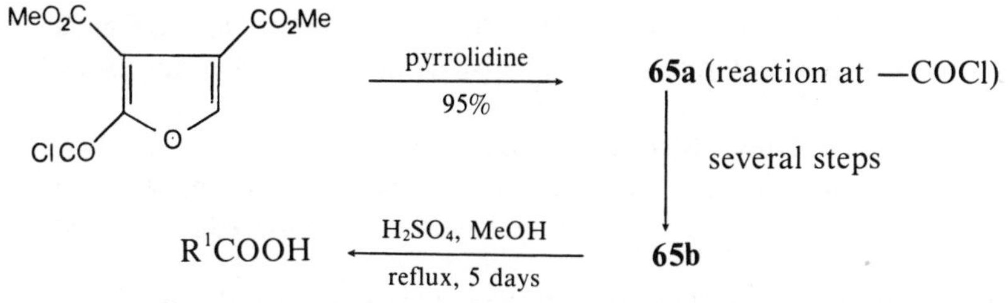

R^1COOH = precursor of DL-camptothecin

[a] A. S. Kende, T. J. Bentley, R. W. Draper, J. K. Jenkins, M. Joyeux, and I. Kubo, *Tetrahedron Lett.*, 1307 (1973).

66. Piperidinamide: RCONR'R", [R'R" = (—CH₂—)₅], 66

Formation[a]

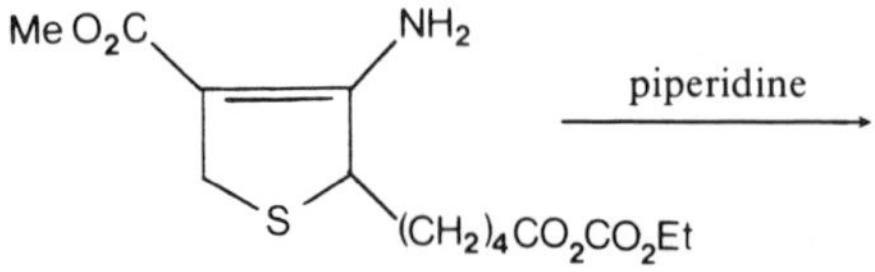

$$\text{piperidine} \longrightarrow 66 \text{ (reaction at } -CO_2CO_2Et)$$

Cleavage[a]

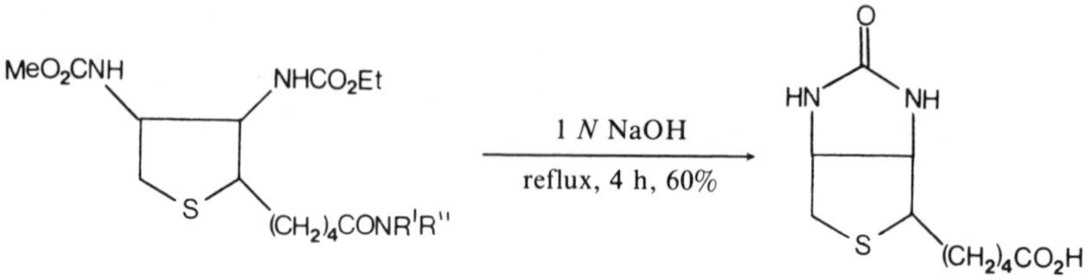

biotin

[a] P. N. Confalone, G. Pizzolato, and M. R. Uskoković, *J. Org. Chem.*, **42**, 1630 (1977).

67. o-Nitroanilide: RCONR'C₆H₄-o-NO₂, R' ≠ H, 67

68. N-7-Nitroindolylamide:

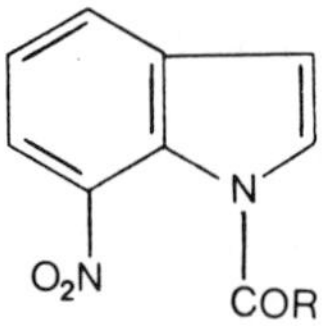

, 68

69. N-8-Nitro-1,2,3,4-tetrahydroquinolylamide:

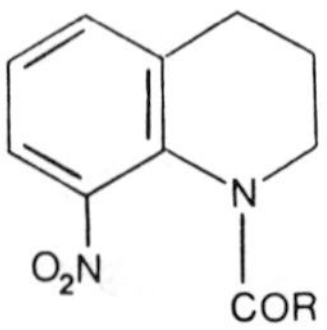

, 69

Amides **67**, (R' = Me, n-Bu, c-C₆H₁₁, Ph, PhCH₂; ≠ H),[a] **68**,[b] and **69**[c] are cleaved in high yields under mild conditions by irradiation at 350 nm (5–10 h).

[a] B. Amit and A. Patchornik, *Tetrahedron Lett.*, 2205 (1973).
[b] B. Amit, D. A. Ben-Efraim, and A. Patchornik, *J. Am. Chem. Soc.*, **98**, 843 (1976).
[c] B. Amit, D. A. Ben-Efraim, and A. Patchornik, *J. Chem. Soc., Perkin Trans. 1*, 57 (1976).

70. p-Ⓟ-Benzenesulfonamide: RCONHSO₂C₆H₄-p-Ⓟ, 70

Compound **70**, prepared from an amino acid activated ester and a polystyrene-sulfonamide, is stable to acidic hydrolysis (CF₃COOH; HBr/HOAc). It is cleaved by the "safety-catch" method shown below.[a]

Cleavage[a]

$$70 \xrightarrow[\text{stable}]{\text{CH}_2\text{N}_2, \text{ Et}_2\text{O–acetone}} \underset{\underset{\text{reactive}}{\overset{|}{\text{Me}}}}{\text{RCONSO}_2\text{C}_6\text{H}_4\text{-}p\text{-}\textcircled{P}} \xrightarrow{0.5 \text{ } N \text{ NaOH}} \text{RCOOH}$$

[a] G. W. Kenner, J. R. McDermott, and R. C. Sheppard, *J. Chem. Soc., Chem. Commun.*, 636 (1971).

HYDRAZIDES

71. Hydrazides: RCONHNH₂, 71

See also pp. 187–188.

Cleavage

(1) $\underset{\overset{|}{\text{NHCO}_2\text{CH}_2\text{Ph}}}{\text{CH}_2\text{CONHCH}_2\text{CONHNH}_2} \xrightarrow[25°, \text{ 10 min, } 74\%]{\text{NBS/H}_2\text{O}^a} \underset{\overset{|}{\text{NHCO}_2\text{CH}_2\text{Ph}}}{\text{CH}_2\text{CONHCH}_2\text{COOH}}$

(2) $\underset{\overset{|}{\text{NHCOPh}}}{\text{RCHCONHNH}_2} \xrightarrow[48°, \text{ 24 h, } 100\%]{60\% \text{ HClO}_4{}^b} \underset{\overset{|}{\text{NHCOPh}}}{\text{RCHCOOH}}$

   71a   **b**

   R = Ph   optically pure

(3) **71a** $\xrightarrow[94\%]{\text{POCl}_3/\text{H}_2\text{O}^b}$ **b**

(4) **71a** $\xrightarrow[\text{HCl/HOAc}^b]{\text{HBr/HOAc or}}$ **b**, 94%

[a] H. T. Cheung and E. R. Blout, *J. Org. Chem.*, **30**, 315 (1965).
[b] J. Schnyder and M. Rottenberg, *Helv. Chim. Acta*, **58**, 521 (1975).

72. *N*-Phenylhydrazide: RCONHNHC₆H₅, 72

See also pp. 187–188.

  Phenylhydrazides have been prepared from amino acid esters and phenyl-hydrazine in 70% yield[a]; they are cleaved by oxidation [Cu(OAc)₂, 95°, 10 min, 67% yield[b]; FeCl₃/1 *N* HCl, 96°, 14 min, 85% yield[c]].

[a] R. B. Kelly, *J. Org. Chem.*, **28**, 453 (1963).
[b] E. W.-Leitz and K. Kühn, *Chem. Ber.*, **84**, 381 (1951).
[c] H. B. Milne, J. E. Halver, D. S. Ho, and M. S. Mason, *J. Am. Chem. Soc.*, **79**, 637 (1957).

73.   N,N'-Diisopropylhydrazide: RCON(i-C₃H₇)NH-i-C₃H₇, 73

See also pp. 187–188.

Compound **73**, prepared to protect penicillin derivatives, is cleaved by oxidation[a]:

$Pb(OAc)_4$/Py, 25°, 10 min, 90% yield
$NaIO_4$/H_2O–THF, H_2SO_4, 20°, 5 min, 89% yield
Aq NBS/THF–Py, 20°, 10 min, 90% yield
CrO_3/HOAc, 25°, 10 min, 65% yield

A number of di- and trisubstituted hydrazides of penicillin and cephalosporin derivatives were prepared to study the effect of N-substitution on ease of oxidative cleavage.[b]

[a] D. H. R. Barton, M. Girijavallabhan, and P. G. Sammes, *J. Chem. Soc., Perkin Trans. 1*, 929 (1972).
[b] D. H. R. Barton et al., *J. Chem. Soc., Perkin Trans. 1*, 1477 (1977).

6

Protection for The Thiol Group

---

*Included in Reactivity Chart 7.

---

*Included in Reactivity Chart 7.

Protection for the thiol group is important in many areas of organic research, particularly in peptide and protein syntheses that often involve the amino acid cysteine, $HSCH_2CH(NH_2)COOH$, CySH. The synthesis[a] of coenzyme A, which converts a carboxylic acid into a thioester, an acyl transfer agent in the biosynthesis or oxidation of fatty acids, also requires the use of thiol protective groups.

A free —SH group can be protected as a thioether or a thioester, or oxidized to a symmetrical disulfide, from which it is regenerated by reduction. Thioethers are in general formed by reaction of the thiol, in a basic solution, with a halide; they are cleaved by reduction with sodium/ammonia, by acid-catalyzed hydrolysis, or by reaction with a heavy metal ion such as silver(I) or mercury(II), followed by hydrogen sulfide. Some groups, including *S*-diphenylmethyl and *S*-triphenyl-methyl thioethers, and *S*-2-tetrahydropyranyl and *S*-isobutoxymethyl hemithio-acetals, can be oxidized by thiocyanogen, $(SCN)_2$, or iodine to a disulfide that is subsequently reduced to the thiol. Thioesters are formed and cleaved in the same way as oxygen esters; they are more reactive to nucleophilic substitution, as indicated by their use as "activated esters." Several miscellaneous protective groups, including thiazolidines, unsymmetrical disulfides, and *S*-sulfenyl derivatives, have been used to a more limited extent. This chapter discusses some synthetically useful thiol protective groups.[b] The more useful groups are included in Reactivity Chart 7.

[a]   J. G. Moffatt and H. G. Khorana, *J. Am. Chem. Soc.*, **83**, 663 (1961).

[b]   See also: Y. Wolman, "Protection of the Thiol Group," in *The Chemistry of the Thiol Group*, S. Patai, Ed., Wiley-Interscience, New York, 1974, Vol. 15/2, pp. 669–684; R. G. Hiskey, V. R. Rao, and W. G. Rhodes, "Protection of Thiols," in *Protective Groups in Organic Chemistry*, J. F. W. McOmie, Ed., Plenum, New York and London, 1973, pp. 235–308; J. F. W. McOmie, "Protective Groups," *Adv. Org. Chem.*, **3**, 251–255 (1963).

THIOETHERS

S-Benzyl and substituted *S*-benzyl derivatives, readily cleaved with sodium/ammonia, are the most frequently used thioethers. *n*-Alkyl thioethers are difficult to cleave and have not been used as protective groups. Alkoxymethyl or alkylthiomethyl hemithio- or dithioacetals ($RSCH_2OR'$ or $RSCH_2SR'$) can be cleaved by acidic hydrolysis, or by reaction with silver or mercury salts, respectively. Mercury(II) salts also cleave dithioacetals, $RS-CH_2SR'$, *S*-tri-phenylmethyl thioethers, $RS-CPh_3$, *S*-diphenylmethyl thioethers, $RS-CHPh_2$, *S*-acetamidomethyl derivatives, $RS-CH_2NHCOCH_3$, and *S*-(*N*-ethylcarba-mates), $RS-CONHEt$. *S*-*t*-Butyl thioethers, RS-*t*-Bu, are cleaved if refluxed with mercury(II); *S*-benzyl thioethers, $RS-CH_2Ph$, are cleaved if refluxed with mercury(II)/1 *N* HCl. Some *β*-substituted *S*-ethyl thioethers are cleaved by reactions associated with the *β*-substituent.

1.   *S*-Benzyl Thioether: $RSCH_2C_6H_5$, 1

Formation[a]

$$CySH + C_6H_5CH_2Cl \xrightarrow[\text{or NH}_3]{\substack{2\ N\ \text{NaOH/EtOH} \\ 30\ \text{min, } 25°}} \mathbf{1},\ 90\%$$

Cleavage

$$(1)\quad \mathbf{1} \xrightarrow[\text{10 min}]{\text{Na/NH}_3{}^{b}} RSH$$

Sodium in boiling butyl alcohol[c] or in boiling ethyl alcohol[d] can be used if the benzyl thioether is insoluble in ammonia.

(2) $1 \xrightarrow[\text{25°, 1 h}]{\text{HF/anisole}^e}$ RSH

Anisole acts as a cation scavenger to prevent polymerization of benzyl fluoride or its attack on a tyrosyl ether. The authors list 15 protective groups that are cleaved by this method, including some branched-chain carbonates and esters, benzyl esters and ethers, the nitro protective group in arginine, and S-benzyl and S-t-butyl thioethers. They report that 12 protective groups are stable under these conditions, including some straight-chain carbonates and esters, N-benzyl derivatives, and S-methyl, S-ethyl, and S-isopropyl thioethers.[e]

(3a) $1 \xrightarrow[\text{NH}_3, \text{ 90 min}]{\text{electrolysis}^f}$ RSH

(b) $C_6H_5CH_2SCH_2CHCO_2H \xrightarrow[\text{DMF/R}_4\text{N}^+\text{X}^-]{-2.8 \text{ V}^g} HSCH_2CHCO_2H, \text{ 82\%}$
$\qquad\qquad\qquad$ | $\qquad\qquad\qquad\qquad\qquad\qquad\qquad$ |
$\qquad\qquad$ NHCOCH$_3$ $\qquad\qquad\qquad\qquad\qquad\qquad$ NHCOCH$_3$

[a] M. Frankel, D. Gertner, H. Jacobson, and A. Zilkha, *J. Chem. Soc.*, 1390 (1960).

[b] J. E. T. Corrie, J. R. Hlubucek, and G. Lowe, *J. Chem. Soc., Perkin Trans. 1*, 1421 (1977).

[c] W. I. Patterson and V. du Vigneaud, *J. Biol. Chem.*, **111**, 393 (1935).

[d] K. Hofmann, A. Bridgwater, and A. E. Axelrod, *J. Am. Chem. Soc.*, **71**, 1253 (1949).

[e] S. Sakakibara, Y. Shimonishi, Y. Kishida, M. Okada, and H. Sugihara, *Bull. Chem. Soc. Jpn.*, **40**, 2164 (1967).

[f] D. A. J. Ives, *Can. J. Chem.*, **47**, 3697 (1969).

[g] V. G. Mairanovsky, *Angew. Chem., Inter. Ed., Engl.*, **15**, 281 (1976).

2. S-4-Methylbenzyl, RSCH$_2$C$_6$H$_4$-4-CH$_3$, 2a, and S-3,4-Dimethylbenzyl Thioether: RSCH$_2$C$_6$H$_3$-3,4-(CH$_3$)$_2$, 2b

The S-4-methyl-[a,b] and S-3,4-dimethylbenzyl[c] thioethers have been used to protect cysteine during solid-phase synthesis. Both compounds are stable to CF_3CO_2H/CH_2Cl_2. The 4-methylbenzyl derivative is completely cleaved by 50% HF/anisole, 0°, 1 h; the 3,4-dimethylbenzyl thioether is completely cleaved by HF/anisole, 0°, 10 min. (The unsubstituted benzyl thioether is completely cleaved by HF/anisole, 20°, 1 h.)[b]

[a] B. W. Erickson and R. B. Merrifield, *J. Am. Chem. Soc.*, **95**, 3750 (1973).

[b] D. H. Live, W. C. Agosta, and D. Cowburn, *J. Org. Chem.*, **42**, 3556 (1977).

[c] D. Yamashiro, R. L. Noble, and C. H. Li, *J. Org. Chem.*, **38**, 3561 (1973).

3. *S*-*p*-Methoxybenzyl Thioether: RSCH$_2$C$_6$H$_4$-*p*-OCH$_3$, 3

Formation[a]

$$\text{CySH} + \text{ClCH}_2\text{C}_6\text{H}_4\text{-}p\text{-OCH}_3 \xrightarrow{\text{NH}_3} \text{3, 78\%}$$

Cleavage

(1) 3 $\xrightarrow[\text{0°, 10–30 min}]{\text{Hg(OAc)}_2/\text{CF}_3\text{CO}_2\text{H}^{b}}$ or $\xrightarrow[\text{20°, 2–3 h}]{\text{Hg(OCOCF}_3)_2/\text{aq HOAc}^{b}}$

$\xrightarrow[\text{HSCH}_2\text{CH}_2\text{OH}]{\text{H}_2\text{S or}}$ CySH, quant

An *S*-*t*-butyl thioether is cleaved in quantitative yield under these conditions.[b]

(2) 3 $\xrightarrow[\text{reflux}]{\text{CF}_3\text{CO}_2\text{H}^{a}}$ RSH

An *S*-*p*-methoxybenzyl thioether is stable to HBr/HOAc.[a]

(3) 3 $\xrightarrow[\text{anisole, 25°, 1 h}]{\text{anhydr HF}^{c}}$ RSH, quant

[a] S. Akabori, S. Sakakibara, Y. Shimonishi, and Y. Nobuhara, *Bull. Chem. Soc. Jpn.*, **37**, 433 (1964).
[b] O. Nishimura, C. Kitada, and M. Fujino, *Chem. Pharm. Bull.*, **26**, 1576 (1978).
[c] S. Sakakibara and Y. Shimonishi, *Bull. Chem. Soc. Jpn.*, **38**, 1412 (1965).

4. *S*-*o*- or *p*-Hydroxy- or Acetoxybenzyl Thioether: RSCH$_2$C$_6$H$_4$-*o*-(or *p*-)-OR′, 4

R′ = H or COCH$_3$

Compounds such as **4** that can be cleaved by base were developed to protect 1-phenyl-5-mercaptotetrazole, **a**, a development restrainer in photographic processes.[a]

Formation (→) / Cleavage (←)[a]

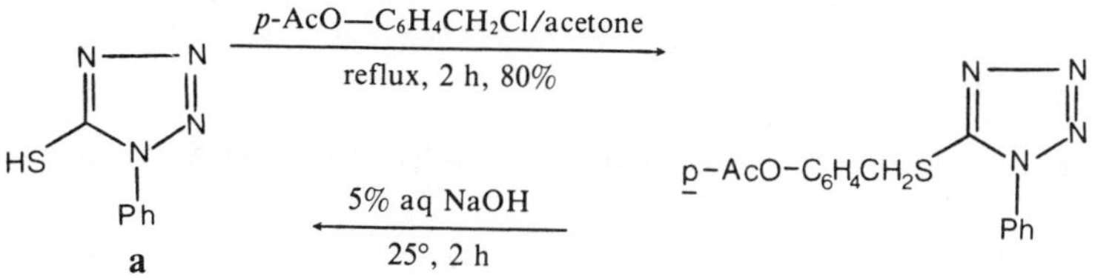

[a] L. D. Taylor, J. M. Grasshoff, and M. Pluhar, *J. Org. Chem.*, **43**, 1197 (1978).

5. *S-p*-Nitrobenzyl Thioether: $RSCH_2C_6H_4\text{-}p\text{-}NO_2$, 5

Formation

$$RSH + ClCH_2C_6H_4\text{-}p\text{-}NO_2 \xrightarrow[\substack{0°, 1\ h\ \to \\ 25°, 0.5\ h}]{1\ N\ NaOH^a} \text{ or } \xrightarrow[C_6H_5CH_3]{NaH^b} \textbf{5},\ 68\%^{\,a}$$

$$RSH = CySH \cdot HCl^a$$
$$= 4\text{-mercapto-}\beta\text{-lactam}^b$$

Cleavage

$$\textbf{5} \xrightarrow[\text{7--8 h, 60--68\%}]{\text{H}_2/\text{Pd--C, HCl or HOAc}} RSCH_2C_6H_4\text{-}p\text{-}NH_2 \xrightarrow[\text{20 h, 60\%}]{HgSO_4/H_2SO_4} \xrightarrow[\text{15 min}]{H_2S}$$

$$RSH,\ 60\%^{\,b} \text{ or RSSR, 76\% after air oxidn}^a$$

[a] M. D. Bachi and K. J. Ross-Petersen, *J. Org. Chem.*, **37**, 3550 (1972).
[b] M. D. Bachi and K. J. Ross-Petersen, *J. Chem. Soc., Chem. Commun.*, 12 (1974).

6. *S*-4-Picolyl Thioether: $RSCH_2\text{-}4\text{-pyridyl}$, 6

Formation (→) / *Cleavage* (←)[a]

$$\xrightarrow[\text{60\%}]{\text{4-picolyl chloride}}$$

$$\text{CySH} \qquad\qquad\qquad\qquad \textbf{6}$$

$$\xleftarrow[\text{0.25 } M\ H_2SO_4,\ 88\%]{\text{Electrolytic redn}}$$

S-4-Picolylcysteine is stable to CF_3CO_2H (7 days), to HBr/HOAc, and to 1 *M* NaOH. References are included for electrolytic reduction of seven other protective groups.[a]

[a] A. Gosden, R. Macrae, and G. T. Young, *J. Chem. Res., Synop.*, 22 (1977).

7. *S*-2-Picolyl *N*-Oxide Thioether: $RSCH_2\text{-}2\text{-pyridyl}$ *N*-Oxide, 7

Formation[a]

$$RSH + 2\text{-picolyl chloride } N\text{-oxide} \xrightarrow{\text{aq NaOH}} \textbf{7},\ \text{moderate yields}$$

Cleavage[a]

$$7 \xrightarrow[\substack{\text{reflux, 7 min} \\ \text{or } 25°, 1.5 \text{ h}}]{\text{Ac}_2\text{O}} \underset{\substack{| \\ \text{OAc}}}{\text{RSCH-2-pyridyl}} \xrightarrow[25°, 3\text{–}12 \text{ h}]{\text{aq NaOH}} \text{RSH, } 79\%$$

[a] Y. Mizuno and K. Ikeda, *Chem. Pharm. Bull.*, **22**, 2889 (1974).

8. *S*-9-Anthrylmethyl Thioether: RSCH$_2$-9-anthryl, 8

Formation (→) / Cleavage (←)[a]

$$\text{RSNa} \quad \xrightarrow[-20°, \text{N}_2]{\text{9-anthrylmethyl chloride/DMF}} \quad \mathbf{8}$$

$$\xleftarrow[0\text{–}25°, 2\text{–}5 \text{ h, } 68\text{–}92\%]{\text{CH}_3\text{SNa/DMF or HMPA}}$$

[a] N. Kornblum and A. Scott, *J. Am. Chem. Soc.*, **96**, 590 (1974).

S-Diphenylmethyl, Substituted *S*-Diphenylmethyl, and *S*-Triphenylmethyl Thioethers

S-Diphenylmethyl, substituted *S*-diphenylmethyl, and *S*-triphenylmethyl thio-ethers have often been formed or cleaved by the same conditions, although some-times in rather different yields. As an effort has been made to avoid repetition in the sections that describe these three protective groups, the reader should glance at all the sections.

9. *S*-Diphenylmethyl Thioether: RSCH(C$_6$H$_5$)$_2$, 9

Formation[a]

$$\text{CySH} + (\text{C}_6\text{H}_5)_2\text{CHOH} \xrightarrow[25°, 15 \text{ min}]{\text{CF}_3\text{CO}_2\text{H}} \text{or} \xrightarrow[50°, 2 \text{ h}]{\text{HBr/HOAc}} \mathbf{9}, >90\%$$

Boron trifluoride etherate (in HOAc, 60–80°, 15 min, high yields)[b] also cata-lyzes formation of *S*-diphenylmethyl and *S*-triphenylmethyl thioethers from aralkylcarbinols.

Yields of thioethers, formed under nonacidic conditions (Ph$_2$CHCl or Ph$_3$CCl, DMF, 80–90°, 2 h, N$_2$), are not as high (RSCHPh$_2$, 50% yield; RSCPh$_3$, 75% yield)[c] as the yields obtained under the acidic conditions described above.

Cleavage

(1) **9** $\xrightarrow[\text{30°, 2 h}]{\text{CF}_3\text{CO}_2\text{H/2.5\% PhOH}^a}$ RSH, 65%

Zervas et al.[a] tried many conditions for the acid-catalyzed formation and removal of the *S*-diphenylmethyl, *S*-4,4'-dimethoxydiphenylmethyl, and *S*-triphenylmethyl thioethers. The best conditions for the *S*-diphenylmethyl thioether are shown above. Phenol or anisole act as cation scavengers.

(2) **9** $\xrightarrow{\text{Na/NH}_3{}^c}$ CySH, 97%

Sodium/ammonia is an efficient but nonselective reagent (RS—Ph, RS—CH$_2$Ph, RS—CPh$_3$, and RS—SR are also cleaved).

(3) **9** $\xrightarrow[\text{HOAc}]{\text{ClSC}_6\text{H}_4\text{-2-NO}_2}$ RSSC$_6$H$_4$-2-NO$_2$ $\xrightarrow[\substack{\text{HS(CH}_2)_2\text{OH} \\ \text{or} \\ \text{dithioerythritol}}]{\text{NaBH}_4 \text{ or}}$ RSH, quant
                                                    ca. quant

S-Triphenylmethyl, *S*-4,4'-dimethoxydiphenylmethyl, and *S*-acetamidomethyl groups are also removed by this method.[d]

[a] I. Photaki, J. T.-Papadimitriou, C. Sakarellos, P. Mazarakis, and L. Zervas, *J. Chem. Soc. C*, 2683 (1970).
[b] R. G. Hiskey and J. B. Adams, Jr., *J. Org. Chem.*, **30**, 1340 (1965).
[c] L. Zervas and I. Photaki, *J. Am. Chem. Soc.*, **84**, 3887 (1962).
[d] A. Fontana, *J. Chem. Soc., Chem. Commun.*, 976 (1975).

10.   *S*-Bis(4-methoxyphenyl)methyl Thioether: RSCH(C$_6$H$_4$-4-OCH$_3$)$_2$, 10

Formation[a]

CySH + ClCH(C$_6$H$_4$-4-OCH$_3$)$_2$ $\xrightarrow[\text{25°, 2 days}]{\text{DMF}}$ **10**, 96%

Cleavage[a]

Also, see the discussion that precedes compound **9**.

10 $\xrightarrow[\text{50–60°, 30 min}]{\text{HBr/HOAc}}$ or $\xrightarrow[\text{reflux, 30 min}]{\text{CF}_3\text{CO}_2\text{H/C}_6\text{H}_5\text{OH}}$ RSH, quant

[a] R. W. Hanson and H. D. Law, *J. Chem. Soc.*, 7285 (1965).

11. *S*-5-Dibenzosuberyl Thioether:

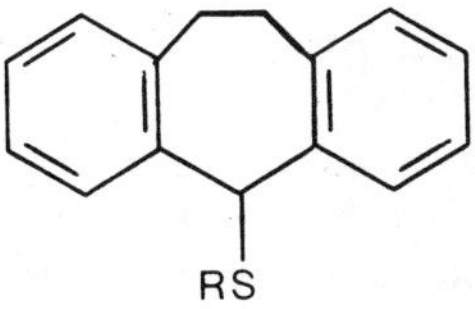

, 11

5-Dibenzosuberyl alcohol reacts in 60% yield with cysteine to give a thioether, **11**, that is cleaved by mercury(II) acetate or oxidized by iodine to cystine. The dibenzosuberyl group has also been used to protect —OH, —NH$_2$, and —COOH groups.[a]

[a] J. Pless, *Helv. Chim. Acta*, **59**, 499 (1976).

12. *S*-Triphenylmethyl Thioether: $RSC(C_6H_5)_3$, 12

S-Triphenylmethyl thioethers have been formed by reaction of the thiol with triphenylmethylcarbinol/anhydr CF$_3$CO$_2$H (85–90% yield) or with triphenylmethyl chloride (75% yield).

Cleavage

Also, see the discussion that precedes compound **9**.

(1) R'SCPh$_3$ $\xrightarrow[\text{90°, 1.5 h}]{\text{HCl/aq HOAc}^a}$ R'SH

(2) R'SCPh$_3$ $\xrightarrow[\substack{\text{reflux, 3 h} \rightarrow \\ \text{25°, 12 h}}]{\text{Hg(OAc)}_2\text{/EtOH}^a}$ $\xrightarrow{\text{H}_2\text{S}}$ R'SH, 61%

(3) R'SCPh$_3$ $\xrightarrow[\text{90°, 1.5 h}]{\text{AgNO}_3\text{/EtOH, Py}^a}$ $\xrightarrow{\text{H}_2\text{S}}$ R'SH, 47%

$$R' = \underset{\underset{Ph_2CHS}{|}}{CBZNHCys} - \underset{|}{CysGlyOEt}$$

An *S*-triphenylmethyl thioether can be selectively cleaved in the presence of an *S*-diphenylmethyl thioether by acidic hydrolysis or by heavy-metal ions.[a]

As a result of the structure of the substrate, the relative yields of cleavage by AgNO$_3$ and Hg(OAc)$_2$ can be reversed.[b]

(4) Thiocyanogen [(SCN)$_2$, 5°, 4 h, 40%] selectively oxidizes an *S*-triphenylmethyl thioether (RS—CPh$_3$) to the disulfide (RSSR) in the presence of an *S*-diphenylmethyl thioether.[c]

(5) *S*-Triphenylmethylcysteine is readily oxidized by iodine (CH$_3$OH, 25°) to cystine.[d] *S*-Benzyl and *S*-*t*-butyl thioethers are stable to the action of iodine.

(6) **12** $\xrightarrow[\text{DMF, R}_4\text{N}^+\text{X}^-]{-2.6\ \text{V}^e}$ RSH

[a] R. G. Hiskey, T. Mizoguchi, and H. Igeta, *J. Org. Chem.*, **31**, 1188 (1966).

[b] R. G. Hiskey and J. B. Adams, *J. Org. Chem.*, **31**, 2178 (1966).

[c] R. G. Hiskey, T. Mizoguchi, and E. L. Smithwick, *J. Org. Chem.*, **32**, 97 (1967).

[d] B. Kamber, *Helv. Chim. Acta*, **54**, 398 (1971).

[e] V. G. Mairanovsky, *Angew. Chem., Inter. Ed., Engl.*, **15**, 281 (1976).

13. *S*-Diphenyl-4-pyridylmethyl Thioether: $RSC(C_6H_5)_2$-4-pyridyl, 13

Formation (→) / Cleavage (←)[a]

$$\text{CySH} + (\text{C}_6\text{H}_5)_2\text{C-4-C}_5\text{H}_5\text{N} \xrightarrow[\text{60°, 48 h}]{\text{BF}_3 \cdot \text{Et}_2\text{O/HOAc}}$$
$$\underset{\text{OH}}{\big|}$$

13

$$\text{CySH} \xleftarrow[\text{25°, 15 min}]{\text{Hg(OAc)}_2/\text{HOAc, pH 4}}$$

Compound **13** is stable to acids (e.g., CF_3COOH, 21°, 48 h; 45% HBr/HOAc, 21°); it is oxidized by iodine to cystine (91% yield).[a]

[a] S. Coyle and G. T. Young, *J. Chem. Soc., Chem. Commun.*, 980 (1976).

14. *S*-Phenyl Thioether: RSC_6H_5, 14

A phenyl alkyl sulfide has been cleaved to the thiol by electrolysis ($-2.7\ \text{V}$, DMF, $R_4N^+X^-$).[a]

[a] V. G. Mairanovsky, *Angew. Chem., Inter. Ed., Engl.*, **15**, 281 (1976).

15. *S*-2,4-Dinitrophenyl Thioether: RSC_6H_3-2,4-$(NO_2)_2$, 15

Formation (→) / Cleavage (←)[a]

$$\xrightarrow[\text{base}]{\text{F-C}_6\text{H}_3\text{-2,4-(NO}_2)_2}$$

RSH   **15**

$$\xleftarrow[\text{22°, 1 h}]{\text{HS(CH}_2)_2\text{OH, pH 8}}$$

$$\text{RSH = CySH, quant}$$

The sulfhydryl group in cysteine can be selectively protected in the presence of the amino group by reaction with 2,4-dinitrophenol at pH 5-6.[b]

[a] S. Shaltiel, *Biochem. Biophys. Res. Commun.*, **29**, 178 (1967).

[b] H. Zahn and K. Traumann, *Z. Naturforsch.*, **9b**, 518 (1954).

16. *S-t*-Butyl Thioether: RSC(CH$_3$)$_3$, 16

Formation[a]

$$CySH + (CH_3)_2C{=}CH_2 \xrightarrow[\text{25}°, \text{ 12 h}]{H_2SO_4/CH_2Cl_2} \textbf{16}, 73\%$$

Compound **16** is stable to HBr/HOAc and to CF$_3$CO$_2$H.

Cleavage

(1) **16** $\xrightarrow[\text{0}°, \text{ 15 min}]{Hg(OAc)_2/CF_3CO_2H/\text{anisole}^b}$ or $\xrightarrow[\text{25}°, \text{ 1 h}]{Hg(OCOCF_3)_2/\text{aq HOAc}^b}$

$\xrightarrow{H_2S}$ RSH (= CySH, ca. quant)

(2) **16** $\xrightarrow[\text{20}°, \text{ 30 min}]{HF/\text{anisole}^c}$ CySH

[a] F. M. Callahan, G. W. Anderson, R. Paul, and J. E. Zimmerman, *J. Am. Chem. Soc.*, **85**, 201 (1963).

[b] O. Nishimura, C. Kitada, and M. Fujino, *Chem. Pharm. Bull.*, **26**, 1576 (1978).

[c] S. Sakakibara, Y. Shimonishi, Y. Kishida, M. Okada, and H. Sugihara, *Bull. Chem. Soc. Jpn.*, **40**, 2164 (1967).

17. *S*-1-Adamantyl Thioether: RS-1-adamantyl, 17

Formation[a]

$$CySH + \text{1-adamantyl alcohol} \xrightarrow[\text{25}°, \text{ 12 h}]{CF_3COOH} \textbf{17}, 90\%$$

Cleavage[a]

17 $\xrightarrow[\text{0}°, \text{ 15 min}]{Hg(OAc)_2/CF_3CO_2H}$ or $\xrightarrow[\text{20}°, \text{ 60 min}]{Hg(OCOCF_3)_2/\text{aq HOAc}}$ RSH, 100%

[a] O. Nishimura, C. Kitada, and M. Fujino, *Chem. Pharm. Bull.*, **26**, 1576 (1978).

Substituted *S*-Methyl Derivatives

HEMITHIO, DITHIO, AND AMINOTHIO ACETALS

18.   *S*-Methoxymethyl Hemithioacetal: RSCH₂OCH₃, 18

Formation

$$\text{ArSH} \xrightarrow[\text{BrCH}_2\text{CO}_2\text{Et}]{\text{Zn/(CH}_3\text{O)}_2\text{CH}_2} \left[\text{ArSZnBr} \right] \xrightarrow{\text{(CH}_3\text{O)}_2\text{CH}_2} \text{ArSCH}_2\text{OCH}_3$$

$$80\text{--}82\%$$

Formation of a methoxymethyl thioether with dimethoxymethane[a] avoids the use of the carcinogen chloromethyl methyl ether.[b]

[a] F. Dardoize, M. Gaudemar, and N. Goasdoue, *Synthesis*, 567 (1977).

[b] T. Fukuyama, S. Nakatsuka, and Y. Kishi, *Tetrahedron Lett.*, 3393 (1976).

19.   *S*-Isobutoxymethyl Hemithioacetal: RSCH₂OCH₂CH(CH₃)₂, 19

Formation (→) / Cleavage (←)[a]

$$\xrightarrow[82\%]{\text{ClCH}_2\text{OCH}_2\text{CH(CH}_3)_2}$$

CySH   19

$$\xleftarrow[\text{rapid}]{2\,N\ \text{HBr/HOAc}}$$

Compound **19** is stable to 2 *N* hydrochloric acid, and to 50% acetic acid; some decomposition occurs in 2 *N* sodium hydroxide.[a]

The hemithioacetal is also stable to 12 *N* hydrochloric acid in acetone (used to remove an *N*-triphenylmethyl group), and to hydrazine hydrate in refluxing ethanol (used to cleave an *N*-phthaloyl group). It is cleaved by boron trifluoride etherate in acetic acid, by silver nitrate in ethanol, and by trifluoroacetic acid. The hemithioacetal is oxidized to a disulfide by thiocyanogen, (SCN)₂.[b]

[a] P. J. E. Brownlee, M. E. Cox, B. O. Handford, J. C. Marsden, and G. T. Young, *J. Chem. Soc.*, 3832 (1964).

[b] R. G. Hiskey and J. T. Sparrow, *J. Org. Chem.*, **35**, 215 (1970).

20. *S*-2-Tetrahydropranyl Hemithioacetal: RS-2-tetrahydropyranyl, 20

Formation[a]

$$\text{RSH} + \text{dihydropyran} \xrightarrow[\text{0°, 0.5 h} \to \text{25°, 1 h}]{\text{BF}_3 \cdot \text{Et}_2\text{O, Et}_2\text{O}} \textbf{20, satisfactory yields}$$

Cleavage

(1) **20** $\xrightarrow[\text{0°, 10 min}]{\text{aq AgNO}_3{}^b}$ CySAg, quant

(2) **20** $\xrightarrow[\text{90 min}]{\text{HBr/CF}_3\text{CO}_2\text{H}^c}$ CySH, 100%

An *S*-tetrahydropyranyl hemithioacetal is stable to 4 *N* HCl/CH$_3$OH, 0° and to reduction with Na/NH$_3$. (An *O*-tetrahydropyranyl acetal is cleaved by 0.1 *N* HCl, 22°, $t_{1/2} = 4$ min.)[d]

Compound **20** is oxidized to a disulfide by iodine[b] or thiocyanogen, $(\text{SCN})_2$.[e]

[a] R. G. Hiskey and W. P. Tucker, *J. Am. Chem. Soc.*, **84**, 4789 (1962).
[b] G. F. Holland and L. A. Cohen, *J. Am. Chem. Soc.*, **80**, 3765 (1958).
[c] K. Hammerström, W. Lunkenheimer, and H. Zahn, *Makromol. Chem.*, **133**, 41 (1970).
[d] B. E. Griffin, M. Jarman, and C. B. Reese, *Tetrahedron*, **24**, 639 (1968).
[e] R. G. Hiskey and W. P. Tucker, *J. Am. Chem. Soc.*, **84**, 4794 (1962).

21. *S*-Benzylthiomethyl Dithioacetal: RSCH$_2$SCH$_2$C$_6$H$_5$, 21

Formation[a]

$$\text{CySH} + \text{ClCH}_2\text{SCH}_2\text{C}_6\text{H}_5 \xrightarrow{\text{NH}_3} \textbf{21}, 91\%$$

Cleavage[a]

$$\textbf{21} \xrightarrow[\text{HS(CH}_2)_2\text{SH, 25°, 5–20 min}]{\text{Hg(OAc)}_2/\text{H}_2\text{O}/80\% \text{ HOAc}} \xrightarrow[\text{2 h}]{\text{H}_2\text{S}} \text{RSH, high yield}$$

The removal of a dithioacetal protective group from compound **21** with mercury(II) acetate avoids certain side reactions that occur when an *S*-benzyl thioether is cleaved with sodium/ammonia. The dithioacetal is stable to hydrogen bromide/acetic acid used to cleave benzyl carbamates.[a]

S-Phenylthiomethyl dithioacetals (RSCH$_2$SC$_6$H$_5$) were prepared and cleaved by similar methods.[a]

The dithioacetal (**21**) is stable to catalytic reduction (H_2/Pd–C, CH_3OH–HOAc, 12 h, e.g., the conditions used to cleave a *p*-nitrobenzyl carbamate).[b]

[a] P. J. E. Brownlee, M. E. Cox, B. O. Handford, J. C. Marsden, and G. T. Young, *J. Chem. Soc.*, 3832 (1964).

[b] R. Camble, R. Purkayastha, and G. T. Young, *J. Chem. Soc.*, C, 1219 (1968).

22. Thiazolidine Derivative:

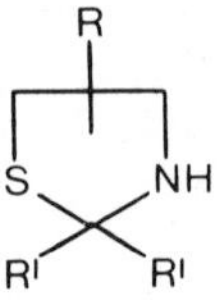

, **22**, (R′ = H, CH_3)

Thiazolidines have been prepared from β-aminothiols, for example cysteine, to protect the —SH and —NH groups during syntheses of peptides including glutathione.[a] Thiazolidines are oxidized to symmetrical disulfides with iodine[b]; they do not react with thiocyanogen in a neutral solution.[c]

Formation[d]

$$CySH \cdot HCl + (CH_3)_2CO \xrightarrow[\text{6 h}]{\text{reflux}}$$

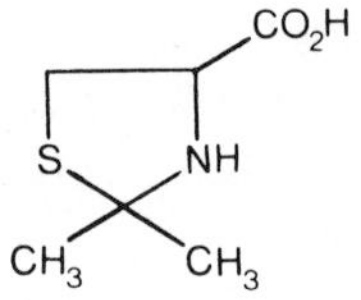

, **22a**, 82%

Cleavage[d]

(1) **22a** $\xrightarrow[\text{25°, 3 days}]{HCl/H_2O/CH_3OH}$ CySH, high yield

(2) **22a** $\xrightarrow[\substack{\text{25°, 2 days} \\ \text{or} \\ \text{60–70°, 15 min}}]{HgCl_2,\ H_2O} \xrightarrow[\text{20 min}]{H_2S}$ CySH, 30–40%

[a] F. E. King, J. W. Clark-Lewis, G. R. Smith, and R. Wade, *J. Chem. Soc.*, 2264 (1959).

[b] S. Ratner and H. T. Clarke, *J. Am. Chem. Soc.*, **59**, 200 (1937).

[c] R. G. Hiskey and W. P. Tucker, *J. Am. Chem. Soc.*, **84**, 4789 (1962).

[d] J. C. Sheehan and D.-D. H. Yang, *J. Am. Chem. Soc.*, **80**, 1158 (1958).

23. *S*-Acetamidomethyl Aminothioacetal: $RSCH_2NHCOCH_3$, 23

Formation[a]

$$CySH \cdot HCl + HOCH_2NHCOCH_3 \xrightarrow[\text{25°, 1–2 days}]{\text{concd HCl, pH 0.5}} \textbf{23}, 52\%$$

Cleavage[a]

(1)   **23** $\xrightarrow[\text{25°, 1 h}]{\text{Hg(OAc)}_2, \text{pH 4}}$ $\xrightarrow{\text{H}_2\text{S}}$ [CySH] $\xrightarrow{\text{air}}$ CySSCy, $> 98\%$

An *S*-acetamidomethyl group is hydrolyzed by the strongly acidic (6 *N* HCl, 110°, 6 h) or strongly basic conditions used to cleave amide bonds. It is stable to anhydrous trifluoracetic acid and to hydrogen fluoride (0°, 1 h; 18°, 1 h, 10% cleaved). It is stable to zinc in acetic acid and to hydrazine in acetic acid or methanol.[a]

(2)   **23** + *o*-NO$_2$—C$_6$H$_4$SCl $\xrightarrow{\text{HOAc}}$ RS—SC$_6$H$_4$-*o*-NO$_2$ $\xrightarrow[\text{NaBH}_4]{\substack{\text{HO(CH}_2)_2\text{SH} \\ \text{or}}}$

RSH, quant[b]

[a] D. F. Veber, J. D. Milkowski, S. L. Varga, R. G. Denkewalter, and R. Hirschmann, *J. Am. Chem. Soc.,* **94,** 5456 (1972); J. D. Milkowski, D. F. Veber, and R. Hirschmann, *Org. Synth.,* **59,** 190 (1980).

[b] L. Moroder, F. Marchiori, G. Borin, and E. Schoffone, *Biopolymers,* **12,** 493 (1973); A. Fontana, *J. Chem. Soc., Chem. Commun.,* 976 (1975).

24. *S*-Benzamidomethyl Aminothioacetal: RSCH$_2$NHCOC$_6$H$_5$, 24

S-Benzamidomethyl-*N*-methylcysteine has been prepared as a crystalline derivative (HOCH$_2$NHCOC$_6$H$_5$, anhydr CF$_3$CO$_2$H, 25°, 45 min, 88% yield as the trifluoroacetate salt) and cleaved (100% yield) by treatment with mercury(II) acetate (pH 4, 25°, 1 h) followed by hydrogen sulfide. Attempted preparation of *S*-acetamidomethyl-*N*-methylcysteine resulted in noncrystalline material, shown by TLC to be a mixture.[a]

[a] P. K. Chakravarty and R. K. Olsen, *J. Org. Chem.,* **43,** 1270 (1978).

25. *S*-Acetyl-, *S*-Carboxy-, and *S*-Cyanomethyl Thioethers: ArSCH$_2$X, 25

25a, X = —COCH$_3$, **25b,** X = —COOH, **25c,** X = —CN

In an attempt to protect thiophenols during electrophilic substitution reactions on the aromatic ring, three thioethers were prepared: **25a, 25b,** and **25c.** After acetylation of the aromatic ring (moderate yields), the protective group was converted to the disulfide in moderate yields, 50–60%, by oxidation with hydrogen peroxide/boiling mineral acid, nitric acid, or acidic potassium permanganate.[a]

[a] D. Walker, *J. Org. Chem.,* **31,** 835 (1966).

Substituted S-Ethyl Derivatives

A thiol (usually as the anion) can undergo Michael addition to an activated double bond, resulting in protection of the sulfhydryl group as a substituted S-ethyl derivative (e.g., compounds **26**, **27**, and **28**).

26. S-2-Nitro-1-phenylethyl Thioether: $RSCH(C_6H_5)CH_2NO_2$, 26

Formation[a]

$$CySH + C_6H_5CH{=}CHNO_2 \xrightarrow[\text{pH 7-8, 10 min}]{\textit{N}\text{-methylmorpholine}} \textbf{26}, 70\%$$

The protective group is removed by mildly alkaline conditions that do not cleave methyl or benzyl esters. This protective group has been used in two new syntheses of glutathione.[a]

[a] G. Jung, H. Fouad, and G. Heusel, *Angew. Chem., Inter. Ed. Engl.*, **14**, 817 (1975).

27. S-2,2-Bis(carboethoxy)ethyl Thioether: $RSCH_2CH(COOC_2H_5)_2$, 27

Formation $(\rightarrow)$ */ Cleavage* $(\leftarrow)$[a]

$$\xrightarrow[74\%]{CH_2{=}C(COOEt)_2, \text{ EtOH, 1 h}}$$

$$CySH \qquad\qquad\qquad\qquad\qquad 27$$

$$\xleftarrow[20°, 5\text{--}10 \text{ min}, 80\%]{1 \textit{ N} \text{ KOH/EtOH}}$$

Compound **27**, stable to acidic reagents such as trifluoracetic acid and hydrogen bromide/acetic acid, has been used in a synthesis of glutathione.[a]

[a] T. Wieland and A. Sieber, *Liebigs Ann. Chem.*, **722**, 222 (1969); **727**, 121 (1969).

28. S-1-m-Nitrophenyl-2-benzoylethyl Thioether:
   $ArSCH(C_6H_4$-m-$NO_2)CH_2COC_6H_5$, 28

Formation[a]

$$C_6H_5SH + C_6H_5COCH{=}CHC_6H_4\text{-}m\text{-}NO_2 \xrightarrow[C_6H_6]{\text{piperidine}} \textbf{28}, 96\%$$

Cleavage[a]

$$28 \xrightarrow[\text{pH 8-10}]{\text{Pb(OAc)}_2/\text{EtOH}} [(C_6H_5S)_2Pb] \xrightarrow{\text{dil HCl}} C_6H_5SH, 77\%$$

Compound **28** was used to protect thiophenols during electrophilic substitution reactions of the benzene ring.

[a] A. H. Herz and D. S. Tarbell, *J. Am. Chem. Soc.*, **75**, 4657 (1953).

THIOESTERS

29. *S*-Acetyl Derivative: RSCOCH₃, 29

30. *S*-Benzoyl Derivative: RSCOC₆H₅, 30

Formation[a]

$$\text{CBZCySH} \xrightarrow{(CH_3CO)_2O/KHCO_3} \textbf{29}, 55\%$$

$$\text{CySH} \xrightarrow[\text{0-5}°, \text{30 min}]{C_6H_5COCl/NaOH/KHCO_3} \textbf{30}, 50\%$$

Cleavage[a]

(1) **29**   CBZCySH, 100%, 2 min

$$\xrightarrow[20°]{0.2 \ N \ \text{NaOH}, \ N_2}$$

     30   CySH, 100%, 15 min

(2) **29**   CBZCySH, 100%, 12 min

$$\xrightarrow[20°]{\text{aq NH}_3, \ N_2}$$

     30   CySH, 95%, 30 min

(3) **29**   CBZCySH, substantial amount

$$\xrightarrow[25°, \ 30 \ \text{min}]{\text{HBr/HOAc}}$$

     30   CySH, 5%

(4) **29**   CBZCySH, 5%

$$\xrightarrow[\text{reflux, 30 min}]{CF_3CO_2H/C_6H_5OH}$$

     30   CySH, 2%

Two disadvantages are associated with the use of S-acetyl or S-benzoyl derivatives in peptide syntheses: (a) base-catalyzed hydrolysis of S-acetyl- and S-benzoylcysteine occurs with β-elimination to give olefinic side products, $CH_2{=}C$-$(NHPG)CO-$[b]; (b) the yields of peptides formed by coupling an unprotected amino group in an S-acylcysteine are low because of prior $S \rightarrow N$ acyl migration.[c]

An S-acetyl group is stable to oxidation of a double bond by ozone ($-20°$, 5.5 h, 73% yield).[d]

[a] L. Zervas, I. Photaki, and N. Ghelis, *J. Am. Chem. Soc.,* **85,** 1337 (1963).

[b] R. G. Hiskey, R. A. Upham, G. M. Beverly, and W. C. Jones, Jr., *J. Org. Chem.,* **35,** 513 (1970).

[c] R. G. Hiskey, T. Mizoguchi, and T. Inui, *J. Org. Chem.,* **31,** 1192 (1966).

[d] I. Ernest, J. Gosteli, C. W. Greengrass, W. Holick, D. E. Jackman, H. R. Pfaendler, and R. B. Woodward, *J. Am. Chem. Soc.,* **100,** 8214 (1978).

31. S-Thiobenzoyl Derivative: $RSCSC_6H_5$, 31

Formation

$$
\begin{array}{ccc}
CH_2SH & & CH_2SCSPh \\
| & & | \\
CHOH & + C_6H_5CSSMe \xrightarrow[\;25°,\ 1.5\ h\;]{\text{cat. NaOCH}_3/\text{abs CH}_3\text{OH}} & CHOH \\
| & & | \\
CH_2OH & & CH_2OH \\
\\
\mathbf{a} & & 54\%
\end{array}
$$

The base-catalyzed reaction of **a** with methyl dithiobenzoate selectively protects a thiol group in the presence of a hydroxyl group.[a]

[a] E. J. Hedgley and N. H. Leon, *J. Chem. Soc. C,* 467 (1970).

Thiocarbonate Derivatives

When cysteine reacts with an alkyl or aryl chloroformate, both the —SH and —NH groups are protected as a thiocarbonate and as a carbamate, respectively. Selective or simultaneous removal of the protective groups is possible (e.g., see the *Formation* and *Cleavage* of compound **34**).

32. S-2,2,2-Trichloroethoxycarbonyl Derivative: $RSCOOCH_2CCl_3$, 32

Cleavage[a]

$$
\mathbf{32} \xrightarrow[\text{LiClO}_4/\text{CH}_3\text{OH}]{-\ 1.50\ \text{V}} \underset{\substack{90\% \quad\ \ 90\%}}{RSH \text{ or } RSSR}
$$

$R = C_6H_5CH_2$; N-acetylcysteine

[a] M. F. Semmelhack and G. E. Heinsohn, *J. Am. Chem. Soc.,* **94,** 5139 (1972).

33.   *S-t*-Butoxycarbonyl Derivative: RSCOOC(CH₃)₃, 33

t-Butyl chloroformate reacts with cysteine to protect both the amine and thiol groups; as with *N,S*-bis(benzyloxycarbonyl)cysteine, selective or simultaneous removal of the *N*- or *S*-protective groups can be effected.[a]

[a] M. Muraki and T. Mizoguchi, *Chem. Pharm. Bull.*, **19**, 1708 (1971).

34.   *S*-Benzyloxycarbonyl Derivative: RSCOOCH₂C₆H₅, 34

Formation[a]

$$CySH + ClCOOCH_2Ph$$

$\xrightarrow[\text{30 min}]{\text{aq NaOH}}$ PhCH₂OCOSCH₂CH(NHCOOCH₂Ph)COOH, 71%

34a

$\xrightarrow[\text{0°, 1 h} \rightarrow \text{10°, 1 h}]{\text{NaHCO}_3}$ PhCH₂OCOSCH₂CH(NH₂)COOH, 67%

34b

Cleavage

(1) **34a** $\xrightarrow[\text{25°, 1 h}]{\text{concd NH}_4\text{OH}^a}$ HSCH₂CHCOOH, 90%

NHCOOCH₂Ph

A

(2) **34b** $\xrightarrow{\text{Na/NH}_3{}^a}$ CySH, 62%

(3) **34b**

$\xrightarrow[\text{30 min}]{0.1\ N\ \text{NaOCH}_3/\text{CH}_3\text{OH}/\text{N}_2{}^b}$

CySH, 100%, 30 min

34a

A, 50%, 30 min

100%, 2–3 h

An *S*-benzoyl group is removed (95–100% yield) in 5–10 minutes.[b]

(4) **34a** $\xrightarrow[\text{reflux, 30 min}]{\text{CF}_3\text{CO}_2\text{H}^b}$ CySH, ca. quant

(5) **34a** $\xrightarrow[\text{25°, 30 min}]{2\ N\ \text{HBr/HOAc}^b}$ CySH, 10–14% + **34b**, ca. quant[c]

(6)   **34** $\xrightarrow[\text{R}_4\text{N}^+\text{X}^-, \text{ DMF}]{-2.6 \text{ V}^d}$ CySH

[a] A. Berger, J. Noguchi, and E. Katchalski, *J. Am. Chem. Soc.*, **78**, 4483 (1956).

[b] L. Zervas, I. Photaki, and N. Ghelis, *J. Am. Chem. Soc.*, **85**, 1337 (1963).

[c] M. Sokolovsky, M. Wilchek, and A. Patchornik, *J. Am. Chem. Soc.*, **86**, 1202 (1964).

[d] V. G. Mairanovsky, *Angew. Chem., Inter. Ed., Engl.*, **15**, 281 (1976).

35.   *S-p*-Methoxybenzyloxycarbonyl Derivative: RSCOOCH$_2$C$_6$H$_4$-*p*-OCH$_3$, 35

S-p-Methoxybenzyloxycarbonylcysteine has been prepared in low yield (30%). It has been used in peptide syntheses, but is very labile to acids and bases.[a]

[a] I. Photaki, *J. Chem. Soc. C*, 2687 (1970).

Thiocarbamate Derivatives

Thiocarbamates, formed by reaction of a thiol with an isocyanate, are stable in acidic and neutral solutions, and are readily cleaved by basic hydrolysis. The β-elimination that can occur when an *S*-acyl group is removed with base from a cysteine derivative does not occur under the conditions needed to cleave a thiocarbamate.[a]

36.   *S*-(*N*-Ethylcarbamate): RSCONHC$_2$H$_5$, 36

Formation[a]

$$\text{CySH} + \text{C}_2\text{H}_5\text{N}{=}\text{C}{=}\text{O} \xrightarrow[\text{20°, 70 h}]{\text{pH 1} \rightarrow \text{pH 6}} \textbf{36, } 67\%$$

Cleavage

(1)   **36** $\xrightarrow[\text{20°, 20 min}]{1 \text{ } N \text{ NaOH}^a}$ RSH, 100%

(2)   **36** $\xrightarrow[\text{20°, 2 h}]{\text{NH}_3/\text{CH}_3\text{OH}^a}$ or $\xrightarrow[\text{20°, 2 h}]{\text{H}_2\text{NNH}_2/\text{CH}_3\text{OH}^a}$ RSH, 100%

(3)   **36** $\xrightarrow[-30°, 3 \text{ min}]{\text{Na/NH}_3{}^a}$ RSH, 100%

This protective group is stable to acidic hydrolysis (4.5 N HBr/HOAc; 1 N HCl; CF$_3$CO$_2$H, reflux). There is no evidence of S $\rightarrow$ N acyl migration in **36** (RS = cysteinyl).[a]

(4)   **36** $\xrightarrow[\text{30 min}]{\text{Hg(OAc)}_2/\text{H}_2\text{O}/\text{CH}_3\text{OH}^b}$ $\xrightarrow[\text{4 h}]{\text{H}_2\text{S}}$ RSH (= Cys—OEt, 79%)

36 $\xrightarrow{\text{AgNO}_3/\text{H}_2\text{O}/\text{CH}_3\text{OH}^b}$ $\xrightarrow[\text{3 h}]{\text{concd HCl}}$ RSH, 62%

Oxidation of compound **36** (RS = cysteinyl) with performic acid yields cysteic acid.[b]

[a] St. Guttmann, *Helv. Chim. Acta,* **49,** 83 (1966).

[b] H. T. Storey, J. Beacham, S. F. Cernosek, F. M. Finn, C. Yanaihara, and K. Hofmann, *J. Am. Chem. Soc.,* **94,** 6170 (1972).

37.   *S*-(*N*-Methoxymethylcarbamate): RSCONHCH₂OCH₃, 37

Formation[a]

$$\text{CySH} + \text{CH}_3\text{OCH}_2\text{N}{=}\text{C}{=}\text{O} \xrightarrow[\text{2 min}]{\text{pH 4–5}} \textbf{37}, 100\%$$

Cleavage[a]

37 $\xrightarrow{\text{pH 9.6}}$ RSH

RSH = CySH, 100%, 30 min

= glutathione, 80%, 30 min

At pH 4–5 the reaction is selective for protection of thiol groups in the presence of α- or ε-amino groups.[a]

[a] H. Tschesche and H. Jering, *Angew. Chem., Inter. Ed. Engl.,* **12,** 756 (1973).

MISCELLANEOUS DERIVATIVES

Disulfides

A thiol can be protected by oxidation (with O_2; H_2O_2; I_2 . . .) to the corresponding symmetrical disulfide, which subsequently can be cleaved by reduction [Sn/HCl; Na/xylene, Et₂O, or NH₃; LiAlH₄; NaBH₄; or thiols such as HO(CH₂)₂SH]. Unsymmetrical disulfides have also been prepared and are discussed.

UNSYMMETRICAL DISULFIDES

38.   S-Ethyl Disulfide: RSSC$_2$H$_5$, 38

Formation[a]

$$\text{RSH} + \text{C}_2\text{H}_5\underset{\underset{\text{O}}{\downarrow}}{\text{SS}}\text{C}_2\text{H}_5 \xrightarrow[\text{1 h}]{-70°} \textbf{38, } 80\text{–}90\%$$

R = alkyl, aryl

Cleavage[b]

$$\textbf{38} \xrightarrow[>50°]{\text{C}_6\text{H}_5\text{SH}} \text{ or } \xrightarrow[45°, \text{ 15 h}]{\text{HSCH}_2\text{CO}_2\text{H}} \text{RSH, ca. quant}$$

Compound **38** is stable to acid-catalyzed hydrolysis (CF$_3$CO$_2$H) of car-bamates and to ammonolysis (25% NH$_3$/CH$_3$OH).[b]

[a] D. A. Armitage, M. J. Clark, and C. C. Tso, *J. Chem. Soc., Perkin Trans. 1,* 680 (1972).
[b] N. Inukai, K. Nakano, and M. Murakami, *Bull. Chem. Soc. Jpn.,* **40,** 2913 (1967).

39.   S-t-Butyl Disulfide: RSSC(CH$_3$)$_3$, 39

Formation[a]

$$\text{RSH} \cdot \text{HCl} \xrightarrow[0\text{–}5°, \text{ 1.5 h}]{\text{ClSCO}_2\text{CH}_3} \underset{94\%}{\text{RSSCO}_2\text{CH}_3 \cdot \text{HCl}} \xrightarrow[5 \text{ days}]{t\text{-BuSH/CH}_3\text{OH}} \underset{\substack{97\% \text{ crude} \\ 46\% \text{ pure}}}{\textbf{39} \cdot \text{HCl}}$$

Cleavage[b]

$$\textbf{39} \xrightarrow{\text{NaBH}_4} \text{RSH}$$

[a] L. Field and R. Ravichandran, *J. Org. Chem.,* **44,** 2624 (1979).
[b] E. Wünsch and R. Spangenberg, in *Peptides, 1969,* E. Schoffone, Ed., North Holland, Amsterdam, p. 1971.

40.   Substituted S-Phenyl Disulfide: RSSC$_6$H$_4$-Y, 40

Three substituted S-phenyl unsymmetrical disulfides have been prepared, **40a**,[a] **40b**,[b] and **40c**,[c]—compounds **40a** and **40b** by reaction of a thiol with a sulfenyl halide, compound **40c** from a thiol and an aryl thiosulfonate (ArSO$_2$SAr). The disulfides are cleaved by reduction (NaBH$_4$) or by treatment with excess thiol (HSCH$_2$CH$_2$OH).

$$RSS—C_6H_3\text{-}o\text{-}NO_2\text{-}p\text{-}R' \qquad RSS—C_6H_4\text{-}o\text{-}N{=}N—C_6H_5$$

40a   **40b**

R' = H, NO$_2$

$$RSS—C_6H_4\text{-}o\text{-}COOH$$

40c

[a] A. Fontana, E. Scoffone, and C. A. Benassi, *Biochemistry*, **7**, 980 (1968); A. Fontana, *J. Chem. Soc., Chem. Commun.*, 976 (1975).

[b] A. Fontana, F. M. Veronese, and E. Scoffone, *Biochemistry*, **7**, 3901 (1968).

[c] L. Field and P. M. Giles, Jr., *J. Org. Chem.*, **36**, 309 (1971).

Sulfenyl Derivatives

41.   *S*-Sulfonate Derivative: RSSO$_3^-$, 41

Formation[a]

$$RSH + Na_2SO_3 + cat.\ CySH \xrightarrow[\text{pH 7–8.5, 1 h}]{O_2} \textbf{41}, quant$$

Cleavage[a]

$$\textbf{41} \xrightarrow[\text{25°, 2 h, 100\%}]{\text{HSCH}_2\text{CH}_2\text{OH, pH 7.5}} or \xrightarrow{\text{NaBH}_4} RSH$$

S-Sulfonates are stable at pH 1–9; they are unstable in hot acidic solutions and in 0.1 *N* sodium hydroxide.[a]

[a] W. W.-C. Chan, *Biochemistry*, **7**, 4247 (1968).

42.   *S*-Sulfenylthiocarbonate Derivative: RSSCOOR', 42

A number of *S*-sulfenylthiocarbonates have been prepared to protect thiols. A benzyl derivative, **42**, R' = CH$_2$Ph, is stable to trifluoroacetic acid (25°, 1 h) and provides satisfactory protection during peptide syntheses[a]; a *t*-butyl derivative, **42**, R' = *t*-Bu, is too labile in base to provide protection.[a] A methyl derivative, **42**, R' = CH$_3$, has been used to protect a cysteine fragment that is subsequently converted to a cystine.[b]

Formation (→) / *Cleavage* (←)[a]

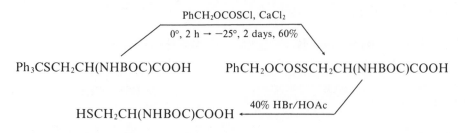

PhCH₂OCOSCl, CaCl₂

0°, 2 h → −25°, 2 days, 60%

Ph₃CSCH₂CH(NHBOC)COOH   PhCH₂OCOSSCH₂CH(NHBOC)COOH

HSCH₂CH(NHBOC)COOH ←——— 40% HBr/HOAc

[a] K. Nokihara and H. Berndt, *J. Org. Chem.*, **43**, 4893 (1978).

[b] R. G. Hiskey, N. Muthukumaraswamy, and R. R. Vunnam, *J. Org. Chem.*, **40**, 950 (1975).

Protection for Dithiols

DITHIO ACETALS AND KETALS

**43.   *S,S'*-Methylene, 43a, *S,S'*-Isopropylidene, 43b, and
S,S'-Benzylidene, 43c, Derivatives**

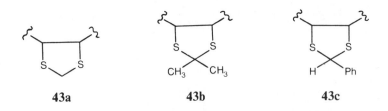

43a   43b   43c

Dithiols, like diols, have been protected as *S,S'*-methylene,[a] *S,S'*-isopropylidene,[b] and *S,S'*-benzylidene[c] derivatives, formed by reaction of the dithiol with formaldehyde, acetone, or benzaldehyde, respectively. The methylene and benzylidene derivatives are cleaved by reduction with sodium/ammonia. The isopropylidene[b] and benzylidene[c] derivatives are cleaved by mercury(II) chloride; with sodium/ammonia the isopropylidene derivative is converted to a monothio ether, HSCH—CHSCHMe₂.[a]

[a] E. D. Brown, S. M. Igbal, and L. N. Owen, *J. Chem. Soc. C*, 415 (1966).

[b] E. P. Adams, F. P. Doyle, W. H. Hunter, and J. H. C. Nayler, *J. Chem. Soc.*, 2674 (1960).

[c] L. W. C. Miles and L. N. Owen, *J. Chem. Soc.*, 2938 (1950).

44. S,S'-p-Methoxybenzylidene Derivative: (RS)$_2$CHC$_6$H$_4$-p-OCH$_3$, 44

Formation[a]

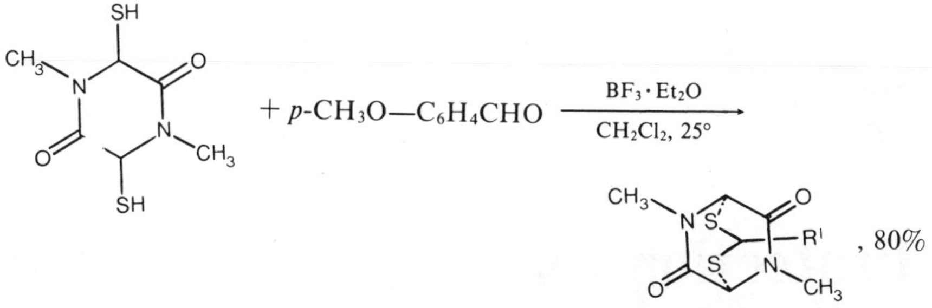

, 80%

44a

R' = p-CH$_3$OC$_6$H$_4$—

"Cleavage"[a]

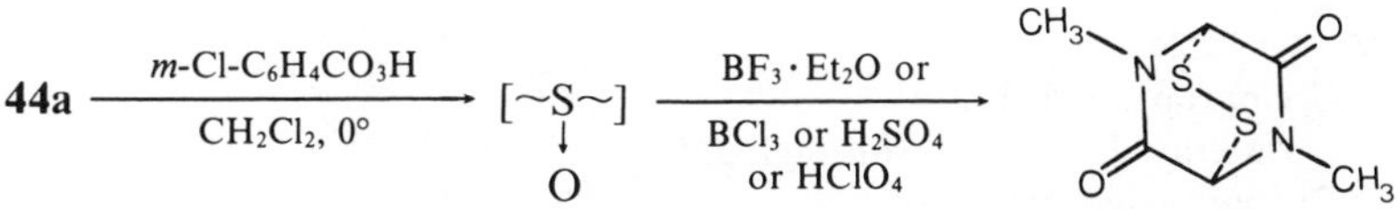

b

The epidithioketopiperazine structure in compound **b** is present in natural products including the gliotoxins and sporidesmins.[a]

[a] Y. Kishi, T. Fukuyama, and S. Nakatsuka, *J. Am. Chem. Soc.,* **95,** 6490 (1973).

7

Protection for
The Amino Group

---

*Included in Reactivity Chart 8.

**Included in Reactivity Chart 9.
***Included in Reactivity Chart 10.

***Included in Reactivity Chart 10.

A great many protective groups have been developed for the amino group, including carbamates ($>NCO_2R$), used for the protection of amino acids in peptide and protein syntheses, and amides ($>NCOR$), used more widely in syntheses of alkaloids and for the protection[a] of the nitrogen bases adenine, cytosine, and guanine in nucleotide syntheses.

Carbamates are formed by reaction of an amine with an azido- or chloroformate or with a carbonate; amides are formed from the acid chloride. n-Alkyl carbamates are cleaved by acid-catalyzed hydrolysis; N-alkylamides are cleaved by acidic or basic hydrolysis at reflux, and by ammonolysis, conditions that cleave peptide bonds.

In this chapter detailed information is provided for the most useful protective groups (in general these are included in Reactivity Charts 8–10); structures and references are given for protective groups that seem to have more limited use. The conventions that are used in this chapter are described on p. xii.[b]

[a]   C. B. Reese, *Tetrahedron*, **34**, 3143 (1978); V. Amarnath and A. D. Broom, *Chem. Rev.*, **77**, 183 (1977).

[b] See also: E. Wünsch, "Blockierung und Schutz der α-Amino-Funktion," in *Methoden der Organischen Chemie* (*Houben-Weyl*), Georg Thieme Verlag, Stuttgart, 1974, Vol. 15/1, pp. 46–305; J. W. Barton, "Protection of N–H Bonds and NR₃," in *Protective Groups in Organic Chemistry,* J. F. W. McOmie, Ed., Plenum Press, New York and London, 1973, pp. 43–93; L. A. Carpino, *Acc. Chem. Res.,* **6,** 191–198 (1973); Y. Wolman, "Protection of the Amino Group," in *The Chemistry of the Amino Group,* S. Patai, Ed., Wiley-Interscience, New York, 1968, Vol. 4, pp. 669–699.

CARBAMATES

Carbamates can be used as protective groups for amino acids to minimize racemization in peptide synthesis. Racemization occurs during the base-catalyzed coupling reaction of an *N*-protected, carboxyl-activated amino acid, and takes place in the intermediate oxazolone that forms readily from an *N*-acyl protected amino acid (R′ = alkyl, aryl) (eq. 1):

oxazolone

To minimize racemization the use of nonpolar solvents, a minimum of base, low reaction temperatures, and carbamate protective groups (R′ = *O*-alkyl or *O*-aryl) is effective. (Recently a carbamate, R′ = O-*t*-Bu, has been reported to form an oxazolone that appears not to racemize during base-catalyzed coupling.)[a]

Many carbamates have been used as protective groups. They are arranged in this chapter in order of increasing complexity of structure. The most useful compounds do not necessarily have the simplest structures, but are *t*-butyl (BOC), readily cleaved by acidic hydrolysis; benzyl (CBZ), cleaved by catalytic hydrogenolysis; 2,4-dichlorobenzyl, stable to the acid-catalyzed hydrolysis of benzyl and *t*-butyl carbamates; 2-(biphenylyl)isopropyl, cleaved more easily than *t*-butyl carbamate by dilute acetic acid; 9-fluorenylmethyl, cleaved by β-elimination with base; isonicotinyl, cleaved by reduction with zinc in acetic acid; 1-adamantyl, readily cleaved by trifluoracetic acid; and 2-phenylisopropyl, slightly more stable than t-butyl carbamate to acidic hydrolysis.

General methods that are used to form carbamates are shown in eq. 2:

$$(2) \quad R^1R^2NH + R'OCOX \xrightarrow{\text{base}} R^1R^2NCOOR'$$

X = Cl: chloroformate

X = N₃: azidoformate

X = −OR': carbonate

X = −OC₆H₄-o-NO₂: mixed carbonate

X = OCOOR': dicarbonate

The t-butyl azidoformate poses some danger because of its heat and shock sensitivity. It would also appear desirable to avoid the use of other azidoformate; when possible.

The more important carbamates are included in Reactivity Chart 8. The conventions that are used in this chapter are described on p. xii.

[a] Work of N. L. Benoiton, reported by J. L. Fox, *Chem. Eng. News,* August 6, 1979, p. 20.

1.   Methyl Carbamate: $R^1R^2NCOOCH_3$, 1

Formation (→) / Cleavage (←)

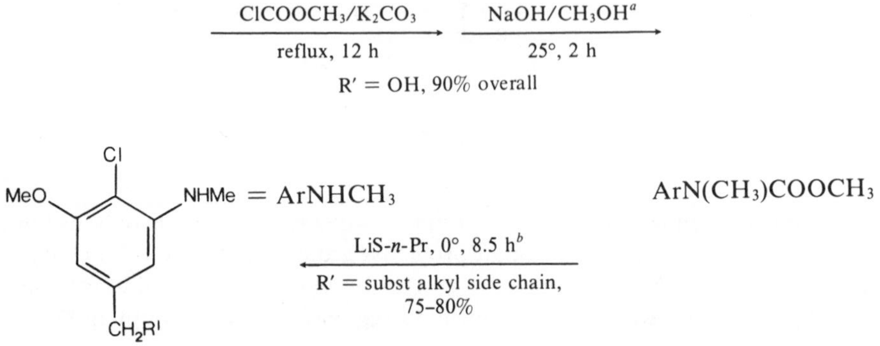

$$\xrightarrow[\text{reflux, 12 h}]{ClCOOCH_3/K_2CO_3} \xrightarrow[\text{25°, 2 h}]{NaOH/CH_3OH^a}$$

R' = OH, 90% overall

NHMe = ArNHCH₃   ArN(CH₃)COOCH₃

$$\xleftarrow[\substack{R' = \text{subst alkyl side chain,} \\ 75-80\%}]{\text{LiS-}n\text{-Pr, 0°, 8.5 h}^b}$$

Methyl carbamates are also cleaved by iodotrimethylsilane (50°, 70% yield),[c] aqueous potassium hydroxide in ethylene glycol (100°, 12 h, 88% yield)[d] or in CH₃OH–H₂O, 25°, 30 min, 100% yield,[e] and hydrogen bromide/acetic acid (25°, 18 h).[f]

[a] E. J. Corey, M. G. Bock, A. P. Kozikowski, A. V. Rama Rao, D. Floyd, and B. Lipshutz, *Tetrahedron Lett.,* 1051 (1978).

[b] E. J. Corey, L. O. Weigel, D. Floyd, and M. G. Bock, *J. Am. Chem. Soc.,* **100**, 2916 (1978).

[c] R. S. Lott, V. S. Chauhan, and C. H. Stammer, *J. Chem. Soc., Chem. Commun.,* 495 (1979).

[d] E. Wenkert, T. Hudlicky, and H. D. H. Showalter, *J. Am. Chem. Soc.,* **100,** 4893 (1978).

[e] M. Natsume and H. Muratake, *Tetrahedron Lett.,* 3477 (1979).

[f] M. C. Wani, H. F. Campbell, G. A. Brine, J. A. Kepler, M. E. Wall, and S. G. Levine, *J. Am. Chem. Soc.,* **94,** 3631 (1972).

2. Cyclopropylmethyl Carbamate: $R^1R^2NCOOCH_2$-c-C_3H_7, 2

See Miscellaneous Carbamates (following compound **72**).

3. 1-Methyl-1-cyclopropylmethyl Carbamate: $R^1R^2NCOOCH(CH_3)$-c-C_3H_7, 3

See Miscellaneous Carbamates (following compound **72**).

4. Diisopropylmethyl Carbamate: $R^1R^2NCOOCH[CH(CH_3)_2]_2$, 4

The ease of removal by acidic hydrolysis (HBr/CH_3NO_2) is R^1R^2N-CO_2-t-Bu $>$ R^1R^2N-CO_2-c-$C_5H_9 > 4.$[a]

[a] F. C. McKay and N. F. Albertson, *J. Am. Chem. Soc.,* **79,** 4686 (1957).

5. 9-Fluorenylmethyl Carbamate:

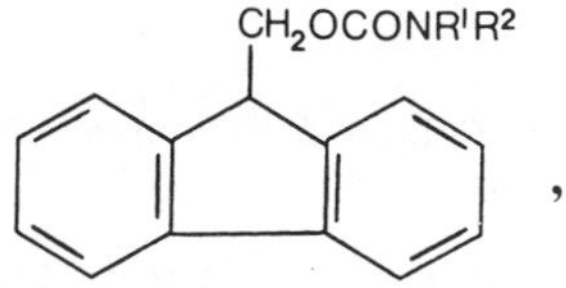

$CH_2OCONR^1R^2$

, 5

Formation[a]

$$R'CHCO_2H + 9\text{-fluorenyl}-CH_2OCOX \xrightarrow{NaHCO_3} 5, 88\text{--}98\%$$
$$\hspace{0.5em}|$$
$$\hspace{0.5em}NH_2$$

$$X = Cl, N_3$$

Cleavage

CH_2

(1) **5** $\xrightarrow{\text{base}^b}$ $R^1R^2NH +$

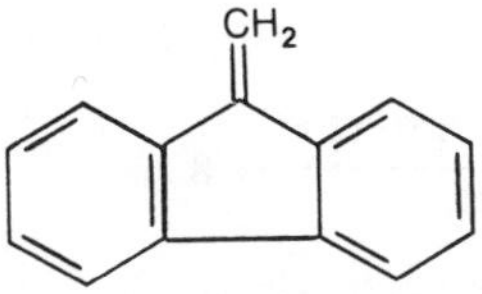

high yields

base = NH_3, 25°, 10–12 h

= piperidine; morpholine; ethanolamine; 25°, few min.

Compound **5** is stable to hydrogen bromide in organic solvents and to trifluoroacetic acid.[a]

The undesired cleavage of compound **5** by amino acids, although a slower reaction than with ammonia or piperidine, can occur during peptide synthesis.[b]

$$(2) \quad 5 \xrightarrow[\text{20\% HOAc-CH}_3\text{OH}]{\text{H}_2/\text{Pd-C}^c} R^1R^2NH$$

$$t_{1/2} = 3\text{--}33 \text{ h}$$

[a] L. A. Carpino and G. Y. Han, *J. Org. Chem.*, **37**, 3404 (1972).
[b] M. Bodanszky, S. S. Deshmane, and J. Martinez, *J. Org. Chem.*, **44**, 1622 (1979).
[c] E. Atherton, C. Bury, R. C. Sheppard, and B. J. Williams, *Tetrahedron Lett.*, 3041 (1979).

6.   9-(2-Sulfo)fluorenylmethyl Carbamate:

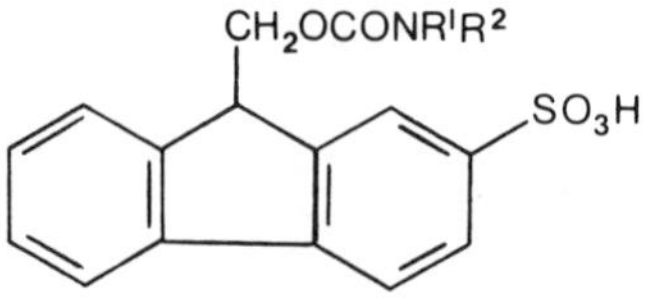

, **6**

Compound **6** is readily cleaved by dilute bases (0.1 N NH$_4$OH; 1% Na$_2$CO$_3$, . . ., 45 min).[a]

[a] R. B. Merrifield and A. E. Bach, *J. Org. Chem.*, **43**, 4808 (1978).

7.   2-Furanylmethyl Carbamate: R^{1}R^{2}NCOOCH$_2$-2-furanyl, 7

Compound **7** is readily cleaved by acidic hydrolysis (HBr/HOAc, 25°, 3 min) and by hydrogenolysis (H$_2$/Pd-C, neutral).[a]

[a] G. Losse and K. Neubert, *Tetrahedron Lett.*, 1267 (1970).

Substituted Ethyl Carbamates

8.   2,2,2-Trichloroethyl Carbamate: R^{1}R^{2}NCOOCH$_2$CCl$_3$, 8

Formation (→) / Cleavage (←)

$$R^1R^2NH \xrightarrow[\text{Py or aq NaOH}]{\text{ClCO}_2\text{CH}_2\text{CCl}_3/25°, 12 \text{ h}^a} 8$$

$$R^1R^2NH \xleftarrow[\text{or pH 5.5--7.2, 18 h, 96\%}]{\text{Zn/aq THF, pH 4.2, 30 min, 86\%}^b} 8$$

$$R^1R^2NH \xleftarrow{-1.70 \text{ V, 0.1 } M \text{ LiClO}_4, 85\%^c} 8$$

A 2,2,2-trichloroethyl carbamate can be used to protect a hindered amine (e.g., adamantylamine), since the protective group can undergo "assisted removal" via attack by zinc at a 2-chloro group; this effect avoids the need to hydrolyze a hindered carbamate.

[a] T. B. Windholz and D. B. R. Johnston, *Tetrahedron Lett.*, 2555 (1967).

[b] G. Just and K. Grozinger, *Synthesis,* 457 (1976).

[c] M. F. Semmelhack and G. E. Heinsohn, *J. Am. Chem. Soc.,* **94,** 5139 (1972).

9. 2-Haloethyl Carbamate: $R^1R^2NCOOCH_2X$, 9

$X = CH_2Br, CH_2Cl, CHBr_2, CCl_3$: cleaved by the supernucleophile cobalt(I) phthalocyanine.[a]

$X = CH_2I, CHCl_2, CCl_3, CBr_3$: cleaved by electrolytic reduction.[b]

[a] H. Eckert and I. Ugi, *Liebigs Ann. Chem.,* 278 (1979).

[b] E. Kasafirek, *Tetrahedron Lett.,* 2021 (1972): M. F. Semmelhack and G. E. Heinsohn, *J. Am. Chem. Soc.,* **94,** 5139 (1972).

10. 2-Iodoethyl Carbamate: $R^1R^2NCOOCH_2CH_2I$, 10

Formation ($\rightarrow$) / Cleavage ($\leftarrow$)[a]

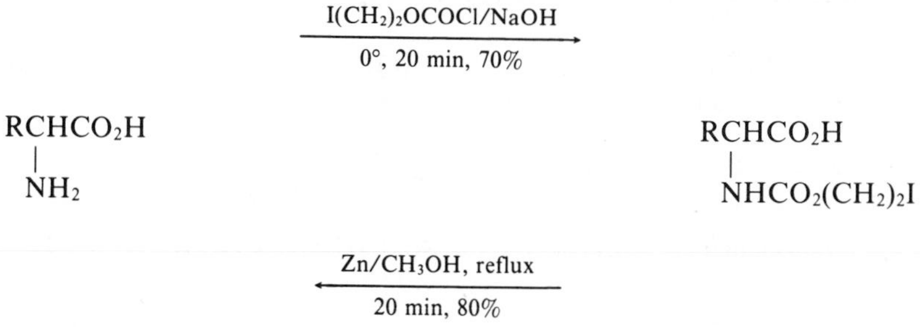

$$\underset{0°,\ 20\ min,\ 70\%}{\xrightarrow{\text{I(CH}_2)_2\text{OCOCl/NaOH}}}$$

RCHCO₂H
|
NH₂

RCHCO₂H
|
NHCO₂(CH₂)₂I

$$\underset{20\ min,\ 80\%}{\xleftarrow{\text{Zn/CH}_3\text{OH, reflux}}}$$

[a] J. Grimshaw, *J. Chem. Soc.,* 7136 (1965).

11. 2-Trimethylsilylethyl Carbamate: $R^1R^2NCOOCH_2CH_2Si(CH_3)_3$, 11

Formation ($\rightarrow$) / Cleavage ($\leftarrow$)

$R^1R^2NH \xrightarrow[\ X = Cl,\ N_3\ ]{Me_3Si(CH_2)_2OCOX^a}$ 11

$R^1R^2NH \xleftarrow[\substack{50°,\ 8\ h,\ 93\%\ or\\ 28°,\ 70\ h,\ 93\%}]{n\text{-Bu}_4N^+F^-/KF\cdot 2H_2O/CH_3CN^b}$ 11

$$R^1R^2NH \xleftarrow{\quad CF_3CO_2H^a \quad} 11$$

[a] L. A. Carpino, J.-H. Tsao, H. Ringsdorf, E. Fell, and G. Hettrich, *J. Chem. Soc., Chem. Commun.*, 358 (1978).

[b] L. A. Carpino and A. C. Sau, *J. Chem. Soc., Chem. Commun.*, 514 (1979).

12. 2-Methylthioethyl Carbamate: $R^1R^2NCOOCH_2CH_2SCH_3$, 12

Compound 12 is cleaved by 0.01 N NaOH after alkylation to $Me_2S^+CH_2CH_2$— or by 0.1 N NaOH after oxidation to $MeSO_2CH_2CH_2$—.[a]

[a] H. Kunz, *Chem. Ber.*, 109, 3693 (1976).

13. 2-Methylsulfonylethyl Carbamate: $R^1R^2NCOOCH_2CH_2SO_2CH_3$, 13

Compound 13 is readily cleaved by base (1 N NaOH/CH_3OH, 5 s); it is stable to catalytic hydrogenolysis and does not poison the catalyst. It is stable to liquid hydrogen fluoride (30 min).[a]

[a] G. I. Tesser and I. C. Balvert-Geers, *Int. J. Pept. Protein Res.*, 7, 295 (1975).

14. 2-(*p*-Toluenesulfonyl)ethyl Carbamate: $R^1R^2NCOOCH_2CH_2SO_2C_6H_4$-*p*-$CH_3$, 14

Compound 14 is stable to acid (HCl/HOAc) and to catalytic hydrogenation; it is cleaved by base (1 M NaOH, < 1 h).[a]

[a] A. T. Kader and C. J. M. Stirling, *J. Chem. Soc.*, 258 (1964).

15. 2-Phosphonioethyl Carbamate: $R^1R^2NCOOCH_2CH_2\overset{+}{P}RR'R''$ X^-, 15

R, R', R″ = CH_3, C_6H_5

Compound 15 is stable to trifluoroacetic acid; it is cleaved by mild bases (pH 8.4; 0.1 N NaOH, 1 min, 100% yield).[a]

[a] H. Kunz, *Angew. Chem., Inter. Ed., Engl.*, 17, 67 (1978).

Substituted Propyl and Isopropyl Carbamates

A number of protective groups can be considered either as α-substituted benzyl derivatives or 2-substituted isopropyl derivatives. Compounds that are generally cleaved by catalytic hydrogenation are included with the benzyl derivatives;

those that are generally cleaved by acidic hydrolysis are included with the iso-propyl derivatives.

16. 1,1-Dimethylpropynyl Carbamate: $R^1R^2NCOOC(CH_3)_2C≡CH$, 16

Formation (→) / Cleavage (←)[a]

$$\underset{50-80\%}{\overset{HC≡CC(CH_3)_2OCOCl}{\xrightarrow{\hspace{3cm}}}}$$

$$\underset{\substack{|\\NH_2}}{RCHCO_2H} \qquad\qquad\qquad \textbf{16}$$

$$\underset{50\%}{\overset{H_2/Pd-C, CH_3OH, 4\ h}{\xleftarrow{\hspace{3cm}}}}$$

S-Protected cysteine derivatives are stable to hydrogenolysis of the propynyl carbamate.[a]

[a] G. L. Southard, B. R. Zaborowsky, and J. M. Pettee, *J. Am. Chem. Soc.*, **93**, 3302 (1971).

17. 1,1-Dimethyl-3-(*N,N*-dimethylcarboxamido)propyl Carbamate: $R^1R^2NCOOC(CH_3)_2CH_2CH_2CON(CH_3)_2$, 17

See Miscellaneous Carbamates (following compound **72**).

18. 1,1-Diphenyl-3-(*N,N*-diethylamino)propyl Carbamate: $R^1R^2NCOOC(Ph)_2CH_2CH_2NEt_2$, 18

See the discussion that accompanies compound **70**.

19. 1-Methyl-1-(1-adamantyl)ethyl Carbamate: $R^1R^2NCOOC(CH_3)_2$-1-adamantyl, 19

Compound **19** is rapidly cleaved by trifluoroacetic acid (3% CF_3CO_2H, 0°, 4–5 min, 10^3 times faster than the *t*-butyl carbamate).[a]

[a] H. Kalbacher and W. Voelter, *Angew. Chem., Inter. Ed., Engl.*, **17**, 944 (1978).

20. 1-Methyl-1-phenylethyl Carbamate: $R^1R^2NCOOC(CH_3)_2C_6H_5$, 20

Compound **20** has some advantages over the chemically similar 2-(biphenylyl)-2-propyl carbamate in terms of expense, characterization, and storage stability.[a]

Formation (→) / *Cleavage* (←)[a]

$$\underset{\text{tetramethylguanidine, 60–70\%}}{\overset{\text{PhC(CH}_3)_2\text{OCO}_2\text{Ph, DMSO, 25°, 2.5 h}}{\longrightarrow}}$$

$$R^1R^2NH \hspace{9cm} 20$$

$$\overset{\text{CF}_3\text{CO}_2\text{H/CH}_2\text{Cl}_2}{\underset{\text{30 min, 90–100\%}}{\longleftarrow}}$$

[a] B. E. B. Sandberg and U. Ragnarsson, *Int. J. Pept. Protein Res.,* **6,** 111 (1974).

21. 1-Methyl-1-(3,5-dimethoxyphenyl)ethyl Carbamate: $R^1R^2NCOOC(CH_3)_2$-3,5-dimethoxyphenyl, 21

Compound **21,** prepared in 80% yield from the azidoformate, is cleaved by photolysis and, as expected, by acidic hydrolysis (CF_3CO_2H, 20°, 8 min, 100% yield).[a]

[a] C. Birr, W. Lochinger, G. Stahnke, and P. Lang, *Liebigs Ann. Chem.,* **763,** 162 (1972).

22. 1-Methyl-1-(4-biphenylyl)ethyl Carbamate: $R^1R^2NCOOC(CH_3)_2$-C_6H_4-p-C_6H_5, 22

Compound **22,** prepared from the mixed carbonate or azidoformate in yields of 35–80%, is readily cleaved by acidic hydrolysis (dil CF_3CO_2H/CH_2Cl_2, 10 min, quant, and 3000 times faster than the *t*-butyl carbamate), conditions that do not cleave benzyl carbamates, esters, or ethers.[a]

[a] R. S. Feinberg and R. B. Merrifield, *Tetrahedron,* **28,** 5865 (1972).

23. 1-Methyl-1-(*p*-phenylazophenyl)ethyl Carbamate: $R^1R^2NCOOC(CH_3)_2$—C_6H_4-p-N=N—C_6H_5, 23

See the discussion that accompanies compound **64.**

24. 1,1-Dimethyl-2-haloethyl Carbamate: $R^1R^2NCOOC(CH_3)_2CH_2X$, 24

$$X = Br, Cl$$

Formation[a]

$$R^1R^2NH + ClCO_2C(CH_3)_2CH_2X \underset{0°, 1.5 \text{ h}}{\overset{\text{H}_2\text{O/CHCl}_3\text{/THF,Et}_3\text{N}}{\longrightarrow}} 24$$

$$41–79\%, X = Br$$
$$60–86\%, X = Cl$$

Cleavage[a]

(1)   **24** $\xrightarrow[\text{reflux, 1 h}]{\text{CH}_3\text{OH}}$ R^1R^2NH

(2)   **24** $\xrightarrow[25°]{\text{BF}_3\cdot\text{Et}_2\text{O}/\text{CF}_3\text{CO}_2\text{H}}$ R^1R^2NH

(3)   **24** $\xrightarrow[25°,\ 1\ h]{4\ N\ \text{HBr/HOAc}}$ R^1R^2NH

(4)   **24** $\xrightarrow{\text{Na/NH}_3}$ R^1R^2NH

A halo-substituted *t*-butyl chloroformate is a more stable compound than the unsubstituted *t*-butyl chloroformate.[a]

[a] T. Ohnishi, H. Sugano, and M. Miyoshi, *Bull. Chem. Soc. Jpn.*, **45**, 2603 (1972).

25.   1,1-Dimethyl-2,2,2-trichloroethyl Carbamate: $R^1R^2NCOOC(CH_3)_2CCl_3$, 25

Compound **25** is stable to the alkaline hydrolysis of methyl esters, and to the acidic hydrolysis of *t*-butyl esters. It is rapidly cleaved by the supernucleophile lithium cobalt(I)-phthalocyanine and by zinc in acetic acid.[a]

[a] H. Eckert, M. Listl, and I. Ugi, *Angew. Chem., Inter. Ed., Engl.*, **17**, 361 (1978).

26.   1,1-Dimethyl-2-cyanoethyl Carbamate: $R^1R^2NCOOC(CH_3)_2CH_2CN$, 26

Formation (→) / Cleavage (←)[a]

$$\xrightarrow[25°,\ \text{few hours, }82\%]{\text{ClCO}_2\text{C(CH}_3)_2\text{CH}_2\text{CN, aq NaOH/THF}}$$

R^1R^2NH **26**

$$\xleftarrow[25°,\ 6\ h,\ 90\%]{\text{K}_2\text{CO}_3/\text{H}_2\text{O or Et}_3\text{N/H}_2\text{O, pH 10}}$$

Compound **26** is stable to trifluoroacetic acid.[a]

[a] E. Wünsch and R. Spangenberg, *Chem. Ber.*, **104**, 2427 (1971).

27. Isobutyl Carbamate: $R^1R^2NCOOCH_2CH(CH_3)_2$, 27

Compound **27**, used to protect the nitrogen bases in nucleoside syntheses, is stable to sodium hydroxide (which cleaves isobutyl carbonates) and to hydrazine (which cleaves amide bonds). Compound **27** is cleaved by concentrated ammonium hydroxide.[a]

[a] R. L. Letsinger and P. S. Miller, *J. Am. Chem. Soc.*, **91**, 3356 (1969).

28. *t*-Butyl Carbamate (BOC Group): $R^1R^2NCOOC(CH_3)_3$, 28

Many reagents are available for formation and removal[a] of the widely used *t*-butyl carbamate. Examples of some efficient, general methods are included below.

Formation

(1) $R^1R^2NH + (CH_3)_3COCO_2CO_2C(CH_3)_3 \xrightarrow[\text{25°, 10–30 min}]{\text{NaOH, H}_2\text{O}^b}$ **28,** 75–95%

(2) $R^1R^2NH + (CH_3)_3COCO_2N{=}C(CN)C_6H_5 \xrightarrow[\text{several hours}]{\text{Et}_3\text{N, 25°}^c}$ **28,** 72–100%

(3) $R^1R^2NH + $

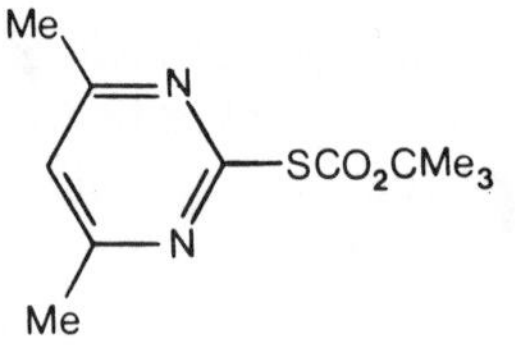

$\xrightarrow[\text{2–24 h}]{\text{Et}_3\text{N, 25°}^d}$ **28,** 80–100%

Cleavage

(1)   **28** $\xrightarrow[\text{25°, 30 min}]{3\ M\ \text{HCl/EtOAc}^e}$ R^1R^2NH, 96%

(2)   **28** $\xrightarrow[\text{20°, 1 h}]{\text{CF}_3\text{CO}_2\text{H/PhSH}^f}$ R^1R^2NH, 100%

Thiophenol acts as a scavenger for *t*-butyl cations, and thus prevents alkylation of methionine or tryptophan.[f]

(3)   **28** $\xrightarrow[\substack{\text{CHCl}_3\text{ or CH}_3\text{CN} \\ \text{25°, 6 min}}]{\text{Me}_3\text{SiI}^g}$ R^1R^2NH, 100%

Iodotrimethylsilane cleaves carbamates, esters, ethers, and acetals under neutral, nonhydrolytic conditions.[g]

(4) **28** $\xrightarrow[\substack{CH_2Cl_2/CH_3NO_2 \\ 0 \to 25°, \, 3\text{--}5 \text{ h}}]{AlCl_3/PhOCH_3{}^h}$ RNH$_2$, 73–88%

[a] For more specialized conditions, see L. A. Carpino, *Acc. Chem. Res.,* **6,** 191 (1973), refs. 12 and 17.

[b] D. S. Tarbell, Y. Yamamoto, and B. M. Pope, *Proc. Natl. Acad. Sci., U.S.A.,* **69,** 730 (1972).

[c] M. Itoh, D. Hagiwara, and T. Kamiya, *Bull. Chem. Soc. Jpn.,* **50,** 718 (1977).

[d] T. Nagasawa, K. Kuroiwa, K. Narita, and Y. Isowa, *Bull. Chem. Soc. Jpn.,* **46,** 1269 (1973).

[e] G. L. Stahl, R. Walter, and C. W. Smith, *J. Org. Chem.,* **43,** 2285 (1978).

[f] B. F. Lundt, N. L. Johansen, A. Vølund, and J. Markussen, *Int. J. Pept. Protein Res.,* **12,** 258 (1978).

[g] R. S. Lott, V. S. Chauhan, and C. H. Stammer, *J. Chem. Soc., Chem. Commun.,* 495 (1979).

[h] T. Tsuji, T. Kataoka, M. Yoshioka, Y. Sendo, Y. Nishitani, S. Hirai, T. Maeda, and W. Nagata, *Tetrahedron Lett.,* 2793 (1979).

29. *t*-Amyl Carbamate: R^{1}R^{2}NCOOC(CH$_3$)$_2$CH$_2$CH$_3$, 29

See Miscellaneous Carbamates (following compound **72**).

30. Cyclobutyl Carbamate: R^{1}R^{2}NCOO-*c*-C$_4$H$_7$, 30

See the discussion that accompanies compound **34**.

31. 1-Methylcyclobutyl Carbamate: R^{1}R^{2}NCO$_2$, 31

See the discussion that accompanies compound **34**.

32. Cyclopentyl Carbamate: R^{1}R^{2}NCOO-*c*-C$_5$H$_9$, 32

See Miscellaneous Carbamates (following compound **72**).

33. Cyclohexyl Carbamate: R^{1}R^{2}NCOO-*c*-C$_6$H$_{11}$, 33

See Miscellaneous Carbamates (following compound **72**).

34. 1-Methylcyclohexyl Carbamate: R^{1}R^{2}NCO$_2$,34

These carbocyclic carbamates, compounds **30,**[a] **31,**[a] and **34,**[a] are stable to 50% aqueous acetic acid, which can cleave *t*-butyl carbamates (3–7%). They are cleaved, at differing rates, by trifluoroacetic acid.

Formation (→) / Cleavage (←)[a]

(1)

$$\xrightarrow[\text{75\%}]{\text{ClCO}_2\text{-}c\text{-C}_4\text{H}_7,\ \text{NaHCO}_3}$$

PhCH₂CHCO₂H   **30**
|
NH₂

$$\xleftarrow[t_{1/2} > 300\ \text{min}]{\text{CF}_3\text{CO}_2\text{H},\ 25°}$$

(2)

$$\xrightarrow{\text{ClCO}_2\text{-}c\text{-C}_4\text{H}_6\text{-1-CH}_3,\ \text{Et}_3\text{N/CHCl}_3,\ 2\ \text{h}}$$

PhCH₂CHCO₂CH₃   **31**
|
NH₂

$$\xleftarrow[t_{1/2} = 180\ \text{min}]{\text{HCOOH},\ 25°}$$

(The amino acid of compound **31** is cleaved by CF_3COOH, 25°, $t_{1/2} = 2$ min.)

(3)

$$\xrightarrow{\text{ClCO}_2\text{-}c\text{-C}_6\text{H}_{10}\text{-1-CH}_3,\ \text{Et}_3\text{N/CHCl}_3,\ 2\ \text{h}}$$

PhCH₂CHCO₂CH₃   **34**
|
NH₂

$$\xleftarrow[t_{1/2} = 1\ \text{min}]{\text{CF}_3\text{CO}_2\text{H},\ 25°}$$

[a] S. F. Brady, R. Hirschmann, and D. F. Veber, *J. Org. Chem.,* **42,** 143 (1977).

35. 1-Adamantyl Carbamate: R^1R^2NCOO-1-adamantyl, 35

Formation (→) / Cleavage (←)[a]

$$\xrightarrow[\text{27–98\%}]{\text{1-adamantyl-OCOCl, NaOH}}$$

R^1R^2NH **35**

$$\xleftarrow[\text{100\%}]{\text{CF}_3\text{CO}_2\text{H},\ 25°,\ 15\ \text{min}}$$

Adamantyl carbamates of amino acids are crystalline; some *t*-butyl carbamates are oils.[a]

[a] W. L. Haas, E. V. Krumkalns, and K. Gerzon, *J. Am. Chem. Soc.,* **88,** 1988 (1966).

36. Isobornyl Carbamate: R^1R^2NCOO-isobornyl, 36

See Miscellaneous Carbamates (following compound **72**).

37.   Vinyl Carbamate: $R^1R^2NCOOCH=CH_2$, 37

The double bond in compound **37** is very reactive to electrophilic reagents, and thus is readily cleaved by reaction with bromine or mercuric acetate.

Formation

(1) $R^1R^2NH + CH_2=CHOCOCl^a$ $\xrightarrow[\text{pH 9-10}]{\text{aq MgO/dioxane}}$ **37**, 90%

(2) $R^1R^2NH + CH_2=CHOCOSPh^b$ $\xrightarrow[\substack{\text{dioxane or DMF, H}_2\text{O} \\ 25°, 16\ h}]{\text{Et}_3\text{N}}$ **37**, 50–80%

Cleavage[a]

(1)   **37** $\xrightarrow[97\%]{\text{anhyd HCl/dioxane, 25°}}$ or $\xrightarrow[94\%]{\text{HBr/HOAc}}$ R^1R^2NH

(2)   **37** $\xrightarrow{\text{Br}_2/\text{CH}_2\text{Cl}_2}$ $\xrightarrow{\text{CH}_3\text{OH}}$ R^1R^2NH, ~95%

(3)   **37** $\xrightarrow[25°]{\text{Hg(OAc)}_2/\text{HOAc-H}_2\text{O(9:1)}}$ R^1R^2NH, ~97%

[a] R. A. Olofson, Y. S. Yamamoto, and D. J. Wancowicz, *Tetrahedron Lett.*, 1563 (1977).
[b] A. J. Duggan and F. E. Roberts, *Tetrahedron Lett.*, 595 (1979).

38.   Allyl Carbamate: $R^1R^2NCOOCH_2CH=CH_2$, 38

Like allyl carbonates prepared to protect hydroxyl groups, allyl carbamates can be cleaved under aprotic conditions by nickel carbonyl.[a]

Formation ($\rightarrow$) / Cleavage ($\leftarrow$)[a]

$$R^1R^2NH \quad \xrightarrow{\text{CH}_2=\text{CHCH}_2\text{OCOCl, Py}} \quad \textbf{38}$$

$$\xleftarrow[55°, 4\ h,\ 83-95\%]{\text{Ni(CO)}_4/\text{DMF-H}_2\text{O(95:5)}}$$

[a] E. J. Corey and J. W. Suggs, *J. Org. Chem.*, **38**, 3223 (1973).

39.   Cinnamyl Carbamate: $R^1R^2NCOOCH_2CH=CHC_6H_5$, 39

The authors suggest that the cinnamyl group, often used to protect a carboxyl group as an ester that may then be cleaved under neutral conditions, can also be used to protect an amine or hydroxyl group.[a]

Cleavage[a]

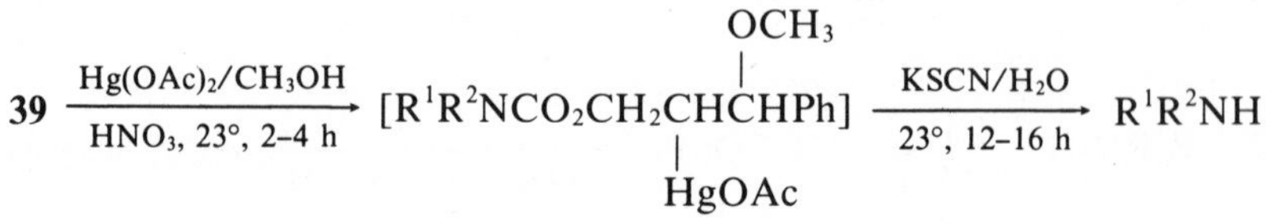

$$39 \xrightarrow[\text{HNO}_3,\ 23°,\ 2\text{-}4\ h]{\text{Hg(OAc)}_2/\text{CH}_3\text{OH}} [R^1R^2NCO_2CH_2\overset{\overset{\displaystyle OCH_3}{|}}{C}H\overset{\overset{\displaystyle }{|}}{C}HPh] \xrightarrow[23°,\ 12\text{-}16\ h]{\text{KSCN/H}_2\text{O}} R^1R^2NH$$

with HgOAc substituent on the CHPh carbon

[a] E. J. Corey and M. A. Tius, *Tetrahedron Lett.*, 2081 (1977).

40.   Phenyl Carbamate: $R^1R^2NCOOC_6H_5$, 40

Formation (→) / Cleavage (←)

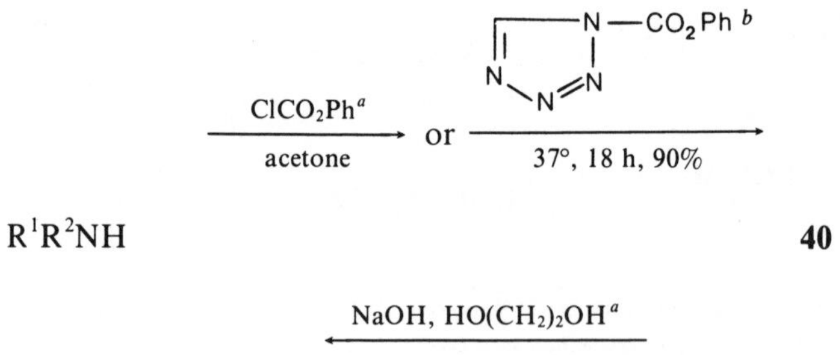

$$R^1R^2NH \xrightarrow[\text{acetone}]{\text{ClCO}_2\text{Ph}^a} \text{or} \xrightarrow[37°,\ 18\ h,\ 90\%]{} \quad 40$$

$$\xleftarrow{\text{NaOH, HO(CH}_2)_2\text{OH}^a}$$

The tetrazolecarbamate is less reactive than the chloroformate and attacks only the primary -NH_2 group in adenine.

[a] J. D. Hobson and J. G. McCluskey, *J. Chem. Soc. C,* 2015 (1967).
[b] R. W. Adamiak and J. Stawinski, *Tetrahedron Lett.*, 1935 (1977).

41.   2,4,6-Tri-*t*-butylphenyl Carbamate: $R^1R^2NCOOC_6H_2$-2,4,6-(t-C_4H_9)$_3$, 41.

See Miscellaneous Carbamates (following compound **72**).

42.   *m*-Nitrophenyl Carbamate: $R^1R^2NCOOC_6H_4$-*m*-NO_2, 42

Compound **42** is cleaved by photolysis ($h\nu > 290$ nm, 2 h, 60% yield), as well as by weak alkali.[a]

[a] Th. Wieland, Ch. Lamperstorfer, and Ch. Birr, *Makromol. Chem.*, **92**, 277 (1966).

43. *S*-Phenyl Thiocarbamate: $R^1R^2NCOSPh$, 43

Compound **43,** prepared from the thiophenyl chloroformate in an organic solvent,[a] is cleaved by oxidation with perbenzoic acid.[b] If compound **43** is prepared
in an aqueous solution, the CO–SPh bond is cleaved by a free amino group to
give a urea derivative.[a]

[a] W. H. Schuller and C. Niemann, *J. Am. Chem. Soc.,* **75,** 3425 (1953).
[b] J. Kollonitsch, A. Hajós, and V. Gábor, *Chem. Ber.,* **89,** 2288 (1956).

44. 8-Quinolyl Carbamate:

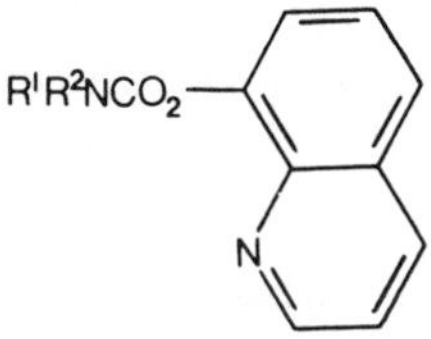

, **44.**

An 8-quinolyl carbamate is cleaved under neutral conditions by Cu(II)- or Ni(II)-
catalyzed hydrolysis.[a]

Formation (→) / Cleavage (←)[a]

$$R^1R^2NH + (8\text{-quinolyl-O})_2CO \xrightarrow[25°, \ 3 \ h, \ 33\%]{Et_3N/THF–H_2O}$$

44

$$R^1R^2NH + \xleftarrow[\substack{25°, \ aq \ acetone, \\ 1 \ h, \ 55–93\%}]{Cu(II) \ or \ Ni(II)}$$

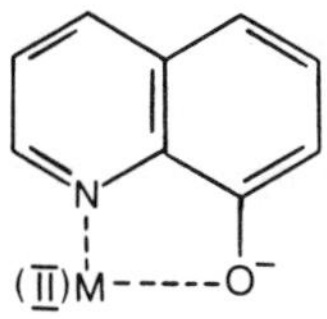

[a] E. J. Corey and R. L. Dawson, *J. Am. Chem. Soc.,* **84,** 4899 (1962).

45. *N*-Hydroxypiperidine Carbamate: $R^1R^2NCO_2N$

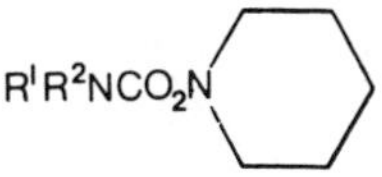

, **45**

A piperidinyl carbamate, stable to aqueous alkali and to cold acid (30% HBr,
25°, several hours) is best cleaved by reduction.[a]

Formation[a]

$$R^1R^2NH + 1\text{-piperidinyl-OCOX} \xrightarrow{Et_3N} \textbf{45,} \ 55–85\%$$

X = 2,4,5-trichlorophenyl, . . .

Cleavage[a]

(1) **45** $\xrightarrow[\text{20°, 30 min}]{\text{H}_2/\text{Pd–C, HOAc}}$ R^{1}R^{2}NH, 95%

(2) **45** $\xrightarrow[\text{1 }N\text{ H}_2\text{SO}_4,\text{ 20°, 90 min}]{\text{Electrolysis, 200 mA}}$ R^{1}R^{2}NH, 90–93%

(3) **45** $\xrightarrow[\text{20°, 5 min, 93%}]{\text{Na}_2\text{S}_2\text{O}_4/\text{HOAc}}$ or $\xrightarrow[\text{20°, 10 min, 94%}]{\text{Zn/HOAc}}$ R^{1}R^{2}NH

[a] D. Stevenson and G. T. Young, *J. Chem. Soc. C,* 2389 (1969).

46.   4-(1,4-Dimethylpiperidinyl) Carbamate: R^{1}R^{2}NCOO— , **46**

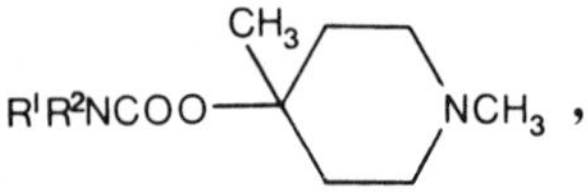

See the discussion that accompanies compound **70.**

47.   4,5-Diphenyl-3-oxazolin-2-one: 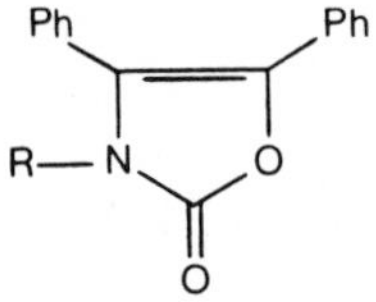 , **47**

Compound **47** is stable to aqueous sodium hydroxide (20°, 48 h), hydrazine (reflux in EtOH, 2 h), and acid (HBr/HOAc, 20°, 24 h, or anhyd CF$_3$CO$_2$H, 20°, 3 h). Racemization does not occur during formation, cleavage, or coupling of compound **47** (R = R$_2$'CHCO—).[a]

Formation[a]

RNH$_2$ + 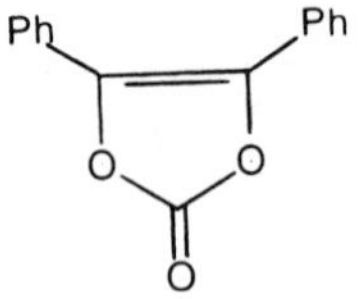 $\xrightarrow[\text{0.5 h}]{\text{DMF}}$ $\xrightarrow{\text{CF}_3\text{CO}_2\text{H}}$ **47**, 67–85%

Cleavage[a]

(1) **47** $\xrightarrow[\text{25° 12 h}]{\text{H}_2/\text{Pd–C, aq HCl}}$ RNH$_2$, quant

(2) **47** $\xrightarrow{\text{Na/NH}_3}$ RNH$_2$, 75–85%

$$(3) \quad 47 \xrightarrow[]{\textit{m}-ClC_6H_4CO_3H} \xrightarrow[]{H_2O} RNH_2, 70\%$$

[a] J. C. Sheehan and F. S. Guziec, *J. Org. Chem.*, **38**, 3034 (1973).

48. Benzyl Carbamate (CBZ Group): $R^1R^2NCOOCH_2C_6H_5$, 48

Benzyl carbamates, widely used in peptide syntheses since 1932, are readily cleaved by the original,[a] selective method of catalytic reduction, as well as by a number of newer methods. They are stable to hydrazine,[b] used to remove *N*-phthaloyl groups, to warm acetic acid, used to remove *N*-triphenylmethyl groups, and to trifluoroacetic acid at room temperature, used to cleave *t*-butyl carbamates and *t*-butyl esters.

Formation

$$R^1R^2NH + ClCO_2CH_2Ph \xrightarrow[0°, 30 \text{ min}]{\text{aq Na}_2\text{CO}_3{}^a} 48, 72\%$$

Cleavage

$$(1) \quad 48 \xrightarrow[]{H_2/Pd-C^a} R^1R^2NH$$

$$(2) \quad 48 \xrightarrow[-33°, 3-8 \text{ h}]{H_2/Pd \text{ black, } NH_3{}^c} R^1R^2NH, \text{ quant}$$

When ammonia is used as the solvent, cysteine or methionine units in a peptide do not poison the catalyst.[c]

$$(3) \quad 48 \xrightarrow[\substack{\text{Hydrogen-donor solvent,} \\ 25° \text{ or reflux in EtOH,} \\ 15 \text{ min to 2 h}}]{\text{Pd-C or Pd black}} R^1R^2NH, 80-100\%$$

Several hydrogen donors including cyclohexene,[d,e] 1,4-cyclohexadiene,[f] and formic acid,[g,h] have been used for catalytic transfer hydrogenation, in general a more rapid reaction than catalytic hydrogenation.

$$(4) \quad 48 \xrightarrow[\text{reflux, 3 h, 80\%}]{Et_3SiH/cat. \ Et_3N/cat. \ PdCl_2{}^i} \text{ or } \xrightarrow[25°, 6 \text{ min, } 100\%]{Me_3SiI/CH_3CN^j} R^1R^2NH$$

S-Benzyl groups are stable to trialkylsilanes; benzyl esters and benzyl ethers are cleaved.[i]

(5)   **48** $\xrightarrow[\text{0–25°, 5 h, 73\%}]{\text{AlCl}_3/\text{PhOCH}_3{}^k}$ or $\xrightarrow[\substack{-10°,\ 1\ h\ \rightarrow\ 25°,\ 2\ h \\ 80\text{–}100\%}]{\text{BBr}_3/\text{CH}_2\text{Cl}_2{}^l}$ R^1R^2NH

Cleavage with a Lewis acid avoids the racemization[l] that accompanies cleavage under basic conditions or the acid-catalyzed cleavage of a β-lactam ring.[k]

Benzyl carbamates of larger peptides can be cleaved by boron tribromide in trifluoroacetic acid,[m] since the peptides are more soluble in acid than in methylene chloride.

(6)   **48** $\xrightarrow[\text{55°, 4 h, aq CH}_3\text{OH}]{\text{253.7 nm}^n}$ $R^1R^2NH,\ 70\%$

Photolytic cleavage can be used, as in this example, for an acid-sensitive, halogen-substituted amino sugar.[n]

(7)   **48** $\xrightarrow[\text{DMF, R}_4\text{N}^+\text{X}^-]{-2.9\ V^o}$ $R^1R^2NH,\ 70\text{–}80\%$

Electrochemical reduction can be used to cleave, selectively, benzyl and β-haloethyl carbamates.[o]

(8)   **48** $\xrightarrow{\text{H}_2\text{NNH}_2{}^p}$ R^1R^2NH

A benzyl carbamate, used to protect the indole nitrogen in tryptophan, can be cleaved by hydrazine, as well as by catalytic hydrogenation, and hydrogen fluoride.[p]

(9)   A number of less selective, acidic reagents can be used to cleave benzyl carbamates: HBr/HOAc[q]; 50% CF_3CO_2H (25°, 14 days, partially cleaved)[r]; 70% HF/Py[s]; CF_3SO_3H[t]; and FSO_3H[u] or CH_3SO_3H.[u]

[a] M. Bergmann and L. Zervas, *Ber.,* **65,** 1192 (1932).

[b] R. A. Boissonnas, *Adv. Org. Chem.,* **3,** 159 (1963), ref. 31: "R. A. Boissonnas and F. Leitner, unpublished results, Ph.D. thesis of F. Leitner, Geneva, 1955."

[c] J. Meienhofer and K. Kuromizu, *Tetrahedron Lett.,* 3259 (1974).

[d] A. E. Jackson and R. A. W. Johnstone, *Synthesis,* 685 (1976).

[e] G. M. Anantharamaiah and K. M. Sivanandaiah, *J. Chem. Soc., Perkin Trans. 1,* 490 (1977).

[f] A. M. Felix, E. P. Heimer, T. J. Lambros, C. Tzougraki, and J. Meienhofer, *J. Org. Chem.,* **43,** 4194 (1978).

[g] K. M. Sivanandaiah and S. Gurusiddappa, *J. Chem. Res. Synop.,* 108 (1979).

[h] B. ElAmin, G. M. Anantharamaiah, G. P. Royer, and G. E. Means, *J. Org. Chem.,* **44,** 3442 (1979).

[i] L. Birkofer, E. Bierwirth, and A. Ritter, *Chem. Ber.,* **94,** 821 (1961).

[j] R. S. Lott, V. S. Chauhan, and C. H. Stammer, *J. Chem. Soc., Chem. Commun.,* 495 (1979).

[k] T. Tsuji, T. Kataoka, M. Yoshioka, Y. Sendo, Y. Nishitani, S. Hirai, T. Maeda, and W. Nagata, *Tetrahedron Lett.*, 2793 (1979).

[l] A. M. Felix, *J. Org. Chem.*, **39**, 1427 (1974).

[m] J. Pless and W. Bauer, *Angew Chem., Inter. Ed., Engl.*, **12**, 147 (1973).

[n] S. Hanessian and R. Masse, *Carbohydr. Res.*, **54**, 142 (1977).

[o] V. G. Mairanovsky, *Angew. Chem., Inter. Ed., Engl.*, **15**, 281 (1976).

[p] M. Chorev and Y. S. Klausner, *J. Chem. Soc., Chem. Commun.*, 596 (1976).

[q] D. Ben-Ishai and A. Berger, *J. Org. Chem.*, **17**, 1564 (1952).

[r] A. R. Mitchell and R. B. Merrifield, *J. Org. Chem.*, **41**, 2015 (1976).

[s] S. Matsuura, C.-H. Niu, and J. S. Cohen, *J. Chem. Soc., Chem. Commun.*, 451 (1976).

[t] H. Yajima, N. Fujii, H. Ogawa, and H. Kawatani, *J. Chem. Soc., Chem. Commun.*, 107 (1974).

[u] H. Yajima, H. Ogawa, and H. Sakurai, *J. Chem. Soc., Chem. Commun.*, 909 (1977).

49. 2,4,6-Trimethylbenzyl Carbamate: $R^1R^2NCOOCH_2C_6H_2$-2,4,6-$(CH_3)_3$, 49

See Miscellaneous Carbamates (following compound **72**).

50. *p*-Methoxybenzyl Carbamate: $R^1R^2NCOOCH_2C_6H_4$-*p*-OCH_3, 50

Compound **50** can be prepared from the azide,[a] carbonate,[b] or thiocarbonate,[c] and cleaved by trifluoroacetic acid at 0°, conditions under which a benzyl carbamate is stable.[a]

Compound **50** is also cleaved by many of the methods described for benzyl carbamates, including hydrolysis by sulfonic acids,[d] and catalytic transfer hydrogenation.

[a] F. Weygand and K. Hunger, *Chem. Ber.*, **95**, 1 (1962).

[b] J. H. Jones and G. T. Young, *Chem. Ind. (London)*, 1722 (1966).

[c] T. Nagasawa, K. Kuroiwa, K. Narita, and Y. Isowa, *Bull. Chem. Soc. Jpn.*, **46**, 1269 (1973).

[d] H. Yajima, H. Ogawa, N. Fujii, and S. Funakoshi, *Chem. Pharm. Bull.*, **25**, 740 (1977).

51. 3,5-Dimethoxybenzyl Carbamate: $R^1R^2NCOOCH_2C_6H_3$-3,5-$(OCH_3)_2$, 51

Compound **51** is readily cleaved by photolysis (H_2O–dioxane, 1 h, 65–85%).[a]

[a] J. W. Chamberlin, *J. Org. Chem.*, **31**, 1658 (1966).

52. *p*-Decyloxybenzyl Carbamate: $R^1R^2NCOOCH_2C_6H_4$-*p*-$OC_{10}H_{21}$, 52

See Miscellaneous Carbamates (following compound **72**).

53. *p*-Nitrobenzyl Carbamate: $R^1R^2NCOOCH_2C_6H_4$-*p*-NO_2, 53

Formation[a]

$$R^1R^2NH \xrightarrow[\text{0°, 1.5 h}]{\text{ClCO}_2\text{CH}_2\text{C}_6\text{H}_4\text{-}p\text{-NO}_2,\text{ base}} \textbf{53},\ 78\%$$

Cleavage

$$53 \xrightarrow[\text{10 h, 87\%}]{\text{H}_2/\text{Pd--C}} \text{ or } \xrightarrow[\text{60°, 2 h, ~68\%}]{\text{4 } N \text{ HBr/HOAc}} \text{R}^1\text{R}^2\text{NH}^a$$

A nitrobenzyl carbamate is more readily cleaved by hydrogenolysis than a benzyl carbamate; it is more stable to acid-catalyzed hydrolysis than a benzyl carbamate, and therefore selective cleavage is possible.[a]

$$(2) \quad 53 \xrightarrow[\text{DMF, R}_4\text{N}^+\text{X}^-]{-1.2 \text{ V}^b} \text{R}^1\text{R}^2\text{NH}$$

[a] J. E. Shields and F. H. Carpenter, *J. Am. Chem. Soc.*, **83**, 3066 (1961).
[b] V. G. Mairanovsky, *Angew. Chem., Inter. Ed. Engl.*, **15**, 281 (1976).

54.   *o*-Nitrobenzyl Carbamate: $\text{R}^1\text{R}^2\text{NCOOCH}_2\text{C}_6\text{H}_4$-*o*-$\text{NO}_2$, 54

55.   3,4-Dimethoxy-6-nitrobenzyl Carbamate:
$\text{R}^1\text{R}^2\text{NCOOCH}_2(\text{3,4-dimethoxy-6-nitrophenyl})$, 55

Compounds **54**[a] and **55**[a] can be prepared from the chloroformates and labile structures such as amino sugars, since the protective groups can be cleaved photolytically ($h\nu = 320$ nm, 25°, 12–16 h, 70–95%).

[a] B. Amit, U. Zehavi, and A. Patchornik, *J. Org. Chem.*, **39**, 192 (1974).

56.   *p*-Bromobenzyl Carbamate: $\text{R}^1\text{R}^2\text{NCOOCH}_2\text{C}_6\text{H}_4$-*p*-Br, 56

See Miscellaneous Carbamates (following compound **72**).

57.   Chlorobenzyl Carbamate: $\text{R}^1\text{R}^2\text{NCOOCH}_2\text{C}_6\text{H}_4$-Cl, 57

Compound **57,** more stable than a benzyl carbamate during acid-catalyzed hydrolsis of an N^α-*t*-butyl carbamate, has been used to protect the N^ϵ-amino group in lysine.[a,b]

[a] K. Noda, S. Terada, and N. Izumiya, *Bull. Chem. Soc. Jpn.*, **43**, 1883 (1970).
[b] B. W. Erickson and R. B. Merrifield, *J. Am. Chem. Soc.*, **95**, 3757 (1973).

58.   2,4-Dichlorobenzyl Carbamate: $\text{R}^1\text{R}^2\text{NCOOCH}_2\text{C}_6\text{H}_3$-2,4-$\text{Cl}_2$, 58

Formation[a]

$$\text{R}^1\text{R}^2\text{NH} + p\text{-O}_2\text{N-C}_6\text{H}_4\text{OCO}_2\text{CH}_2\text{C}_6\text{H}_3\text{-2,4-Cl}_2 \xrightarrow[\substack{\text{18-crown-6} \\ i\text{-Pr}_2\text{NEt, CH}_3\text{CN}}]{\text{KF}} \textbf{58}, 60\%$$

Cleavage[a]

$$58 \quad \xrightarrow[\substack{\text{EtOH, 4 h} \\ 78\%}]{\text{H}_2/\text{Pd–C}} \quad \text{or} \quad \xrightarrow[\substack{0°, 45 \text{ min} \\ 25\%}]{\text{HF}} \quad R^1R^2NH$$

2,4-Dichlorobenzyl carbamate is 80 times more stable to acid than the unsubstituted benzyl carbamate; it is stable to the acid-catalyzed removal of *t*-butyl carbamates. Potassium fluoride catalysis during the formation eliminates the base-catalyzed racemization that occurs when sodium hydride is used.[a]

[a] Y. S. Klausner and M. Chorev, *J. Chem. Soc., Perkin Trans. 1,* 627 (1977).

59. *p*-Cyanobenzyl Carbamate: $R^1R^2NCOOCH_2C_6H_4$-*p*-CN, 59

Compound **59** is similar to the *m*-chlorobenzyl carbamate, **57,** developed by the same chemists. It is cleaved by strong acid (36% HBr/HOAc, 25°, 4 h, 100% yield).[a]

[a] K. Noda, S. Terada, and N. Izumiya, *Bull. Chem. Soc. Jpn.,* **43,** 1883 (1970).

60. *o*-(*N,N*-Dimethylcarboxamido)benzyl Carbamate: $R^1R^2NCOOCH_2C_6H_4$-*o*-CON(CH_3)_2, 60

Compound **60** can be used to provide increased solubility of long chain peptides. It is cleaved by strong acid (45% HBr/HOAc, 20°, 3 h), but is stable to trifluoroacetic acid (20°, 3 h).[a]

[a] S. Coyle, O. Keller, and G. T. Young, *J. Chem. Soc., Perkin Trans. 1,* 1459 (1979).

61. *m*-Chloro-*p*-acyloxybenzyl Carbamate[a,b]:

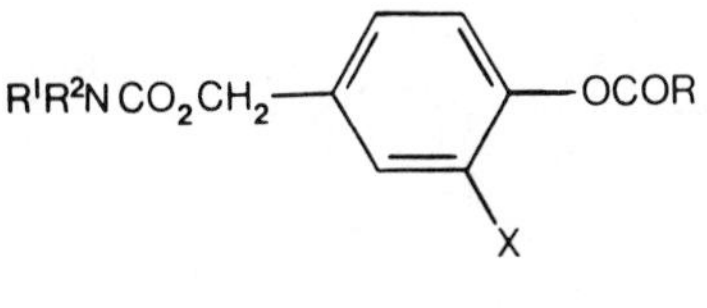

$$R^1R^2N\,CO_2CH_2 - \underset{X}{\bigcirc} - OCOR \quad , \textbf{61, X = Cl, H}$$

Cleavage

$$(1) \quad \textbf{61} \quad \xrightarrow[\substack{\text{NaHSO}_3, 1 \text{ h} \\ X = \text{Cl}}]{\text{NaHCO}_3/\text{Na}_2\text{CO}_3 \text{ or } \text{H}_2\text{O}_2/\text{NH}_3} \quad \text{or} \quad \xrightarrow[\substack{10 \text{ min, } 100\% \\ X = \text{Cl}}]{0.1 \ N \ \text{NaOH}}$$

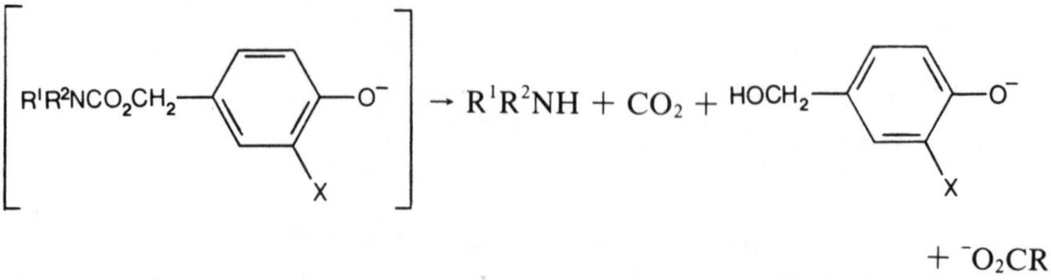

$$\left[R^1R^2NCO_2CH_2 - \overset{X}{\underset{}{\bigcirc}} - O^- \right] \rightarrow R^1R^2NH + CO_2 + HOCH_2 - \overset{X}{\underset{}{\bigcirc}} - O^-$$

$$+ \ ^-O_2CR$$

(2) **61** $\xrightarrow{\text{H}_2/\text{Pd-C}}$ or $\xrightarrow{\text{HBr/HOAc}}$ R^1R^2NH

[a] M. Wakselman and E. G.-Jampel, *J. Chem. Soc., Chem. Commun.*, 593 (1973).
[b] G. Le Corre, E. G.-Jampel, and M. Wakselman, *Tetrahedron*, **34**, 3105 (1978).

62. *p*-(Dihydroxyboryl)benzyl Carbamate: $R^1R^2NCOOCH_2C_6H_4$-*p*-$B(OH)_2$, 62

The dihydroxyboryl group modifies the behavior of a benzyl carbamate in several respects: solubility and affinity properties can be changed by complex formation with catechols and similar diols; the protective group can be cleaved under mild conditions by oxidation with hydrogen peroxide.[a]

This useful protective group has not been included in a Reactivity Chart because the dihydroxyboryl group, easily protected during a synthesis, would appear to be reactive to many of the reagents in the charts.

Formation[a]

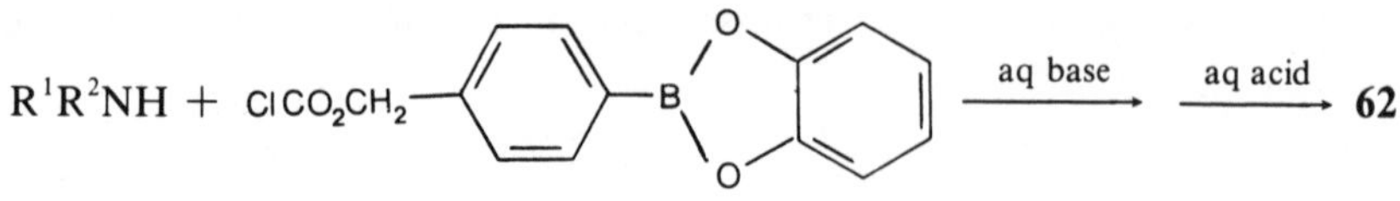

$$R^1R^2NH + ClCO_2CH_2 - \bigcirc - B \overset{O}{\underset{O}{\diagdown}} \bigcirc \xrightarrow{\text{aq base}} \xrightarrow{\text{aq acid}} \textbf{62}$$

Cleavage[a]

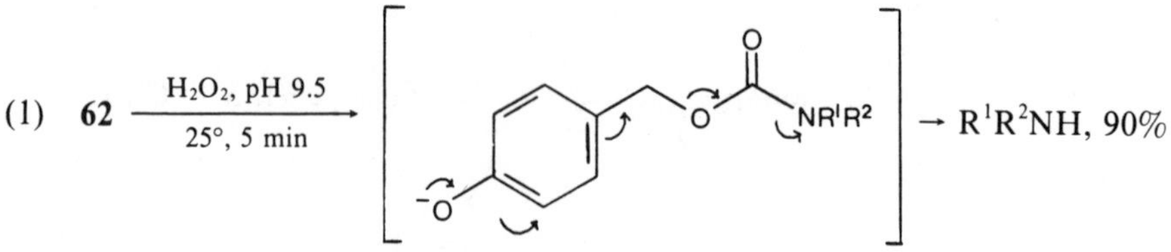

(1) **62** $\xrightarrow[\text{25°, 5 min}]{\text{H}_2\text{O}_2, \text{ pH } 9.5}$ $\left[\ \right] \rightarrow R^1R^2NH, 90\%$

(2) **62** $\xrightarrow{\text{H}_2/\text{Pd-C}}$ or $\xrightarrow{\text{HBr/HOAc}}$ R^1R^2NH

[a] D. S. Kemp and D. C. Roberts, *Tetrahedron Lett.*, 4629 (1975).

63. *p*-(Phenylazo)benzyl Carbamate: $R^1R^2NCOOCH_2C_6H_4$-*p*-N=N—C_6H_5, 63

64. *p*-(*p*′-Methoxyphenylazo)benzyl Carbamate: $R^1R^2NCOOCH_2C_6H_4$-*p*-N=N—C_6H_4-*p*-OCH_3, 64

Compounds **23,**[a] **63,**[b] and **64**[b] are colored, easily detectable substances. Compound **23,** prepared in 50–80% yield from the azidoformate or mixed carbonate, is readily cleaved by acidic hydrolysis (CF_3CO_2H/CH_2Cl_2, 25°, 5 min, 100% yield). Compounds **63** and **64** are cleaved by catalytic hydrogenation or acidic hydrolysis (HBr/HOAc).

[a] A. T.-Kyi and R. Schwyzer, *Helv. Chim. Acta,* **59,** 1642 (1976).

[b] R. Schwyzer, P. Sieber, and K. Zatskó, *Helv. Chim. Acta,* **41,** 491 (1958).

65. 5-Benzisoxazolylmethyl Carbamate: 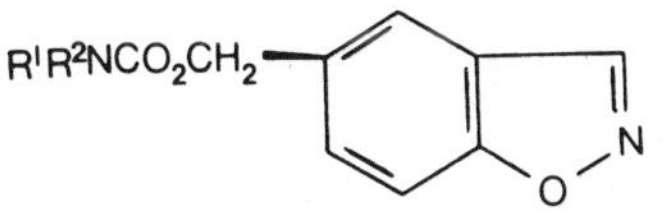, 65

Formation[a]

$$R^1R^2NH + ClCO_2CH_2\text{-5-benzisoxazole} \xrightarrow[0°,\ 1\ h]{pH\ 8.5\text{–}9,\ CH_3CN} \textbf{65},\ 63\%$$

Cleavage[a]

(1) **65** $\xrightarrow[25°,\ 30\ min]{Et_3N/CH_3CN\ or\ DMF}$ 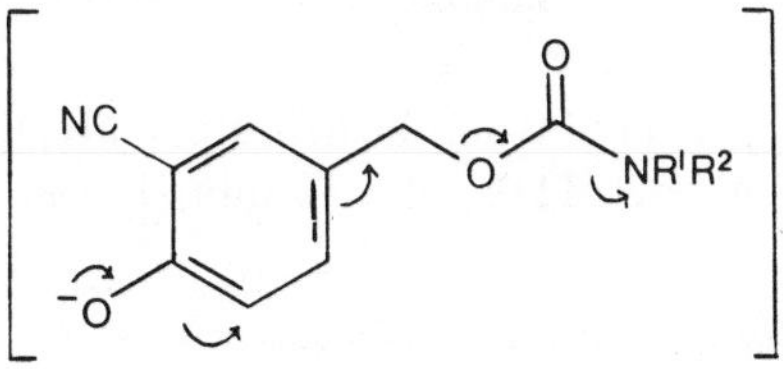 $\xrightarrow[40°,\ 3\ h,\ pH\ 7,\ 92\%]{Na_2SO_3/aq\ EtOH}$

or $\xrightarrow[95\%]{CF_3CO_2H,\ 90\ min}$ R^1R^2NH

(2) **65** $\xrightarrow{H_2/Pd\text{–}C}$ or $\xrightarrow{HBr/HOAc}$ R^1R^2NH

Compound **65** is stable to trifluoroacetic acid.[a]

[a] D. S. Kemp and C. F. Hoyng, *Tetrahedron Lett.,* 4625 (1975).

$R^1R^2NCO_2CH_2$

66. 9-Anthrylmethyl Carbamate: , **66**

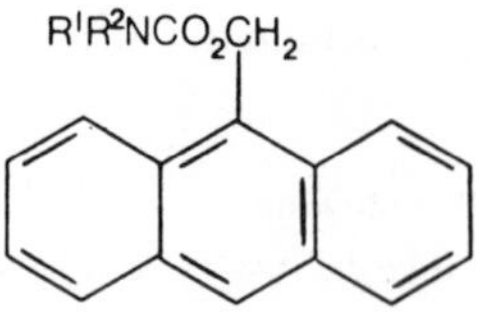

Formation[a]

$$R^1R^2NH + 9\text{-anthryl-}CH_2OCOC_6H_4\text{-}p\text{-}NO_2 \xrightarrow{\text{DMF, }25°} \textbf{66}, 86\%$$

Cleavage[a]

(1) **66** $\xrightarrow[\substack{-20°, 1\text{-}7\text{ h, }77\text{-}91\% \text{ or} \\ 25°, 4\text{ min, }86\%}]{\text{CH}_3\text{SNa, DMF}}$ R^1R^2NH

(2) **66** $\xrightarrow[0°, 5\text{ min}]{\text{CF}_3\text{CO}_2\text{H/CH}_2\text{Cl}_2}$ R^1R^2NH, 88–92%

Compound **66** is stable to 0.01 N lithium hydroxide (25°, 6 h), to 0.1 N sulfuric acid (25°, 1 h), and to 1 M trifluoroacetic acid (25°, 1 h, dioxane).[a]

[a] N. Kornblum and A. Scott, *J. Org. Chem.*, **42**, 399 (1977).

67. Diphenylmethyl Carbamate: $R^1R^2NCOOCH(C_6H_5)_2$, 67

Compound **67,** prepared from the azidoformate, is readily cleaved by mild acid hydrolysis (1.7 N HCl/THF, 65°, 10 min, 100%).[a]

[a] R. G. Hiskey and J. B. Adams, *J. Am. Chem. Soc.*, **87**, 3969 (1965).

68. Phenyl(o-nitrophenyl)methyl Carbamate: $R^1R^2NCOOCH(C_6H_5)C_6H_4\text{-}o\text{-}NO_2$, 68

Compound **68** is cleaved in good yield by irradiation in benzene for 3 hours.[a]

[a] J. A. Barltrop, P. J. Plant, and P. Schofield, *J. Chem. Soc., Chem. Commun.*, 822 (1966).

69. Di(2-pyridyl)methyl Carbamate: $R^1R^2NCOOCH(2\text{-pyridyl})_2$, 69

70.   1-Methyl-1-(4-pyridyl)ethyl Carbamate: $R^1R^2NCOOC(CH_3)_2$-4-pyridyl, 70

Compounds 18,[a] 46,[a] 69,[a] and 70[a] were designed with a basic site to provide increased acid stability. Compound 18 is cleaved by CF_3CO_2H (20°, 1 h, 100%); the analogous 1,1-diphenylpropyl carbamate is cleaved by much milder conditions (80% HOAc, 22°, $t_{1/2} = 52$ min). Compound 46 is stable to CF_3CO_2H (25°, 1 h); the corresponding 1-methylcyclohexyl carbamate is cleaved in 1 min. Compound 69 is stable to 45% HBr/HOAc (20°, 48 h); the corresponding diphenyl carbamate is cleaved by CF_3CO_2H (0°, 5 min) and by HCl/HOAc, 20°. Compound 70 is stable to CF_3CO_2H (20°, 18 h) and to 2 M HBr/HOAc (20°, 48 h). The analogous 2-phenylisopropyl carbamate is cleaved by 1% CF_3CO_2H/CH_2Cl_2 in a few minutes. Compound 70 is cleaved by chemical and electrolytic reduction: H_2/Pd–C, EtOH, 1.75 h, 100%, and 50 mA, 0.05 M H_2SO_4, 0°, 1 h, 91% yield.

[a] S. Coyle, O. Keller, and G. T. Young, *J. Chem. Soc., Perkin Trans. 1,* 1459 (1979).

71.   Isonicotinyl Carbamate: $R^1R^2NCOOCH_2$-4-pyridyl, 71

Compound 71 can be used to protect the ϵ-amino group in lysine. It is stable to acids (e.g., CF_3CO_2H, 20°, 1 h), and can be cleaved reductively (Zn/HOAc, 25°, 1.5 h, 100%, or H_2/Pd–C, HOAc, 10 min, 100%).[a]

[a] D. F. Veber, W. J. Paleveda, Y. C. Lee, and R. Hirschmann, *J. Org. Chem.,* 42, 3286 (1977).

72.   *S*-Benzyl Thiocarbamate: $R^1R^2NCOSCH_2Ph$, 72

Formation[a]

$$R^1R^2NH + PhCH_2SCOCl \xrightarrow[\text{0°, 20 min}]{\text{Et}_3\text{N/CHCl}_3} \textbf{72},\ 35\text{–}75\%$$

Cleavage[a]

$$\textbf{72} \xrightarrow[\text{76\%}]{\text{3 } N \text{ NaOH, 80°, 45 min}} \text{or} \xrightarrow[\text{25\%}]{\text{Na/NH}_3} R^1R^2NH$$

Compound 72 is more stable in a weakly basic solution than is an *S*-phenyl thiocarbamate.[a]

[a] H. B. Milne, S. L. Razniak, R. P. Bayer, and D. W. Fish, *J. Am. Chem. Soc.,* 82, 4582 (1960).

Miscellaneous Carbamates

The carbamates listed below have in general been prepared and cleaved by the usual methods.

	R^1R^2NCOOR'	
Compound number	R'	Reference
2	—CH_2-c-C_3H_5	a
3	—$CH(CH_3)$-c-C_3H_5	a
17	—$C(CH_3)_2CH_2CH_2CON(CH_3)_2$	b
29	—$C(CH_3)_2CH_2CH_3$	c
32	-c-C_5H_9	d
33	-c-C_6H_{11}	d
36	Isobornyl	e
41	2,4,6-Tri-t-butylphenyl	f
49	2,4,6-Trimethylbenzyl	g
52	p-Decyloxybenzyl	h
56	—$CH_2C_6H_4$-p-Br	i

[a] S. F. Brady, R. Hirschmann, and D. F. Veber, J. Org. Chem., **42**, 143 (1977).

[b] S. Coyle, O. Keller, and G. T. Young, J. Chem. Soc., Perkin Trans. 1, 1459 (1979).

[c] S. Sakakibara, I. Honda, K. Takada, M. Miyoshi, T. Ohnishi, and K. Okumura, Bull. Chem. Soc. Jpn., **42**, 809 (1969); S. Matsuura, C.-H. Niu, and J. S. Cohen, J. Chem. Soc., Chem. Commun., 451 (1976).

[d] F. C. McKay and N. F. Albertson, J. Am. Chem. Soc., **79**, 4686 (1957); E. Munekata, Y. Masui, T. Shiba, and T. Kaneko, Bull. Chem. Soc. Jpn., **46**, 3187 (1973).

[e] M. Fujino, S. Shinagawa, O. Nishimura, and T. Fukuda, Chem. Pharm. Bull., **20**, 1017 (1972).

[f] D. Seebach and T. Hassel, Angew. Chem., Inter. Ed., Engl., **17**, 274 (1978).

[g] Y. Isowa, M. Ohmori, M. Sato, and K. Mori, Bull. Chem. Soc. Jpn., **50**, 2766 (1977).

[h] H. Brechbühler, H. Büchi, E. Hatz, J. Schreiber, and A. Eschenmoser, Helv. Chim. Acta, **48**, 1746 (1965).

[i] D. M. Channing, P. B. Turner, and G. T. Young, Nature, **167**, 487 (1951).

Urea-Type Derivatives

73.   N'-Piperidinylcarbonyl Derivative: $R^1R^2NCON(-CH_2-)_5$, 73

Compound **73**, prepared in 68–100% yield from N-chloroformamide, and cleaved by hydrazine (25°, 24 h, 80% yield) or base (2 N NaOH, 25°, 2 h, 84% yield), is used to protect the —NH group of histidine.[a]

[a] G. Jäger, R. Geiger, and W. Siedel, Chem. Ber., **101**, 3537 (1968).

74.   N'-p-Toluenesulfonylaminocarbonyl Derivative: R^{1}R^{2}NCONHSO$_2$C$_6$H$_4$-p-CH$_3$, 74

Compound **74,** prepared from an amino acid and p-tosyl isocyanate in 20–80% yield, is cleaved by alcohols (95% aq EtOH, n-PrOH, or n-BuOH, 100°, 1 h, 95% yield). It is stable to dilute base, to acids (HBr/HOAc or cold CF$_3$CO$_2$H), and to hydrazine.[a]

[a]   B. Weinstein, T. N.-S. Ho, R. T. Fukura, and E. C. Angell, *Synth. Commun.,* **6,** 17 (1976).

75.   N'-Phenylaminothiocarbonyl Derivative: R^{1}R^{2}NCSNHC$_6$H$_5$, 75

Compound **75,** prepared from an amino acid and phenyl isothiocyanate,[a] is cleaved by anhydrous trifluoracetic acid (an N—COCF$_3$ group is stable),[b] and by oxidation (m-ClC$_6$H$_4$CO$_3$H, 0°, 1.5 h, 73% yield; H$_2$O$_2$/HOAc, 80°, 80 min, 44% yield).[c]

[a]   P. Edman, *Acta Chem. Scand.,* **4,** 277 (1950).

[b]   F. Borrás and R. E. Offord, *Nature,* **227,** 716 (1970).

[c]   J. Kollonitsch, A. Hajós, and V. Gábor, *Chem. Ber.,* **89,** 2288 (1956).

AMIDES

Amides, readily prepared from an amine and an acid chloride or anhydride, are relatively stable compounds that, classically, are cleaved by heating in strongly acidic or basic solutions. A number of substrates, including peptides, nucleo-sides, and amino sugars, are unstable to these conditions, and other methods of cleavage have been developed. These include "assisted removal" of haloacetyl, acetoacetyl, and o-nitro-, -amino- or -azophenyl or -benzyl derivatives, oxida-tion of formyl and 3-(p-hydroxyphenyl)propionyl derivatives, reduction of for-myl derivatives, and electrolysis of benzoyl and 3-(p-hydroxyphenyl)propionyl derivatives.

Many N-acyl derivatives have been prepared to protect the —NH group. Among simple amides, hydrolytic stability increases from formyl to acetyl to benzoyl. Lability of the haloacetyl derivatives to mild acid hydrolysis increases with substitution: acetyl < chloroacetyl < dichloroacetyl < trichloroacetyl < trifluoroacetyl.[a] During phosphorylation reactions in nucleotide syntheses, the amino groups in cytosine, adenine, and guanine are protected by the p-methoxy-benzoyl, benzoyl, and isobutyryl or 2-methylbutyryl groups, respectively; these protective groups are removed by ammonolysis.[b] Primary amines can be pro-tected as imides, which prevents activated N-acyl amino acids from racemization via azlactone intermediates. Several review articles discuss amides as —NH pro-tective groups.[c–e] The most commonly used derivatives are included in Reactivity Chart 9; the conventions used in this section are described on p. xii.

[a] R. S. Goody and R. T. Walker, *Tetrahedron Lett.*, 289 (1967).

[b] C. B. Reese, *Tetrahedron*, **34**, 3143–3179 (1978), see p. 3150.

[c] E. Wünsch, "Blockierung und Schutz der α-Amino-Function," in *Methoden der Organischen Chemie (Houben-Weyl)*, Georg Thieme Verlag, Stuttgart, 1974, Vol. 15/1, pp. 164–203, 250–264.

[d] J. W. Barton, "Protection of N—H Bonds and NR₃," in *Protective Groups in Organic Chemistry*, J. F. W. McOmie, Ed., Plenum, New York and London, 1973, pp. 46–56.

[e] Y. Wolman, "Protection of the Amino Group," in *The Chemistry of the Amino Group*, S. Patai, Ed., Wiley-Interscience, New York, 1968, Vol. 4, pp. 669–682.

1.   Formamide: R^1R^2NCHO, 1

Formation

(1)

$$\xrightarrow[\text{25°, 1 h, 78–90\%}]{\text{98\% HCO}_2\text{H, Ac}_2\text{O}^{a}}$$

RCHCO—
 |
 NH$_2$

RCHCO—
 |
 NHCHO

$$\xleftarrow[\substack{\text{25°, 48 h or} \\ \text{reflux, 1 h, 80–95\%}}]{\text{HCl/H}_2\text{O–dioxane}^{a}}$$

A formamide is stable to saponification of an ester group.[a]

(2)   RCHCO$_2$-t-Bu $\xrightarrow[\text{0°, 4 h}]{\text{HCOOH, DCC, Py}^{b}}$ RCHCO$_2$-t-Bu, 87–90%
          | |
         NH$_2$ NHCHO

This method forms *N*-formyl derivatives of *t*-butyl amino acid esters with a minimum of racemization.[b]

(3)   RCHCO$_2$R′ + [HCO$_2$H + EtN=C=N(CH$_2$)$_3$NMe$_2$·HCl $\xrightarrow[\text{15 min}]{0°}$
          |
         NH$_2$·HX

(HCO)$_2$O] $\xrightarrow[\text{5°, 20 h}]{\text{N-methylmorpholine}}$ RCHCO$_2$R′, 65–96%[c]
                                                                    |
                                                                  NHCHO

This method can be used with amine hydrochlorides.[c]

Cleavage

(1)   RCHCO— $\xrightarrow[\text{60°, 4 h}]{\text{H}_2\text{NNH}_2\text{/EtOH}}$ RCHCO—, 60–80%[d]
          | |
         NHCHO   NH$_2$

(2) RCHCO— $\xrightarrow[\text{25°, 5-7 h}]{\text{H}_2/\text{Pd-C, THF-HCl}}$ RCHCO—, quant^e
 | |
 NHCHO   NH$_2$

(3) RCHCO— $\xrightarrow[\text{60°, 2 h}]{\text{15\% H}_2\text{O}_2/\text{H}_2\text{O}}$ RCHCO—, 80%^{f}
 | |
 NHCHO   NH$_2$

(4) RCHCO— $\xrightarrow[\substack{\text{20°, 24 h or} \\ \text{60°, 3 h}}]{\text{CH}_3\text{COCl/PhCH}_2\text{OH}}$ RCHCO—, good yields^g
 | |
 NHCHO   NH$_2$

(5) ArNHCHO $\xrightarrow{\text{254 nm, CH}_3\text{CN}}$ ArNH$_2$, 100%^{h}

(6) R^{1}R^{2}NCHO $\xrightarrow[\text{reflux, 18 h}]{\text{NaOH/H}_2\text{O}}$ R^{1}R^{2}NH, 85%^{i}

During some syntheses of racemic tryptophans, an *N*-formyl malonic acid ester was converted to an α-amino acid.^{i}

^{a} J. C. Sheehan and D.-D. H. Yang, *J. Am. Chem. Soc.*, **80**, 1154 (1958).

^{b} M. Waki and J. Meienhofer, *J. Org. Chem.*, **42**, 2019 (1977).

^{c} F. M. F. Chen and N. L. Benoiton, *Synthesis*, 709 (1979).

^{d} R. Geiger and W. Siedel, *Chem. Ber.*, **101**, 3386 (1968).

^{e} G. Losse and D. Nadolski, *J. Prakt. Chem.*, **24**, 118 (1964).

^{f} G. Losse and W. Zönnchen, *Liebigs Ann. Chem.*, **636**, 140 (1960).

^{g} J. O. Thomas, *Tetrahedron Lett.*, 335 (1967).

^{h} B. K. Barnett and T. D. Roberts, *J. Chem. Soc., Chem. Commun.*, 758 (1972).

^{i} U. Hengartner, A. D. Batcho, J. F. Blount, W. Leimgruber, M. E. Larscheid, and J. W. Scott, *J. Org. Chem.*, **44**, 3748 (1979).

2. Acetamide: R^{1}R^{2}NCOCH$_3$, 2

Formation

N-Acetyl derivatives are formed in base from an amine and acetic anhydride or acetyl chloride. Some other methods are described below.

(1) $\xrightarrow[\text{25°, 2-12 h}]{\text{CH}_3\text{CO}_2\text{C}_6\text{F}_5/\text{DMF}^a}$ HO(CH$_2$)$_n$NHCOCH$_3$, 78-91%

HO(CH$_2$)$_n$NH$_2$

$\xrightarrow[\text{80°, 1-3 h}]{\text{CH}_3\text{CO}_2\text{C}_6\text{F}_5/\text{Et}_3\text{N}^a}$ CH$_3$CO$_2$(CH$_2$)$_n$NHCOCH$_3$, 75-87%

(2) $RNH_2 + R^1R^2NH + (CH_3CO)_2O \xrightarrow[\text{Et}_3N]{\text{18-crown-6}^b} R^1R^2NCOCH_3, 98\%$

Complex formation of a primary amine with 18-crown-6 allows selective acylation of a secondary amine.[b]

(3) $HO—\cdots—NH_2 \xrightarrow[\substack{\text{cat. 1-hydroxybenzotriazole,} \\ 25°, 18-24 \text{ h}}]{RCO_2C_6H_4\text{-}p\text{-}NO_2/\text{DMF}^c} HO—\cdots—NHCOR, 85–98\%$
      cytidine $R = CH_3, Ph$
      derivative

Cleavage

(1) $2 \xrightarrow[\text{reflux, 9 h}]{1.2 \ N \ \text{HCl}} R^1R^2NH, 61–77\%^d$

During the synthesis of penicillamine an *N*-acetyl group was cleaved by acidic hydrolysis.[d]

(2) $2 \xrightarrow[70°, 15 \text{ h}]{85\% \ H_2NNH_2} R^1R^2NH, 68\%^e$
                        amino sugar

(3) $2 \xrightarrow[25°, 1-2 \text{ h}, 90\%]{Et_3O^+BF_4^-/CH_2Cl_2} \left[\begin{array}{c} R^1R^2\overset{+}{N}H\!\!=\!\!CCH_3 \\ | \\ ^-BF_4 \quad OEt \end{array} \right] \xrightarrow[\text{cold}]{\text{aq NaHCO}_3}$

R^1R^2NH, satisfactory yields[f]

Triethyloxonium fluoroborate selectively cleaves *N*-acetyl groups in sugars in the presence of *O*-acetyl groups.[f]

(4) $RNHCOCH_3 \xrightarrow[t_{1/2} = 1.5 \text{ min}]{9 \ N \ NH_3}$ or $\xrightarrow[t_{1/2} = 3-6 \text{ min}]{\text{aq NaOH}} RNH_2{}^g$

RNH_2 = cytidine derivatives

The order of cleavage of *N*-acyl derivatives in base is *N*-acetyl > *N*-benzoyl > *N*-anisoyl.[g]

Note that alkaline hydrolysis of *N*-acyl cytidine derivatives is much faster than hydrolysis of simple amides; for example, compare with (6) on p. 251.

[a] L. Kisfaludy, T. Mohacsi, M. Low, and F. Drexler, *J. Org. Chem.*, **44**, 654 (1979).
[b] A. G. M. Barrett and J. C. A. Lana, *J. Chem. Soc., Chem. Commun.*, 471 (1978).
[c] A. S. Steinfeld, F. Naider, and J. M. Becker, *J. Chem. Res., Synop.*, 129 (1979).
[d] G. A. Dilbeck, L. Field, A. A. Gallo, and R. J. Gargiulo, *J. Org. Chem.*, **43**, 4593 (1978).
[e] D. D. Keith, J. A. Tortora, and R. Yang, *J. Org. Chem.*, **43**, 3711 (1978).
[f] S. Hanessian, *Tetrahedron Lett.*, 1549 (1967).
[g] H. G. Khorana, A. F. Turner, and J. P. Vizsolyi, *J. Am. Chem. Soc.*, **83**, 686 (1961).

3.   Chloroacetamide: $R^1R^2NCOCH_2Cl$, 3

4.   Dichloroacetamide: $R^1R^2NCOCHCl_2$, 4

Haloacetamides, such as compounds **3**, **4**, **5**, and **6**, are more easily cleaved by acid (0.1 *N* HCl) and base (0.01 *N* KOH or 15% NH_3/CH_3OH) than unsubstituted acetamides; rates are reported for *N*-6-acylated cytosines.[a]

Monochloroacetamides, **3**, are cleaved (by "assisted removal") by reagents that contain two nucleophilic groups (e.g., *o*-phenylenediamine,[b] thiourea,[c] 1-piperidinethiocarboxamide,[d] 3-nitropyridine-2-thione,[e] and 2-aminothiophenol[f]):

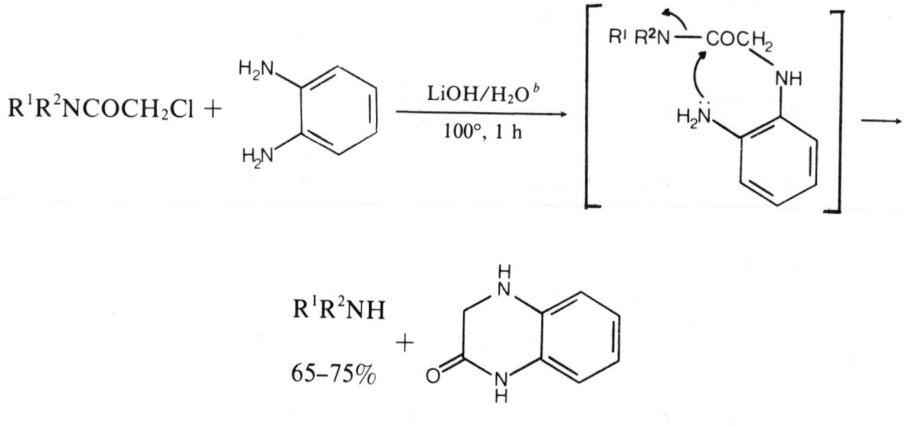

[a] R. S. Goody and R. T. Walker, *Tetrahedron Lett.*, 289 (1967).
[b] R. W. Holley and A. D. Holley, *J. Am. Chem. Soc.*, **74**, 3069 (1952).
[c] M. Masaki, T. Kitahara, H. Kurita, and M. Ohta, *J. Am. Chem. Soc.*, **90**, 4508 (1968).
[d] W. Steglich and H.-G. Batz, *Angew. Chem., Inter. Ed., Engl.*, **10**, 75 (1971).
[e] K. Undheim and P. E. Fjeldstad, *J. Chem. Soc., Perkin Trans. 1*, 829 (1973).
[f] J. D. Glass, M. Pelzig, and C. S. Pande, *Int. J. Pept. Protein Res.*, **13**, 28 (1979).

5.   Trichloroacetamide: $R^1R^2NCOCCl_3$, 5

Formation (→) / *Cleavage* (←)

$$\xrightarrow[\text{65°, 90 min, 65–97\%}]{Cl_3CCOCCl_3/\text{hexane}^a}$$

$$R^1R^2NH \hspace{6cm} 5$$

$$\xleftarrow[\text{1 h, 65\%}]{NaBH_4/EtOH^b}$$

R^1R^2NH = peptide,[b] nucleoside[c]

[a] B. Sukornick, *Org. Synth.*, *Coll. Vol. V*, 1074 (1973).
[b] F. Weygand and E. Frauendorfer, *Chem. Ber.*, **103**, 2437 (1970).
[c] R. S. Goody and R. T. Walker, *Tetrahedron Lett.*, 289 (1967).

6.   Trifluoroacetamide: $R^1R^2NCOCF_3$, 6

Formation

Two methods for the preparation of trifluoroacetamides are described here; references to five other methods are reported by Curphey.[a]

(1) $R^1R^2NH + CF_3CO_2Et \xrightarrow[\text{25°, 15–45 h}]{Et_3N/CH_3OH^a}$ **6**, 75–95%

(2) $RNH_2 + R^1R^2NH + (CF_3CO)_2O \xrightarrow[\text{Et}_3N]{\text{18-crown-6}^b} R^1R^2NCOCF_3$, ca. 95%

Complex formation of a primary amine with 18-crown-6 allows selective acylation of a secondary amine.[b]

Cleavage

(1)   **6** $\xrightarrow[\substack{\text{25°, 4.5 h, facile}^c \\ \text{or 55°, 2 weeks, 53\%}^d}]{K_2CO_3/CH_3OH\text{-}H_2O}$ or $\xrightarrow[\text{95\%}]{Na_2CO_3/CH_3OH\text{-}H_2O^e}$

or $\xrightarrow[\text{CH}_3OH]{NH_3{}^f}$ or $\xrightarrow[\text{25°, 2 h, 79\%}]{0.2\ N\ Ba(OH)_2/CH_3OH^g} R^1R^2NH$

(2)   **6** $\xrightarrow[\text{20° or 60°, 1 h}]{NaBH_4/EtOH} R^1R^2NH$, 60–100%[h]

[a] T. J. Curphey, *J. Org. Chem.*, **44**, 2805 (1979).

[b] A. G. M. Barrett and J. C. A. Lana, *J. Chem. Soc., Chem. Commun.*, 471 (1978).

[c] H. Newman, *J. Org. Chem.*, **30**, 1287 (1965).

[d] J. Quick and C. Meltz, *J. Org. Chem.*, **44**, 573 (1979).

[e] M. A. Schwartz, B. F. Rose, and B. Vishnuvajjala, *J. Am. Chem. Soc.*, **95**, 612 (1973).

[f] M. Imazawa and F. Eckstein, *J. Org. Chem.*, **44**, 2039 (1979).

[g] F. Weygand and W. Swodenk, *Chem. Ber.*, **90**, 639 (1957).

[h] F. Weygand and E. Frauendorfer, *Chem. Ber.*, **103**, 2437 (1970).

7. *o*-Nitrophenylacetamide: $R^1R^2NCOCH_2C_6H_4$-*o*-NO_2, 7

Formation (→) / *Cleavage* (←)[a]

$$\text{RCHCO—}\underset{\displaystyle\text{NH}_2}{|}$$

$$\xrightarrow[\text{Et}_3\text{N, }25°\text{, 3 h, 75–98\%}]{o\text{-NO}_2\text{-C}_6\text{H}_4\text{CH}_2\text{CO}_2\text{C}_6\text{H}_4\text{-}p\text{-NO}_2}$$

7

$$\xleftarrow[\text{quant}]{\text{H}_2/\text{Pd–C, HOAc}}$$

Compounds **7,**[a] **8, 14, 15, 16,** and **19** are cleaved via intramolecular cyclization that results from attack by the amino group following its reduction from the nitro group.

[a] F. Cuiban, *Rev. Roum. Chim.*, **18**, 449 (1973).

8. *o*-Nitrophenoxyacetamide: $R^1R^2NCOCH_2OC_6H_4$-*o*-NO_2, 8

Formation[a]

$$\text{RCHCO—}\underset{\displaystyle\text{NH}_2}{|} \xrightarrow[0°\text{, 20 min}]{o\text{-NO}_2\text{C}_6\text{H}_4\text{OCH}_2\text{COCl/NaOH}} \textbf{8}, 76\%$$

Cleavage[a]

$$\textbf{8} \xrightarrow[\substack{\text{aq NaHCO}_3\\30\text{ min}}]{\text{H}_2/\text{Pt}} \xrightarrow[100°]{\text{H}_2\text{O}} \text{RCHCO—}\underset{\displaystyle\underset{65–75\%}{\text{NH}_2}}{|} \quad + \quad \text{(benzoxazinone)}$$

See the discussion that accompanies compound **7**.

[a] R. W. Holley and A. D. Holley, *J. Am. Chem. Soc.*, **74**, 3069 (1952).

9. Acetoacetamide: $R^1R^2NCOCH_2COCH_3$, 9

Diketene reacts selectively with an —NH group in the presence of an —OH group to form compound 9.[a]

Formation[a]

$$R^1R^2NH + \quad \underset{\underset{\displaystyle CH_2=C\!\!-\!\!O}{|\qquad\quad|}}{CH_2\!\!-\!\!C\!\!=\!\!O} \quad \xrightarrow[0°]{\text{NaOH or Et}_3N/\text{EtOH}} \quad 9, 50\%$$

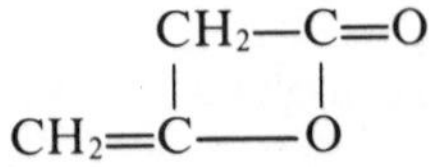

Cleavage[a]

$$9 \xrightarrow[\text{HONH}_2/\text{H}_3\text{O}^+]{\text{PhNHNH}_2 \text{ or}} [R^1R^2NCOCH_2C(CH_3)\!\!=\!\!NXH] \rightarrow R^1R^2NH +$$

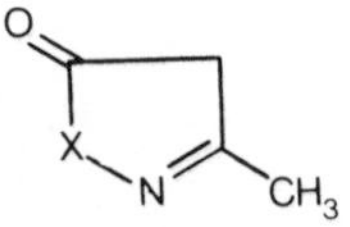

$$X = PhN<, \quad —O—$$

[a] C. Di Bello, F. Filira, V. Giormani, and F. D'Angeli, *J. Chem. Soc. C*, 350 (1969).

10. Pyridiniumacetamide: $R^1R^2NCOCH_2\!\!-\!\!\overset{+}{N}C_5H_5Cl^-$, 10

Compound 10 is readily hydrolyzed by dilute alkali.[a]

Formation[a]

$$R^1R^2NH \rightarrow R^1R^2NCOCH_2Cl \xrightarrow[1\ h]{\text{Py, }90°} 10, 70\text{--}90\%$$

Cleavage[a]

$$10 \xrightarrow[\substack{0.05\ N\ \text{NaOH, }37° \\ t_{1/2}=45\ \text{min}}]{0.1\ N\ \text{NaOH, }25°\ \text{or}} R^1R^2NH$$

[a] C. H. Gaozza, B. C. Cieri, and S. Lamdan, *J. Heterocycl. Chem.*, **8**, 1079 (1971).

11. (N'-Dithiobenzyloxycarbonylamino)acetamide: $R^1R^2NCOCH_2NHCS_2CH_2C_6H_5$, 11

Since compound 11 is readily cleaved by trifluoroacetic acid, it can be used in syntheses of peptides and cephalosporins.[a]

Formation (→) / Cleavage (←)[a]

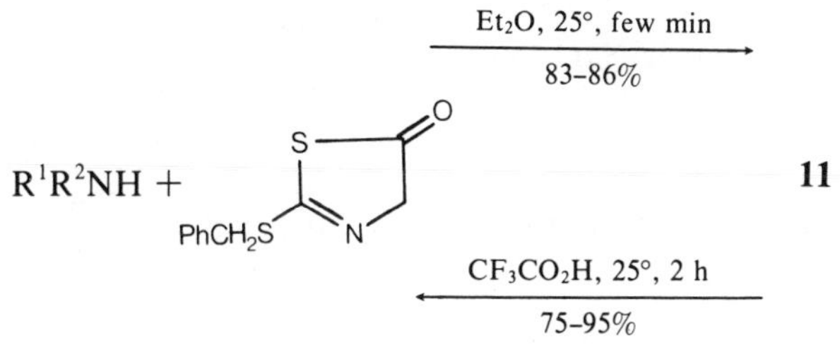

$$R^1R^2NH +$$

$$\xrightarrow[\text{83–86\%}]{\text{Et}_2\text{O, 25°, few min}}$$

$$\xleftarrow[\text{75–95\%}]{\text{CF}_3\text{CO}_2\text{H, 25°, 2 h}}$$

11

[a] F. E. Roberts, *Tetrahedron Lett.*, 325 (1979).

Substituted *N*-Propionyl Derivatives

12. 3-Phenylpropanamide: $R^1R^2NCOCH_2CH_2C_6H_5$, 12

When compound **12** is prepared from a nucleoside, it is hydrolyzed under mild conditions by α-chymotrypsin (37°, pH 7, 2–12 h).[a]

[a] H. S. Sachdev and N. A. Starkovsky, *Tetrahedron Lett.*, 733 (1969).

13. 3-(*p*-Hydroxyphenyl)propanamide: $R^1R^2NCOCH_2CH_2C_6H_4$-*p*-OH, 13

Compound **13**, cleaved by chemical[a] or electrolytic[b] oxidation, has been used in , peptide syntheses.

Cleavage

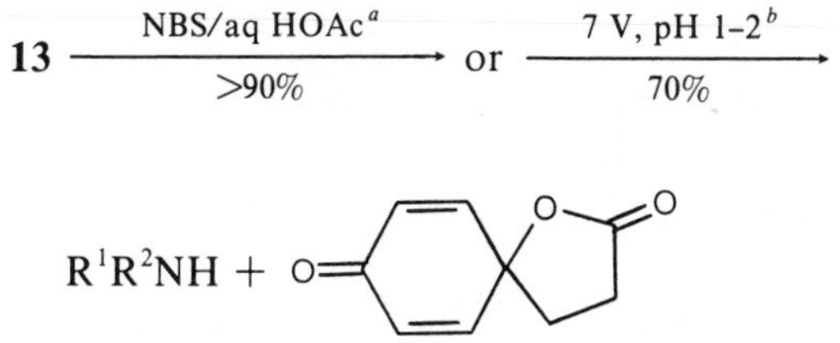

$$\textbf{13} \xrightarrow[\text{>90\%}]{\text{NBS/aq HOAc}^a} \text{ or } \xrightarrow[\text{70\%}]{\text{7 V, pH 1–2}^b}$$

$$R^1R^2NH + O=$$

[a] G. L. Schmir and L. A. Cohen, *J. Am. Chem. Soc.*, **83**, 723 (1961).
[b] L. Farber and L. A. Cohen, *Biochemistry*, **5**, 1027 (1966).

14. 3-(*o*-Nitrophenyl)propanamide: $R^1R^2NCOCH_2CH_2C_6H_4$-*o*-NO$_2$, 14

Compound **14**, used to protect an amino acid, can be cleaved by catalytic transfer hydrogenation.[a] (See also the discussion that accompanies compound **7**.)

Cleavage[a]

$$14 + \text{cyclohexene or NaH}_2\text{PO}_2 \xrightarrow[\text{15 min to 6 h}]{10\% \text{ Pd-C}}$$

$$[R^1R^2NCO(CH_2)_2C_6H_4\text{-}o\text{-}NH_2] \rightarrow R^1R^2NH + \\ 60\text{–}95\%$$

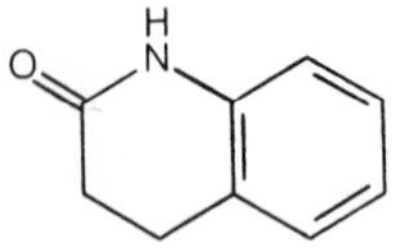

[a] I. D. Entwistle, *Tetrahedron Lett.*, 555 (1979).

15. 2-Methyl-2-(*o*-nitrophenoxy)propanamide: $R^1R^2NCOC(CH_3)_2OC_6H_4\text{-}o\text{-}NO_2$, 15

See the discussion that accompanies compound **7**.

Formation[a]

$$\underset{\underset{NH_2}{|}}{RCHCO-} + p\text{-}NO_2\text{—}C_6H_4OCOC(CH_3)_2OC_6H_4\text{-}o\text{-}NO_2$$

$$\xrightarrow[25°, 5\text{ h}]{Et_3N} \textbf{15}, 60\text{–}85\%$$

Cleavage

$$\textbf{15} \xrightarrow[\text{quant}]{H_2/Pd\text{-}C, HOAc^a} \text{ or } \xrightarrow[THF]{Zn/NH_4Cl} \xrightarrow{HCl/EtOH^b}$$

$$\underset{\underset{NH_2 \cdot HCl}{|}}{RCHCO-} + \\ 45\text{–}75\%$$

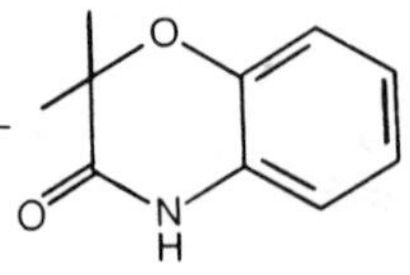

[a] F. Cubian, *Rev. Roum. Chim.*, **18**, 449 (1973).
[b] C. A. Panetta, *J. Org. Chem.*, **34**, 2773 (1969).

16. 2-Methyl-2-(*o*-phenylazophenoxy)propanamide: $R^1R^2NCOC(CH_3)_2OC_6H_4\text{-}o\text{-}N\text{=}NC_6H_5$, 16

See the discussion that accompanies compound **7**.

Formation[a]

$$RCHCO- + HO_2CC(CH_3)_2OC_6H_4\text{-}o\text{-}N=NPh \xrightarrow[25°,\ 5\ h]{EEDQ/THF} \textbf{16},\ 82\text{-}97\%$$
$$|$$
$$NH_2$$

EEDQ =

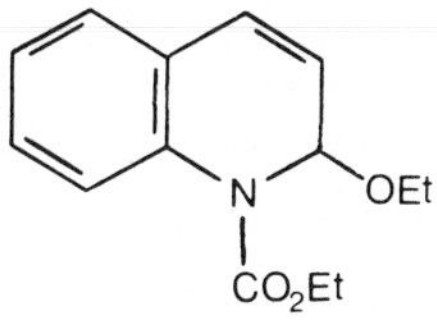

Cleavage[a]

$$\textbf{16} \xrightarrow[1\ h]{KBH_4/Pd\text{-}C} \xrightarrow{pH\ 1} \underset{48\text{-}89\%}{\underset{NH_2}{\overset{RCHCO-}{|}}} + \ \text{(structure)}$$

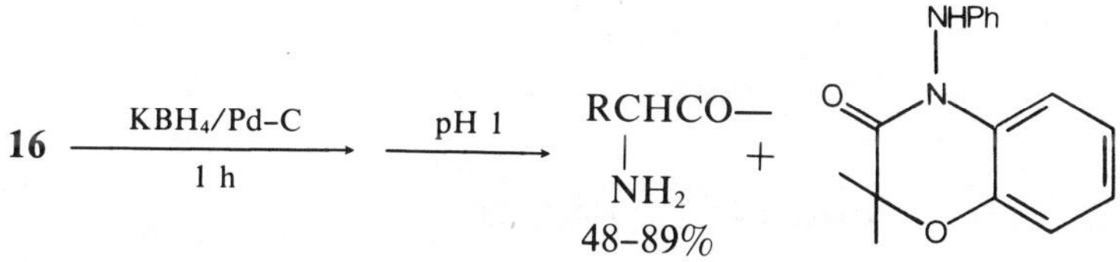

[a] C. A. Panetta and A.-U.-Rahman, *J. Org. Chem.*, **36**, 2250 (1971).

17. 4-Chlorobutanamide: $RNHCO(CH_2)_3Cl$, 17

Compound **17** can be cleaved under mild conditions by hydrolysis of the intermediate iminolactone.[a]

Cleavage[a]

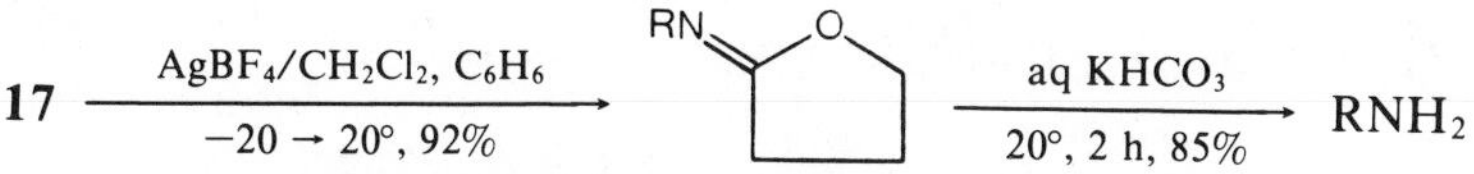

$$\textbf{17} \xrightarrow[-20\ \rightarrow\ 20°,\ 92\%]{AgBF_4/CH_2Cl_2,\ C_6H_6} \ \text{(structure)} \xrightarrow[20°,\ 2\ h,\ 85\%]{aq\ KHCO_3} RNH_2$$

[a] H. Peter, M. Brugger, J. Schreiber, and A. Eschenmoser, *Helv. Chim. Acta*, **46**, 577 (1963).

18. Isobutanamide: $R^1R^2NCOCH(CH_3)_2$, 18

N-Isobutyryl derivatives are used to protect the amino group in guanine; they are more stable to base (1 *N* NaOH) than *N*-acetyl derivatives.[a]

Formation[a]

$$R^1R^2NH + (Me_2CHCO)_2O \xrightarrow[25°,\ 48\ h,\ dark]{Et_4N^+OH^-/Py} \textbf{18}$$

Cleavage[a]

$$18 \xrightarrow[\text{C}_4\text{H}_9\text{NH}_2/\text{CH}_3\text{OH}]{\text{concd NH}_3 \text{ or}} \text{R}^1\text{R}^2\text{NH}$$

[a] H. Büchi and H. G. Khorana, *J. Mol. Biol.*, **72**, 251 (1972).

19. *o*-Nitrocinnamide: $\text{R}^1\text{R}^2\text{NCOCH}{=}\text{CHC}_6\text{H}_4\text{-}o\text{-NO}_2$, 19

Compound **19**, prepared from an amino acid and the acid chloride, is cleaved by reduction (H_2/cat., HOAc, 82% yield).[a] See also the discussion that accompanies compound **7**.

[a] G. Just and G. Rosebery, *Synth. Commun.*, **3**, 447 (1973).

20. Picolinamide: $\text{R}^1\text{R}^2\text{NCO}$—2—pyridyl, 20

Compound **20** is prepared in 95% yield from picolinic acid/DCC and an amino acid, and hydrolyzed in 75% yield by aqueous Cu(OAc)_2.[a]

[a] A. K. Koul, B. Prashad, J. M. Bachhawat, N. S. Ramegowda, and N. K. Mathur, *Synth. Commun.*, **2**, 383 (1972).

Miscellaneous Derivatives

21. *N*-Acetylmethionine Derivative: RNHCOCHNHCOCH_3, 21
$$\overset{|}{}\text{CH}_2\text{CH}_2\text{SCH}_3$$

A methionyl group is selectively attacked by iodoacetamide[a] or cyanogen bromide (CNBr)[b]; hydrolysis liberates the amino acid, in 80 and >90% yields, respectively.

Formation

$$\text{RNH}_2 + \underset{\underset{\text{CH}_2\text{CH}_2\text{SCH}_3}{|}}{\text{HO}_2\text{CCHNHCOCH}_3} \xrightarrow[-15°]{\text{ClCO}_2\text{Et/Et}_3\text{N}^a} \text{21, 62\%}$$

$\text{RNH}_2 = $ amino acid

Cleavage

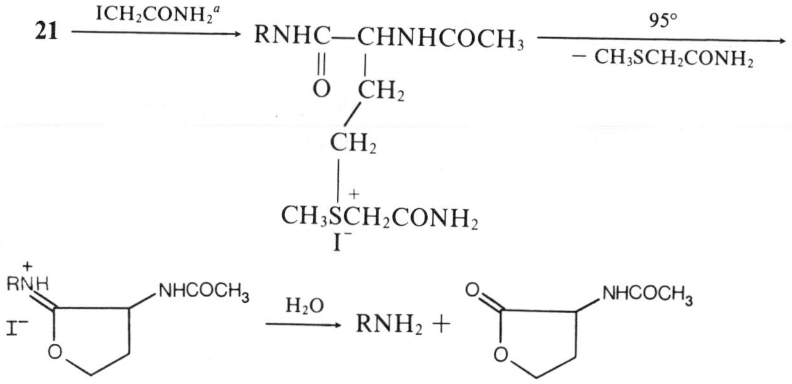

$$21 \xrightarrow{\text{ICH}_2\text{CONH}_2{}^a} \text{RNHC} - \text{CHNHCOCH}_3 \xrightarrow[- \text{CH}_3\text{SCH}_2\text{CONH}_2]{95°}$$

with structure:

RNHC—CHNHCOCH₃ where the C bears =O (shown as ‖ O) and the CH bears CH₂—CH₂—CH₃S⁺CH₂CONH₂ I⁻

Second line structures with H₂O:

$$\xrightarrow{\text{H}_2\text{O}} \text{RNH}_2 +$$

[a] W. B. Lawson, E. Gross, C. M. Foltz, and B. Witkop, *J. Am. Chem Soc.*, **84**, 1715 (1962).
[b] E. Gross and B. Witkop, *J. Biol. Chem.*, **237**, 1856 (1962).

22. *N*-Benzoylphenylalanyl Derivative: $R^1R^2\text{NCOCH(NHCOC}_6\text{H}_5)\text{CH}_2\text{C}_6\text{H}_5$, 22

Compound **22**, prepared from an amino acid and the acyl azide, is selectively cleaved in 80% yield by chymotrypsin.[a]

[a] R. W. Holley, *J. Am. Chem. Soc.*, **77**, 2552 (1955).

23. Benzamide: $R^1R^2\text{NCOC}_6\text{H}_5$, 23

Formation

(1) $R^1R^2\text{NH} + \text{PhCOCl} \xrightarrow{\text{Py, 0°}}$ **23**, high yield[a]

(2) $\text{RNH}_2 + R^1R^2\text{NH} + \text{PhCOCl} \xrightarrow[\text{Et}_3\text{N}]{\text{18-crown-6}} R^1R^2\text{NCOPh}$, 94%[b]

(3) $R^1R^2\text{NH} + \text{PhCOCN} \xrightarrow[25°, \text{ few min}]{\text{cat. Et}_3\text{N}}$ **23**, quant[c]

   $R^1R^2\text{NH} = \text{nucleoside}$

(4) $R^1R^2\text{NH} + \text{PhCOCF(CF}_3)_2 \xrightarrow[25°, 30 \text{ min}]{\text{TMEDA}}$ **23**, high yield[d]

   $\text{TMEDA} = \text{Me}_2\text{N(CH}_2)_2\text{NMe}_2$

(5)   Cytidine + $RCO_2C_6H_4$-p-NO_2 $\xrightarrow[25°, 18\text{-}24\text{ h}]{HOBT/DMF}$ **23**, 85–98%[e]

R = Ph, CH_3; HOBT = 1-hydroxybenzotriazole

Cytidine reacts under mild conditions with an active ester and a catalytic amount of HOBT to give selective —NH protection.[e]

(6) R^1R^2NH +

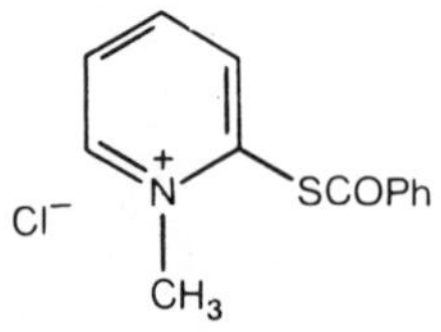

$\xrightarrow[\text{aq NaOH}]{\text{aq } NaHCO_3 \text{ or}}$ **23**, good yields[f]

Compound **a** benzoylates —NH (and —OH) groups under aqueous conditions.[f]

Cleavage

(1) R^1R^2NCOPh $\xrightarrow[\text{reflux, 48 h}]{6\ N\ HCl}$ or $\xrightarrow[25°, 72\text{ h}]{HBr/HOAc}$ R^1R^2NH, 80%[g]

(2) $\begin{array}{c} RCHCO— \\ | \\ NHCOPh \end{array}$ $\xrightarrow[25°, 60\text{ min}]{(HF)_n/Py}$ $\begin{array}{c} RCHCO—, 100\%^h \\ | \\ NH_2 \end{array}$

Polyhydrogen fluoride/pyridine cleaves most of the protective groups used in peptide syntheses, without side reactions.[h]

(3)   **23** $\xrightarrow[t_{1/2} = 16\text{ min}]{9\ N\ NH_3}$ or $\xrightarrow[t_{1/2} = 2.5\text{ h}]{NaOH,\ pH\ 13}$ RNH_2[i]

RNH_2 = cytidine derivative; see the note on p. 253.

(4)   **23** $\xrightarrow[130°, 1\text{-}24\text{ h}]{ArOH^j}$ R^1R^2NH, 76–90%

Fusion of an N,O-dibenzoyl adenosine or cytidine with a phenol selectively removes the N-benzoyl group.[j]

(5)   **23** $\xrightarrow[\substack{Me_4N^+X^-,\ CH_3OH \\ 70\text{ min}}]{-2.3\ V^k}$ R^1R^2NH, 60–90%

Peptide bonds are not cleaved by electrolysis.[k]

(6)   **23** $\xrightarrow[-78°]{\text{(Me}_2\text{CHCH}_2)_2\text{AlH/PhCH}_3{}^l}$ R^1R^2NH, 80%

Since the N-benzoyl group in this substrate, an alkaloid precursor, could not be removed by hydrolysis, a less selective reaction, reductive cleavage with diisobutylaluminum hydride, was used.[l]

[a] E. White, *Org. Synth., Collect. Vol. V*, 336 (1973).

[b] A. G. M. Barrett and J. C. A. Lana, *J. Chem. Soc., Chem. Commun.*, 471 (1978).

[c] A. Holy and M. Soucek, *Tetrahedron Lett.*, 185 (1971).

[d] N. Ishikawa and S. Shin-ya, *Chem. Lett.*, 673 (1976).

[e] A. S. Steinfeld, F. Naider, and J. M. Becker, *J. Chem. Res., Synop.*, 129 (1979).

[f] M. Yamada, Y. Watabe, T. Sakakibara, and R. Sudoh, *J. Chem. Soc., Chem. Commun.*, 179 (1979).

[g] D. Ben-Ishai, J. Altman, and N. Peled, *Tetrahedron*, **33**, 2715 (1977).

[h] S. Matsuura, C.-H. Niu, and J. S. Cohen, *J. Chem. Soc., Chem. Commun.*, 451 (1976).

[i] H. G. Khorana, A. F. Turner, and J. P. Vizsolyi, *J. Am. Chem. Soc.*, **83**, 686 (1961).

[j] Y. Ishido, N. Nakazaki, and N. Sakairi, *J. Chem. Soc., Perkin Trans. 1*, 657 (1977).

[k] L. Horner and H. Neumann, *Chem. Ber.*, **98**, 3462 (1965).

[l] J. Gutzwiller and M. Uskokovic, *J. Am. Chem. Soc.*, **92**, 204 (1970).

24.   *p*-Phenylbenzamide: $R^1R^2NCOC_6H_4$-*p*-C_6H_5, 24

Formation ($\rightarrow$) / Cleavage ($\leftarrow$)[a]

$$R^1R^2NH \qquad \xrightarrow[86\%]{\text{ClCOC}_6\text{H}_4\text{-}p\text{-C}_6\text{H}_5/\text{Et}_3\text{N}} \qquad \mathbf{24}$$

$$\xleftarrow[25°, 4\text{ h}, 81\%]{3\%\ \text{Na(Hg)/CH}_3\text{OH}}$$

p-Phenylbenzamides are usually crystalline; the corresponding N-acetyl and N-benzoyl derivatives are often oils. In this example a hindered amide, stable to the usual hydrolysis conditions, is cleaved by sodium amalgam. Most amides react only slowly with sodium amalgam.[a]

[a] R. M. Scribner, *Tetrahedron Lett.*, 3853 (1976).

25.   *p*-Methoxybenzamide: $R^1R^2NCOC_6H_4$-*p*-OCH_3, 25

Formation ($\rightarrow$) / Cleavage ($\leftarrow$)[a]

$$RNH_2 \qquad \xrightarrow[25°, 2\text{ h}, 84\%]{\text{ClCOC}_6\text{H}_4\text{-}p\text{-OCH}_3/\text{Py}} \qquad \mathbf{25}$$

$$\xleftarrow[t_{1/2} = 64\text{ min}]{9\ M\ \text{NH}_3}$$

$$RNH_2 \xleftarrow[t_{1/2} > 5 \text{ h}]{1 \, N \text{ NaOH}} \textbf{25}$$

RNH_2 = cytidine derivatives; see the note on p. 253.

[a] H. G. Khorana, A. F. Turner, and J. P. Vizsolyi, *J. Am. Chem. Soc.*, **83**, 686 (1961).

26. *o*-Nitrobenzamide: $R^1R^2NCOC_6H_4$-*o*-NO_2, 26

Formation[a]

$$\underset{\underset{NH_2}{|}}{RCHCO_2R'} \xrightarrow[\text{N,N'-carbonyldiimidazole}]{HO_2CC_6H_4\text{-}o\text{-}NO_2/THF, \, 25°, \, 3 \text{ h}} \textbf{26}, 90\%$$

Cleavage[a]

$$\textbf{26} \xrightarrow[\text{30 min, 85\%}]{H_2/PtO_2, \text{ NaHCO}_3} \xrightarrow[\text{25°, 30 min}]{Cu(OAc)_2/H_3O^+} \underset{\underset{NH_2}{|}}{RCHCO_2R'}, 80\%$$

[a] A. K. Koul, J. M. Bachhawat, B. Prashad, N. S. Ramegowda, A. K. Mathur, and N. K. Mathur, *Tetrahedron*, **29**, 625 (1973).

27. *o*-(Benzoyloxymethyl)benzamide: $R^1R^2NCOC_6H_4$-*o*-$CH_2OCOC_6H_5$, 27

Formation[a]

$$R^1R^2NH \xrightarrow[\text{Py, 0°, 78–91\%}]{ClCOC_6H_4\text{-}o\text{-}CH_2OCOC_6H_5} \textbf{27}$$

Cleavage[a]

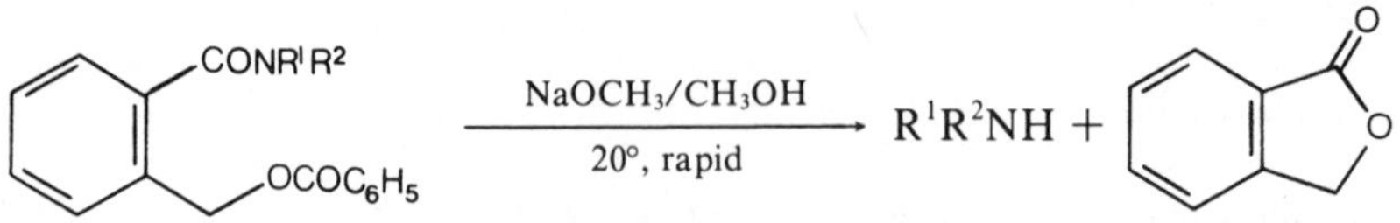

27

[a] B. F. Cain, *J. Org. Chem.*, **41**, 2029 (1976).

28. *p*-Ⓟ-Benzamide: $R^1R^2NCOC_6H_4$-*p*-Ⓟ, 28

A polymer-supported reagent (88% styrene:12% *p*-vinylbenzoic acid:0.2% *p*-divinylbenzene) protects amino groups during oligonucleotide syntheses; the benzamide is cleaved by base (0.4 M NaOH/dioxane–EtOH–H$_2$O, 25°, 36 h, 100% yield).[a]

[a] R. L. Letsinger and V. Mahadevan, *J. Am. Chem. Soc.*, **88**, 5319 (1966).

Cyclic Imide Derivatives

Primary amines can be protected as cyclic imides; acyclic imides have proved to be too reactive for use as protective groups.

29. Phthalimide:

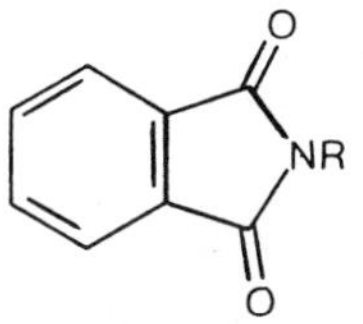

, **29**

Formation

(1)   RNH$_2$ + phthalic anhydride $\xrightarrow[\text{70°, 4 h}]{\text{CHCl}_3{}^a}$ **29**, 85–93%

　　RNH$_2$ = nucleoside

(2)   RNH$_2$ + o-(CH$_3$OOC)C$_6$H$_4$COCl $\xrightarrow[\text{0°, 2 h}]{\text{Et}_3\text{N/THF}^b}$ **29**, 90–95%

　　RNH$_2$ = amino acid

(3)   RNH$_2$ + N—(CO$_2$Et)phthalimide $\xrightarrow[\text{25°, 10–15 min}]{\text{aq Na}_2\text{CO}_3{}^c}$ **29**, 85–95%

　　RNH$_2$ = amino acid

Cleavage

(1) **29** $\xrightarrow[\text{25°, 12 h, 76%}]{\text{H}_2\text{NNH}_2\text{/EtOH}}$ $\xrightarrow{\text{H}_3\text{O}^{+a}}$ or $\xrightarrow[\text{reflux, 2 h. 83%}]{\text{PhNHNH}_2/n\text{-Bu}_3\text{N}^d}$ RNH$_2$

　　RNH$_2$ = peptide^d

(2) **29** $\xrightarrow[\substack{\text{H}_2\text{O/THF}\\68\text{–}90\%}]{\text{Na}_2\text{S}\cdot9\,\text{H}_2\text{O}}$ $\xrightarrow[\text{67–97%}]{\text{DCC (}-\text{H}_2\text{O)}}$ $\xrightarrow{\text{H}_2\text{NNH}_2}$ $\xrightarrow{\text{dil HCl}}$ RNH$_2^e$, 55–95%

This method (eq. 2) is used to cleave N-phthalimido penicillins; hydrazine attacks an intermediate phthalisoimide instead of the azetidinone ring.[e]

[a] T. Sasaki, K. Minamoto, and H. Itoh, *J. Org. Chem.*, **43**, 2320 (1978).

[b] D. A. Hoogwater, D. N. Reinhoudt, T. S. Lie, J. J. Gunneweg, and H. C. Beyerman, *Recl. Trav. Chim. Pays-Bas*, **92**, 819 (1973).

[c] G. H. L. Nefkens, G. I. Tesser, and R. J. F. Nivard, *Recl. Trav. Chim. Pays-Bas*, **79**, 688 (1960).

[d] I. Schumann and R. A. Boissonnas, *Helv. Chim. Acta*, **35**, 2235 (1952).

[e] S. Kukolja and S. R. Lammert, *J. Am. Chem. Soc.*, **97**, 5582 (1975).

30. 2,3-Diphenylmaleimide: 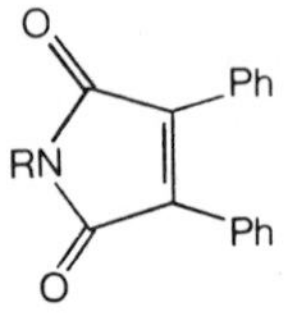 , **30**

Compound **30**, like a phthalimide, is prepared from the anhydride, 33–87% yield, and cleaved by hydrazinolysis, 65–75% yield. It is stable to acid (HBr/HOAc, 48 h) and to mercuric cyanide. It is colored and easily located during chromatography, and has been prepared to protect steroidal amines and amino sugars.[a]

[a] U. Zehavi, *J. Org. Chem.*, **42**, 2819 (1977).

31. Dithiasuccinimide: 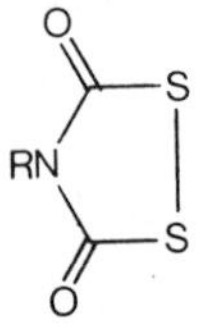 , **31**

Formation (→) / Cleavage (←)[a]

$$
\text{RNH}_2 \xrightarrow[\substack{\text{EtOCS}_2\text{CSOEt} \\ \sim \text{quant}}]{\text{EtOCS}_2\text{CH}_2\text{CO}_2\text{H or}} \text{[RNHCSOEt]} \xrightarrow[\text{70–90\%}]{\text{ClSCOCl, 0–45°}} \textbf{31}
$$

$$
\xleftarrow[\text{25°, 5 min}]{\text{HO(CH}_2)_2\text{SH/Et}_3\text{N}}
$$

Compound **31**, stable to acidic cleavage of *t*-butyl carbamates (12 *N* HCl/HOAc, reflux; HBr/HOAc), to mild base (NaHCO₃), and to photolytic cleavage of *o*-nitrobenzyl carbamates, can be used in orthogonal schemes for protection of peptides. Merrifield defines an orthogonal system as a set of completely independent classes of protective groups. In such a system each class of protective groups can be removed in any order and in the presence of all other classes.[a]

[a] G. Barany and R. B. Merrifield, *J. Am. Chem. Soc.*, **99**, 7363 (1977).

SPECIAL —NH PROTECTIVE GROUPS

The many special derivatives used to protect an —NH group belong to one of three classes: N-alkyl or N-aryl derivatives; imines from carbonyl compounds and enamines from β-dicarbonyl compounds; and N-hetero atom derivatives.

Like other protective groups, these derivatives must be easily and selectively formed, stable during the synthetic sequence, and readily removed. Although compounds in this section, like carbamates and amides, have been used to protect amino acids and nucleosides, they have been used more to protect isolated primary and secondary amines, including indole and imidazole —NH groups. Many of the compounds described here can be removed under quite specialized and mild conditions. N-Alkyl and N-benzyl derivatives are prepared by reaction of the amine with an alkyl or aryl halide in base. To facilitate removal, simple N-alkyl derivatives must contain another functional group (double bond, carbonyl, or aminoacetal group). N-Benzyl derivatives are cleaved by reduction; N-triphenylmethyl derivatives, cleaved by reduction, are also cleaved by acid hydrolysis.

A primary amine reacts with an aromatic aldehyde or an aromatic or aliphatic ketone in neutral or basic solution to form a relatively stable imine; imines formed from aliphatic aldehydes often polymerize. Enamines are formed from β-dicarbonyl compounds. Imines and enamines formed from acyclic β-dicarbonyl compounds are readily cleaved by acid hydrolysis; enamines prepared from cyclic β-dicarbonyl compounds are more stable to acid hydrolysis because of intramolecular hydrogen bonding, and must first be converted to acid-labile imines. A short review describes the use of aldimines, ketimines, and enamines in peptide synthesis.[a]

N-Hetero atom derivatives have a wide range of properties. Silyl, phosphinyl, phosphoryl, sulfenyl, and sulfonyl derivatives are prepared by reaction of the amine and appropriate halide in the presence of base. Silyl derivatives are readily cleaved by heating in water. Phosphinamides and some sulfenamides are cleaved by mild acid hydrolysis; alkylsulfonamides are cleaved by strong acid hydrolysis. Copper chelates, used to protect vicinal -NH/-OH groups, and borane derivatives, used to protect tertiary amines, are removed by mild base hydrolysis. Arylsulfonamides, arylsulfenamides, N-nitro derivatives (used primarily to protect the guanidino group in arginine), N-nitroso derivatives, and N-oxides (used to protect tertiary amines) are removed by reduction.

This section describes some special derivatives that have been prepared to protect the -NH group[b,c]; the more interesting and synthetically useful derivatives are included in Reactivity Chart 10. Conventions that are used in this section are described on p. xii.

---

[a] B. Halpern, in *Chemistry and Biochemistry of Amino Acids, Peptides, and Proteins,* B. Weinstein, Ed., Dekker, New York, 1978, Vol. 5, pp. 95–115.

[b] See also: E. Wünsch, "Blockierung und Schutz der α-Amino-Funktion," in *Methoden der Organischen Chemie (Houben-Weyl),* George Thieme Verlag, Stuttgart, 1974, Vol. 15/1, pp. 265–305.

[c] J. W. Barton, "Protection of N-H Bonds and NR₃", in *Protective Groups in Organic Chemistry*, J. F. W. McOmie, Ed., Plenum, New York and London, 1973, pp. 61–83.

N-Alkyl Derivatives

1.   *N*-Allylamine: $R^1R^2NCH_2CH{=}CH_2$, 1

Formation

$$R^1R^2NH + CH_2{=}CHCH_2Br \xrightarrow{K_2CO_3} 1^a$$

Cleavage

(1) $1 \xrightarrow[\text{DMSO}]{\text{KO-}t\text{-Bu}} R^1R^2NCH{=}CHCH_3 \xrightarrow[\text{0.5 } N \text{ NaOH}]{\text{KMnO}_4} R^1R^2NH, 64\%^a$

  82%

R^1R^2NH = imidazole ring in purines

(2) $1 \xrightarrow[\text{C}_6\text{H}_6/\text{H}_2\text{O}]{(\text{Ph}_3\text{P})_3\text{RhCl}} R^1R^2NCH{=}CHCH_3 \xrightarrow[\text{22}°, \text{15 min}]{1\ N\ \text{HCl/CH}_3\text{OH}} R^1R^2NH^b$

  $\sim100\%$

[a] J. A. Montgomery and H. J. Thomas, *J. Org. Chem.*, **30**, 3235 (1965).
[b] B. Moreau, S. Lavielle, and A. Marquet, *Tetrahedron Lett.*, 2591 (1977).

2.   *N*-Phenacylamine: $R^1R^2NCH_2COC_6H_5$, 2

Formation ($\rightarrow$) */ Cleavage* ($\leftarrow$)^a

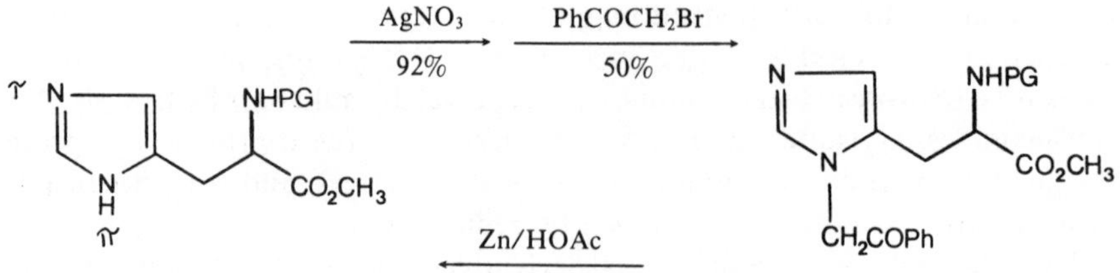

Protection of the π-nitrogen in histidine (defined as the N-atom closest to the methylene group) with a phenacyl group minimizes racemization during peptide syntheses.[a]

[a] A. R. Fletcher, J. H. Jones, W. I. Ramage, and A. V. Stachulski, *J. Chem. Soc., Perkin Trans. 1*, 2261 (1979).

3. *N*-3-Acetoxypropylamine: $R^1R^2NCH_2CH_2CH_2OCOCH_3$, 3

Formation[a]

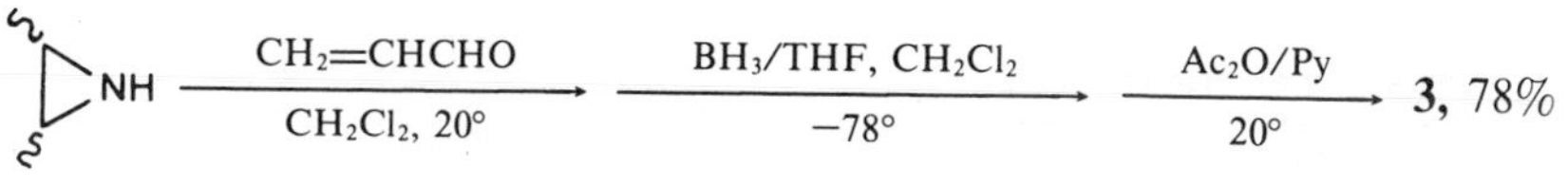

$$\underset{\substack{\\ \\}}{\text{aziridine NH}} \xrightarrow[\text{CH}_2\text{Cl}_2,\ 20^\circ]{\text{CH}_2=\text{CHCHO}} \xrightarrow[-78^\circ]{\text{BH}_3/\text{THF, CH}_2\text{Cl}_2} \xrightarrow[20^\circ]{\text{Ac}_2\text{O/Py}} \text{3, 78\%}$$

Cleavage[a]

$$\text{3} \xrightarrow[20^\circ]{\text{NaOCH}_3/\text{CH}_3\text{OH}} \xrightarrow[\text{CF}_3\text{CO}_2\text{H/Py, 20}^\circ]{\text{DMSO/DCC}} \xrightarrow[20^\circ]{\text{HClO}_4/\text{PhNMe}_2} \text{aziridine NH , 35\%}$$

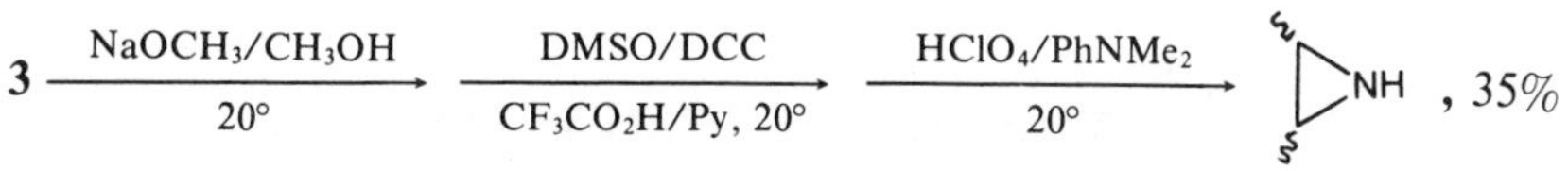

A 3-acetoxypropyl group was used to protect an aziridine —NH group during the synthesis of mitomycins A and C; acetyl, benzoyl, ethoxycarbonyl, and methoxymethyl groups were unsatisfactory.[a]

[a] T. Fukuyama, F. Nakatsubo, A. J. Cocuzza, and Y. Kishi, *Tetrahedron Lett.*, 4295 (1977).

4. *N*-4-Nitro-1-cyclohexyl-2-oxo-3-pyrrolin-3-ylamine:

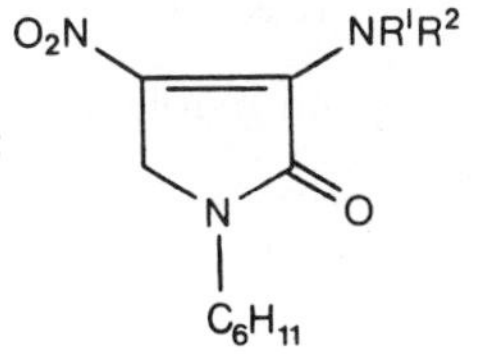

, 4

Amino acids react with the 3-ethoxypyrrolinone at pH 10 to give compound 4 in high yields; the amino acid is regenerated on treatment with ammonia at room temperature (73–88% yield).[a]

[a] P. L. Southwick, R. F. Dufresne, and J. J. Lindsey, *J. Org. Chem.*, **39**, 3351 (1974).

5. Quaternary Ammonium Salts: $R^1R^2R^3N^+CH_3I^-$, 5

Formation[a]

$$R^1R^2R^3N \xrightarrow[20^\circ,\ 24\ h]{\text{CH}_3\text{I/CH}_3\text{OH, KHCO}_3} \text{5, 85–95\%}$$

Compound 5 is generally used to protect tertiary amines during oxidation reactions. The conditions cited above form quaternary salts from primary, secondary, or tertiary amines, including amino acids, in the presence of hydroxyl or phenol groups.[a]

Cleavage[b]

$$5 \xrightarrow[\text{reflux, 24–36 h}]{\text{PhSNa/CH}_3\text{COCH}_2\text{CH}_3} R^1R^2R^3N, 85\%$$

[a] F. C. M. Chen and N. L. Benoiton, *Can. J. Chem.,* **54,** 3310 (1976).

[b] M. Shamma, N. C. Deno, and J. F. Remar, *Tetrahedron Lett.,* 1375 (1966).

AMINO ACETAL DERIVATIVES

6.   N-Methoxymethylamine: $R^1R^2NCH_2OCH_3$, 6

Formation (→) / Cleavage (←)[a]

$$\begin{array}{ccc}
& \xrightarrow[\text{0°, 0.5 h}]{\text{DMSO/NaOH}} & \xrightarrow[\text{22°, 0.5 h, 90\%}]{\text{CH}_3\text{OCH}_2\text{Cl}} \\
\end{array}$$

R^{1}R^{2}NH   6

an indole

$$\xleftarrow[\text{20°, 48 h, 86\%}]{\text{BF}_3 \cdot \text{Et}_2\text{O, Ac}_2\text{O, LiBr}}$$

[a] R. J. Sundberg and H. F. Russell, *J. Org. Chem.,* **38,** 3324 (1973).

7.   N-2-Chloroethoxymethylamine: $R^1R^2NCH_2OCH_2CH_2Cl$, 7

Compound 7 has been prepared from an indole, the chloromethyl ether, and potassium hydride in 50% yield; it is cleaved in 84% yield by potassium cyanide/18-crown-6 in refluxing acetonitrile.[a]

[a] A. J. Hutchison and Y. Kishi, *J. Am. Chem. Soc.,* **101,** 6786 (1979).

8.   N-Benzyloxymethylamine: $R^1R^2NCH_2OCH_2C_6H_5$, 8

Compound 8 is prepared from an indole, the chloromethyl ether, and sodium hydride in 80–90% yield. It is cleaved in 92% yield by catalytic reduction followed by basic hydrolysis.[a]

[a] H. J. Anderson and J. K. Groves, *Tetrahedron Lett.,* 3165 (1971).

9. N-Pivaloyloxymethylamine: $R^1R^2NCH_2OCOC(CH_3)_3$, 9

Compound **9** can be prepared in moderate yield from an indole or imidazole —NH group by reaction with the chloromethyl ether and potassium carbonate. It is cleaved by methanolic ammonia (25°, 4 h, 30–80% yield).[a]

[a] M. Rasmussen and N. J. Leonard, *J. Am. Chem. Soc.,* **89**, 5439 (1967).

10. N-[1-(Alkoxycarbonylamino)-2,2,2-trifluoro]ethylamine:
$R^1R^2NCH(CF_3)NHCO_2R$, 10

$R = CH_2C_6H_5$, t-C_4H_9

Compound **10** has been prepared to protect the imidazole —NH group in histidine.

Formation[a]

$$R^1R^2NH + ClCH(CF_3)NHCO_2R \xrightarrow[\text{2 h}]{\text{Et}_3\text{N/THF}} 10$$

$R = CH_2Ph$, 90%; $R = t$-C_4H_9, 80%

Cleavage[a]

$$R^1R^2NCH(CF_3)NHCO_2CH_2Ph \xrightarrow[\text{6 h}]{\text{H}_2/\text{Pd–C, CH}_3\text{OH}} R^1R^2NH,\ 73\%$$

$$R^1R^2NCH(CF_3)NHCO_2\text{-}t\text{-Bu} \xrightarrow[\text{22°, 15 min}]{\text{aq CF}_3\text{CO}_2\text{H}} R^1R^2NH,\ 77\%$$

[a] F. Weygand, W. Steglich, and P. Pietta, *Chem. Ber.,* **100**, 3841 (1967).

11. N-[1-Trifluoromethyl-1-(p-chlorophenoxymethoxy)-2,2,2-trifluoro]-
ethylamine: $R^1R^2NC(CF_3)_2OCH_2OC_6H_4$-$p$-Cl, 11

The imidazole —NH group in N-benzyloxycarbonylhistidine has been protected, in 70% yield, to prevent it from reaction with activated esters during peptide synthesis. Compound **11** is hydrolyzed by acid (2.4 M HCl/20% HOAc, 20°, 0.5 h, 65% yield).[a]

[a] H. H. Seltzman and T. M. Chapman, *Tetrahedron Lett.,* 2637 (1974).

12.   N-2-Tetrahydropyranylamine: R^1R^2N-2-tetrahydropyranyl, 12

Compound **12** has been prepared from the imidazole nitrogen in purines (dihydropyran, TsOH, 55°, 1.5 h, 50–85% yield); it is cleaved by acid hydrolysis.[a]

[a] R. K. Robins, E. F. Godefroi, E. C. Taylor, L. R. Lewis, and A. Jackson, *J. Am. Chem. Soc.*, **83**, 2574 (1961).

13.   N-2,4-Dinitrophenylamine: $R^1R^2NC_6H_3$-2,4-$(NO_2)_2$, 13

Compound **13** has been prepared from amino sugars (35% yield)[a] and from the imidazole —NH group in histidines (45% yield)[b] by reaction with 2,4-dinitrofluorobenzene and potassium carbonate. It is stable to trifluoroacetic acid, and is cleaved by an anion exchange resin (47% yield).[a] Imidazole —NH groups, but not α-amino acid groups, are quantitatively regenerated from compound **13** by reaction with 2-mercaptoethanol (22°, pH 8, 1 h).[c]

[a] P. F. Lloyd and M. Stacey, *Tetrahedron*, **9**, 116 (1960).
[b] E. Siepmann and H. Zahn, *Biochim. Biophys. Acta*, **82**, 412 (1964).
[c] S. Shaltiel, *Biochem. Biophys. Res. Commun.*, **29**, 178 (1967).

N-Benzyl Derivatives

14.   N-Benzylamine: $R^1R^2NCH_2C_6H_5$, 14

Formation

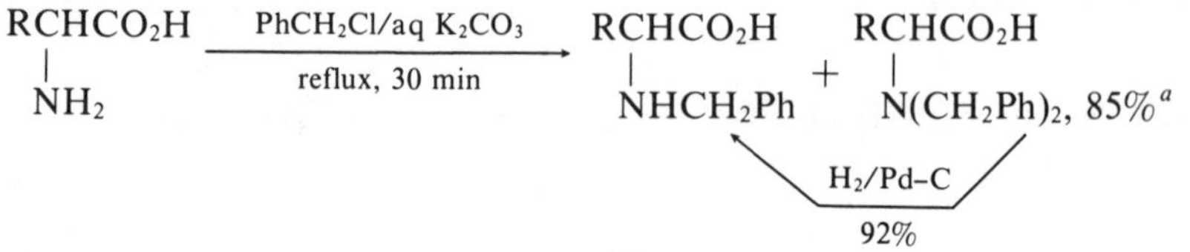

Cleavage

(1)   **14** $\xrightarrow[\substack{4.4\% \text{ HCOOH/CH}_3\text{OH} \\ 25°, 10 \text{ h}}]{\text{Pd–C}}$ R^1R^2NH, 80%[b]

(2)   **14** $\xrightarrow{\text{Na/NH}_3}$ R^1R^2NH, excellent yields[c]

(3)   **14** $\xrightarrow[\text{slow}]{\text{H}_2\text{Pd–C}}$ R^1R^2NH[d]

[a] L. Velluz, G. Amiard, and R. Heymes, *Bull. Soc. Chim. Fr.*, 1012 (1954).

[b] B. ElAmin, G. M. Anantharamaiah, G. P. Royer, and G. E. Means, *J. Org. Chem.*, **44**, 3442 (1979).

[c] V. du Vigneaud and O. K. Behrens, *J. Biol. Chem.*, **117**, 27 (1937).

[d] W. H. Hartung and R. Simonoff, *Org. Reactions*, **VII**, 263 (1953).

15. *N*-3,4-Dimethoxybenzylamine: $R^1R^2NCH_2C_6H_3$-3,4-$(OCH_3)_2$, 15

A 3,4-dimethoxybenzyl derivative, cleaved by acid (concd H_2SO_4/anhyd CF_3CO_2H/anisole), has been used to protect a pyrrole —NH group during the synthesis of a tetrapyrrole pigment precursor. Neither an *N*-benzyl nor an *N*-*p*-methoxybenzyl derivative could be cleaved satisfactorily. Hydrogenolysis of the benzyl derivatives led to cyclohexyl compounds; acidic cleavage resulted in migration of the benzyl groups to the free α-position.[a]

[a] M. I. Jones, C. Froussios, and D. A. Evans, *J. Chem. Soc., Chem. Commun.*, 472 (1976).

16. *N*-*o*-Nitrobenzylamine: $R^1R^2NCH_2C_6H_4$-*o*-NO_2, 16

Compound **16** is cleaved by irradiation (320 nm, 1 h, 100% yield from imidazole —NH protected histidine).[a]

[a] S. M. Kalbag and R. W. Roeske, *J. Am. Chem. Soc.*, **97**, 440 (1975).

17. *N*-Di(*p*-methoxyphenyl)methylamine: $R^1R^2NCH(C_6H_4$-*p*-$OCH_3)_2$, 17

Formation ($\rightarrow$) / *Cleavage* ($\leftarrow$)[a]

$$\text{RCHCO}-\underset{|}{\overset{}{\underset{NH_2}{}}}$$

$$\xrightarrow[\text{0} \rightarrow \text{20°, 20 h, 67\%}]{(p\text{-}CH_3OC_6H_4)_2CHCl, Et_3N}$$

$$\xleftarrow[\text{73\%}]{\text{aq HOAc, 80°, 5 min}}$$

17

[a] R. W. Hanson and H. D. Law, *J. Chem. Soc.*, 7285 (1965).

18. *N*-Triphenylmethylamine: $R^1R^2NC(C_6H_5)_3$, 18

The bulky triphenylmethyl group, readily removed by mild acid hydrolysis, has been used to protect —NH groups in peptide and penicillin/cephalosporin syntheses. Esters of *N*-triphenylmethyl α-amino acids are shielded from hydrolysis, and require forcing conditions for cleavage.

Formation

$$R^1R^2NH + (Ph)_3CCl \xrightarrow[\text{25°, 4 h}]{Et_3N} \textbf{18}^a$$

Cleavage

(1) **18** $\xrightarrow[\text{25°, 3 h}]{\text{HCl/acetone}}$ R^1R^2NH, 80%[a]

(2) **18** $\xrightarrow[\text{EtOH, 4–5 h}]{\text{H}_2/\text{Pd black}}$ R^1R^2NH, 92%[b]

(3) **18** $\xrightarrow{\text{Na/NH}_3}$ R^1R^2NH[c]

[a] H. E. Applegate, C. M. Cimarusti, J. E. Dolfini, P. T. Funke, W. H. Koster, M. S. Puar, W. A. Slusarchyk, and M. G. Young, *J. Org. Chem.,* **44,** 811 (1979).

[b] L. Zervas and D. M. Theodoropoulos, *J. Am. Chem. Soc.,* **78,** 1359 (1956).

[c] H. Nesvadba and H. Roth, *Monatsh. Chem.,* **98,** 1432 (1967).

19. *N*-[(*p*-Methoxyphenyl)diphenylmethyl]amine: $R^1R^2NC(C_6H_5)_2C_6H_4$-*p*-OCH$_3$, 19

Compound **19** is prepared in high yield from amino acids, and is readily cleaved by acid hydrolysis (5% CCl_3CO_2H, 4°, 5 min, 100%).[a]

[a] Y. Lapidot, N. de Groot, M. Weiss, R. Peled, and Y. Wolman, *Biochim. Biophys. Acta,* **138,** 241 (1967).

20. *N*-(Diphenyl-4-pyridylmethyl)amine: $R^1R^2NC(C_6H_5)_2$-C_5H_4N, 20

Compound **20**, prepared from the imidazole —NH group in histidine, and stable to acids (CF_3CO_2H, 21°, 48 h; 45% HBr/HOAc, 1 h) is cleaved by chemical and electrolytic reduction (H_2/Pd–C, 91% yield; Zn/HOAc, 1.5 h, 91% yield; electrolysis, 0°, 2.5 h, 87% yield).[a]

[a] S. Coyle and G. T. Young, *J. Chem. Soc., Chem. Commun.,* 980 (1976).

21. *N*-2-Picolylamine N′-Oxide: $R^1R^2NCH_2$-2-pyridyl *N*-Oxide, 21

Compound **21**, used in oligonucleotide syntheses, is cleaved by acetic anhydride at 22° followed by methanolic ammonia (85–95% yield).[a]

[a] Y. Mizuno, T. Endo, T. Miyaoka, and K. Ikeda, *J. Org. Chem.,* **39,** 1250 (1974).

NR¹R²

22. *N*-5-Dibenzosuberylamine: 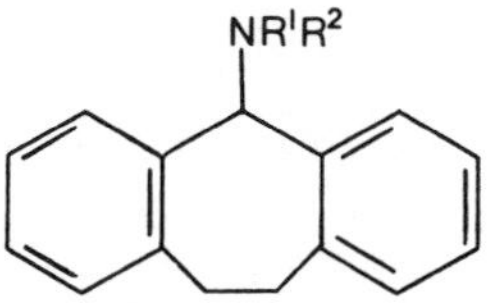 , **22**

Compound **22** is prepared in quantitative yield from an amine or amino acid and the suberyl chloride; this chloride has also been used to protect —OH, —SH, and —COOH groups. Although compound **22** is stable to 5 N HCl/dioxane (22°, 16 h) and to refluxing HBr (1 h), it is completely cleaved by some acids (HCOOH/CH_2Cl_2, 22°, 2 h; CF_3CO_2H/CH_2Cl_2, 22°, 0.5 h; BBr_3/CH_2Cl_2, 22°, 0.5 h; 4 N HBr/HOAc, 22°, 1 h; 60% HOAc, reflux, 1 h) and by reduction (H_2/Pd–C, CH_3OH, 22°, 1 h, 100% cleaved).[a]

[a] J. Pless, *Helv. Chim. Acta,* **59,** 499 (1976).

Imine Derivatives

23. *N*-(*N′,N′*-Dimethylaminomethylene)amine: RN═CHN(CH_3)$_2$, 23

Formation

$$RNH_2 + (CH_3O)_2CHN(CH_3)_2 \xrightarrow[22°, 15\ h]{DMF} \textbf{23, high yields}$$

RNH_2 = nucleosides[a]; amino acids (during C-alkylation)[b]

Cleavage

(1) **23** $\xrightarrow[22°, 6\ h]{dil\ NH_3}$ RNH_2, high yield[a]

(2) **23** $\xrightarrow[reflux,\ 18–72\ h]{concd\ HCl}$ R¹R²CCO$_2$H, 65–90%[b]
$\quad\quad\quad\quad\quad\quad\quad\quad\quad\quad\quad$ |
$\quad\quad\quad\quad\quad\quad\quad\quad\quad\quad NH_2$

[a] J. Zemlicka, S. Chládek, A. Holý, and J. Smrt, *Collect. Czech. Chem. Commun.,* **31,** 3198 (1966).
[b] J. J. Fitt and H. W. Gschwend, *J. Org. Chem.,* **42,** 2639 (1977).

24. *N,N′*-Isopropylidenediamine: RN-------NR¹ , **24**

Compound **24** has been prepared to protect dipeptides during peptide syntheses.[a]

Formation (→) / Cleavage (←)[a]

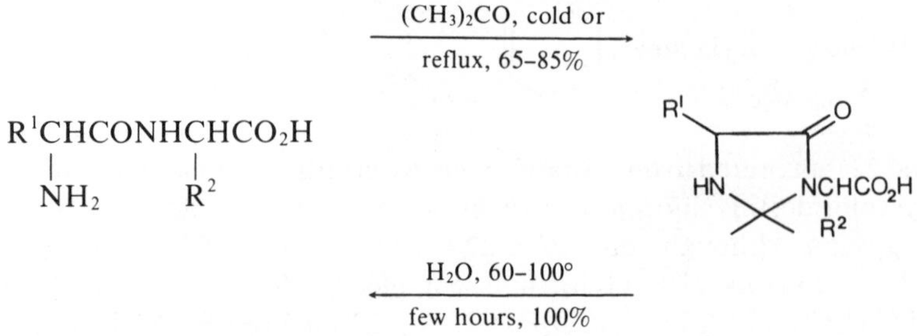
(CH3)2CO, cold or
reflux, 65–85%
R1CHCONHCHCO2H
NH2
R2
R1
O
HN
NCHCO2H
R2
H2O, 60–100°
few hours, 100%

[a] P. M. Hardy and D. J. Samworth, *J. Chem. Soc., Perkin Trans. 1,* 1954 (1977).

25. *N*-Benzylideneamine: $RN{=}CHC_6H_5$, 25

Formation

$$\underset{\displaystyle NH_2}{RCHCO_2CH_3} \xrightarrow{PhCHO/Et_3N} \underset{\displaystyle N{=}CHPh}{RCHCO_2CH_3},\ 80\text{–}90\%^{a}$$

Cleavage

(1) $$\underset{\displaystyle N{=}CHPh}{RR'CCO_2CH_3} \xrightarrow[25°,\ 1\ h]{1\ N\ HCl} \underset{\displaystyle NH_2}{RR'CCO_2CH_3}{}^{a}$$

(2) $$\mathbf{25} \xrightarrow{H_2/Pd\text{–}C,\ CH_3OH} RNH_2 = \text{amino steroid}^{b}$$

(3) 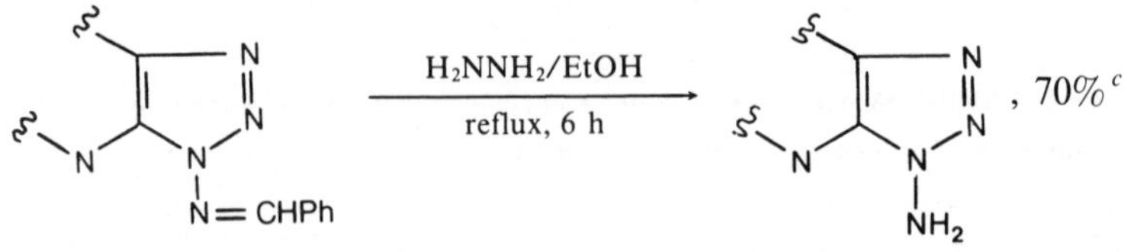
N
N
N
N
N= CHPh
H2NNH2/EtOH
reflux, 6 h
N
N
N
N
NH2
, 70%c

[a] P. Bey and J. P. Vevert, *Tetrahedron Lett.,* 1455 (1977).

[b] R. A. Lucas, D. F. Dickel, R. L. Dziemian, M. J. Ceglowski, B. L. Hensle, and H. B. MacPhillamy, *J. Am. Chem. Soc.,* **82,** 5688 (1960).

[c] G. W. J. Fleet and I. Fleming, *J. Chem. Soc. C,* 1758 (1969).

26. *N-p*-Methoxybenzylideneamine: $RN=CHC_6H_4$-*p*-OCH_3, 26

Compound 26, prepared in 85% yield, has been used to protect glucosamines; it is cleaved in 90% yield by 5 *N* HCl.[a]

[a] M. Bergmann and L. Zervas, *Ber.*, 64, 975 (1931).

27. *N-p*-Nitrobenzylideneamine: $RN=CHC_6H_4$-*p*-NO_2, 27

Compound 27 has been used to protect the 7-amino group in cephalosporins; it is prepared in high yield, and cleaved by acid (*p*-TsOH/EtOAc–H_2O, 22°, 0.5 h, 53–90% yield).[a]

[a] J. L. Douglas, D. E. Horning, and T. T. Conway, *Can. J. Chem.*, 56, 2879 (1978).

28. *N*-Salicylideneamine: $RN=CHC_6H_4$-*o*-OH, 28

Compound 28 has been prepared to protect the ϵ-NH_2 group in lysine; it is readily cleaved by acid (1 *N* HCl, 60–70°, few minutes). Because of hydrogen bonding, compound 28, unlike a benzylidene derivative, is not cleaved by mild catalytic hydrogenation.[a]

[a] J. N. Williams and R. M. Jacobs, *Biochem. Biophys. Res. Commun.*, 22, 695 (1966).

29. *N*-5-Chlorosalicylideneamine: $RN=CHC_6H_3$-2-OH-5-Cl, 29

Compound 29 has been prepared in 40–85% yield; it is cleaved with acid (dilute HCl/CH_3OH, 22°, 3 h, 88% yield). Like compound 28, it is not cleaved by hydrogenation.[a]

[a] J. C. Sheehan and V. J. Grenada, *J. Am. Chem. Soc.*, 84, 2417 (1962).

30. *N*-Diphenylmethyleneamine: $RN=C(C_6H_5)_2$, 30

Compound 30 is prepared in 82% yield from glycine and benzophenone (cat. $BF_3 \cdot Et_2O$/xylene, reflux).

It is stable to phase transfer alkylation, and is cleaved by acid (concd HCl, reflux, 6 h, or aq citric acid, 12 h).[a]

[a] M. J. O'Donnell, J. M. Boniece, and S. E. Earp, *Tetrahedron Lett.*, 2641 (1978).

31. *N*-(5-Chloro-2-hydroxyphenyl)phenylmethyleneamine:
   $RN=C(C_6H_5)C_6H_3$-2-OH-5-Cl, 31

Compound 31, used to protect amino acids, is cleaved by acid (80% HOAc, 80°, 20 min or 25°, 10 h, 69% yield; 1 *N* HCl/EtOH, 15 min, 70% yield; TsOH/Et_2O,

25°, 4 h).[a] If the amino acid —COOH group is protected by a resin during solid phase peptide synthesis, the imine is more stable to acid hydrolysis (e.g., stable to 80% HOAc, 25°, 24 h; cleaved by 0.4 N HCl/THF, 2.5 h).[b]

[a] B. Halpern and A. P. Hope, *Aust. J. Chem.*, **27**, 2047 (1974).
[b] A. Abdipranoto, A. P. Hope, and B. Halpern, *Aust. J. Chem.*, **30**, 2711 (1977).

Enamine Derivatives

32. N-(Acylvinyl)amine: $R^1R^2NC(CH_3)$=$CCOR'$, 32
|

Formation

$$RCHCO_2H + CH_3COCH_2COR' \xrightarrow{KOH/CH_3OH} \textbf{32},\ 65\text{-}87\%\ ^a$$
|
NH_2

$R' = CH_3, C_6H_5, OC_2H_5$

Cleavage

$$\textbf{32} \xrightarrow{\text{2 }N\text{ HCl}} \text{or} \xrightarrow{\text{HOAc}} RCHCO_2H,\ \text{high yields}^a$$
|
NH_2

[a] E. Dane, F. Drees, P. Konrad, and T. Dockner, *Angew. Chem., Inter. Ed. Engl.*, **1**, 658 (1962).

33. N-(5,5-Dimethyl-3-oxo-1-cyclohexenyl)amine: , 33

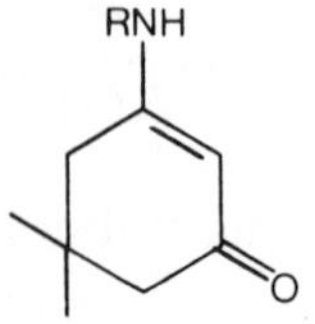

Compound **33**, $R = R'CHCO_2R''$, has been prepared in 70% yield to protect
|
amino acid esters.[a] It is cleaved by treatment with either aqueous bromine[a] or nitrous acid (90% yield).[b]

[a] B. Halpern and L. B. James, *Aust. J. Chem.*, **17**, 1282 (1964).
[b] B. Halpern and A. D. Cross, *Chem. Ind. (London)*, 1183 (1965).

N-Hetero Atom Derivatives

Six categories of N-hetero atom derivatives are considered: N–M (M = boron, copper); N–N (e.g., N-nitro, N-nitroso); N-oxides (used to protect tertiary amines); N–P (e.g., phosphinamides, phosphonamides); N–SiR$_3$ (R = CH$_3$), and N–S (e.g., sulfonamides, sulfenamides). These derivatives provide specialized protection. Borane derivatives protect tertiary amines; copper forms a chelate to protect adjacent —NH$_2$ and —OH groups. *N*-Nitro derivatives protect the guanidino group of arginine; *N*-nitroso derivatives protect secondary amines. Acid-labile phosphinamides and phosphonamides are used in peptide and nucleoside syntheses. Silyl derivatives protect amino-substituted aromatic halides during organometallic reactions. Sulfenamides, readily cleaved by acid, have been used in syntheses of peptides, penicillins, and nucleosides. The sulfonamides are the most versatile of these derivatives. They are used in syntheses of glycosides and nucleosides, to protect aromatic amines during nitration or Friedel-Crafts acylation of the aromatic ring, and to protect the guanidino group of arginine.

N-METAL DERIVATIVES

34. *N*-Borane Derivatives: R^{1}R^{2}R^{3}N$^+$BH$_3^-$, 34

Compound **34** can be prepared from diborane to protect a tertiary amine during oxidation; it is cleaved by refluxing in ethanol[a] or methanolic sodium carbonate.[b]

[a] A. Picot and X. Lusinchi, *Bull. Soc. Chim. Fr.,* 1227 (1977).
[b] M. A. Schwartz, B. F. Rose, and B. Vishnuvajjala, *J. Am. Chem. Soc.,* **95,** 612 (1973).

35. *N*-[Phenyl(pentacarbonylchromium- or - tungsten)carbenyl]amine: R^{1}R^{2}N$\cdots$C$\cdots$M(CO)$_5$, 35
$\quad$ |
$\quad$ R

R = C$_6$H$_5$ or CH$_3$; M = Cr or W

Compound **35,** prepared in 66–97% yield from amino acid esters, is cleaved by acid hydrolysis (CF$_3$CO$_2$H, 20°, 80% yield; 80% HOAc; M = W: BBr$_3$, −25°).[a]

[a] K. Weiss and E. O. Fischer, *Chem. Ber.,* **109,** 1868 (1976).

36. *N*-Copper or *N*-Zinc Chelate: RNH$_2$ --- M --- OH, 36

$\quad$ M = Cu(II), Zn(II)

Formation / Cleavage

(1) H$_2$N(CH$_2$)$_4$CHCO$_2$H $\xrightarrow{\text{aq Cu(II)}}$ H$_2$N(CH$_2$)$_4$CHCO
$\qquad\qquad$ |$\qquad\qquad\qquad\qquad\qquad\qquad$ |$\quad$ |
$\qquad\qquad$ NH$_2$ $\qquad\qquad\qquad\qquad\qquad$ H$_2$N$\quad$OH
$\qquad\qquad\qquad\qquad\qquad\qquad\qquad\qquad\qquad$ \ /
$\qquad\qquad\qquad\qquad\qquad\qquad\qquad\qquad\qquad$ Cu(II)

A copper chelate selectivity protects the α-NH_2 group in lysine. The chelate is cleaved by 2 N HCl or by EDTA: $(HO_2CCH_2)_2NCH_2CH_2N(CH_2CO_2H)_2$.[a]

(2)   In an aminoglycoside a vicinal amino hydroxy group can be protected as a Cu(II) chelate. After acylation of other amine groups, the chelate is cleaved by aqueous ammonia.[b]

(3)   After examination of the complexing ability of Ca(II), Cr(III), Mn(II), Fe(III), Co(II), Ni(II), Cu(II), Zn(II), Ru(III), Ag(I), and Sn(IV), the authors decided that Zn(II) provides the best protection for vicinal amino hydroxy groups during trifluoroacetylation of other amino groups in the course of some syntheses of kanamycin derivatives.[c]

[a]   R. Ledger and F. H. C. Stewart, *Aust. J. Chem.*, **18**, 933 (1965).
[b]   S. Hanessian and G. Patil, *Tetrahedron Lett.*, 1035 (1978).
[c]   T. Tsuchiya, Y. Takagi, and S. Umezawa, *Tetrahedron Lett.*, 4951 (1979).

N–N DERIVATIVES

37.   *N*-Nitroamine: $R^1R^2NNO_2$, 37

Formation[a]

$$H_2NCNH(CH_2)_3CHCO_2H \xrightarrow[0°,\ 1\ h,\ 80\%]{HNO_3/H_2SO_4} H_2NCNH(CH_2)_3CHCO_2H$$

(left structure: $\overset{\|}{NH}$ ... $\underset{NHPG}{|}$; right structure: $\overset{\|}{N-NO_2}$ ... $\underset{NHPG}{|}$)

An N-nitro derivative is used primarily to protect the guanidino group in arginine; it is cleaved by reduction: H_2/Pd–C, HOAc/CH_3OH, ~80% yield[a]; 10% Pd–C/cyclohexadiene, 25°, 2 h, good yields[b]; Pd–C/4% HCOOH–CH_3OH, 5 h, 100% yield[c]; $TiCl_3$/pH 6, 25°, 45 min, 70–98% yield[d]; $SnCl_2$/60% HCO_2H, 63% yield[e]; electrolysis, 1 N H_2SO_4, 1–6 h, 85–95% yield.[f]

[a]   K. Hofmann, W. D. Peckham, and A. Rheiner, *J. Am. Chem. Soc.*, **78**, 238 (1956).
[b]   A. M. Felix, E. P. Heimer, T. J. Lambros, C. Tzougraki, and J. Meienhofer, *J. Org. Chem.*, **43**, 4194 (1978).
[c]   B. ElAmin, G. M. Anantharamaiah, G. P. Royer, and G. E. Means, *J. Org. Chem.*, **44**, 3442 (1979).
[d]   R. M. Freidinger, R. Hirschmann, and D. F. Veber, *J. Org. Chem.*, **43**, 4800 (1978).
[e]   T. Hayakawa, Y. Fujiwara, and J. Noguchi, *Bull. Chem. Soc. Jpn.*, **40**, 1205 (1967).
[f]   P. M. Scopes, K. B. Walshaw, M. Welford, and G. T. Young, *J. Chem. Soc.*, **782** (1965).

38.   *N*-Nitrosoamine: R^1R^2NNO, 38

N-Nitroso derivatives, prepared from secondary amines and nitrous acid, are cleaved by reduction [H_2/Raney Ni, EtOH, 28°, 3.5 h[a]; CuCl/concd HCl[b]]. Since many *N*-nitroso compounds are carcinogens, and because some racemization and cyclodehydration of *N*-nitroso derivatives of *N*-alkyl amino acids occur during peptide syntheses,[c,d] *N*-nitroso derivatives are of limited value as protective groups.

[a] M. Harfenist and E. Magnein, *J. Am. Chem. Soc.,* **79**, 2215 (1957).

[b] C. F. Koelsch, *J. Am. Chem. Soc.,* **68**, 146 (1946).

[c] P. Quitt, R. O. Studer, and K. Vogler, *Helv. Chim. Acta,* **47**, 166 (1964).

[d] F. H. C. Stewart, *Aust. J. Chem.,* **22**, 2451 (1969).

39. Amine *N*-Oxide: $R^1R^2R^3N \rightarrow O$, 39

Amine oxides, prepared to protect tertiary amines during methylation[a,b] and to prevent their protonation in diazotized aminopyridines,[c] can be cleaved by reduction (e.g., SO_2/H_2O, 1 h, 22°, 63% yield[a]; H_2/Pd–C, HOAc–Ac_2O, 7 h, 91% yield[b]; Zn/HCl, 30% yield).[c] Photolytic reduction of an aromatic amine oxide has been reported [i.e., 4-nitropyridine *N*-oxide, 300 nm, $(MeO)_3PO/CH_2Cl_2$, 15 min, 85–90% yield].[d]

[a] F. N. H. Chang, J. F. Oneto, P. P. T. Sah, B. M. Tolbert, and H. Rapoport, *J. Org. Chem.,* **15**, 634 (1950).

[b] J. A. Berson and T. Cohen, *J. Org. Chem.,* **20**, 1461 (1955).

[c] F. Koniuszy, P. F. Wiley, and K. Folkers, *J. Am. Chem. Soc.,* **71**, 875 (1949).

[d] C. Kaneko, A. Yamamoto, and M. Gomi, *Heterocycles,* **12**, 227 (1979).

N–P DERIVATIVES

40. Diphenylphosphinamide: $R^1R^2NPO(C_6H_5)_2$, 40

Phosphinamides are stable to catalytic hydrogenation, used to cleave benzyl esters, and to hydrazine.[a]

Formation[a]

$$RCHCOOR' + Ph_2POCl \xrightarrow{\text{\textit{N}-methylmorpholine}} RCHCOOR'$$

RCHCOOR' + Ph₂POCl → RCHCOOR'
  | |
 NH₂   NHPOPh₂

40a

Cleavage[a]

Compound **40a** is hydrolyzed to the amino acid by the following acidic conditions: $HOAc/HCOOH/H_2O$, 24 h, 100%; 80% CF_3CO_2H, ca. quant; 0.4 *M*

HCl/90% CF_3CH_2OH, ca. quant; p-TsOH/H_2O–CH_3OH, ca. quant; 80% HOAc, 3 days, not completely cleaved.

[a] G. W. Kenner, G. A. Moore, and R. Ramage, *Tetrahedron Lett.*, 3623 (1976).

41, 42. Dimethyl- and Diphenylthiophosphinamide: $R^1R^2NPS(CH_3)_2$, 41, $R^1R^2NPS(C_6H_5)_2$, 42

Compounds **41** and **42** can be prepared from an amino acid and the thiophosphinyl chloride (Me_2PSCl or Ph_2PSCl, respectively). They are cleaved by acid hydrolysis at a rate similar to that for t-butyl carbamates.[a] In solid-phase synthesis they are cleaved more rapidly than t-butyl carbamates by triphenylphosphine dihydrochloride.[b]

[a] S. Ikeda, F. Tonegawa, E. Shikano, K. Shinozaki, and M. Ueki, *Bull. Chem. Soc. Jpn.*, **52**, 1431 (1979).

[b] M. Ueki, T. Inazu, and S. Ikeda, *Bull. Chem. Soc. Jpn.*, **52**, 2424 (1979).

43. Diethyl Phosphoramidate: $R^1R^2NPO(OC_2H_5)_2$, 43

Formation ($\rightarrow$) / Cleavage ($\leftarrow$)

$$\underset{0°, 1\text{ h} \rightarrow 22°, 1\text{ h}, 75\text{-}90\%}{\xrightarrow{(EtO)_2POH,\ CCl_4,\ aq\ NaOH,\ PhCH_2\overset{+}{N}Et_3Cl^{-\,a}}}$$

R^1R^2NH **43**

$$\underset{80\text{-}95\%}{\xleftarrow{g.\ HCl/THF,\ 22°,\ 12\ h^{\,b}}}$$

[a] A. Zwierzak, *Synthesis*, 507 (1975).

[b] A. Zwierzak and J. B.-Piotrowicz, *Angew. Chem., Inter. Ed. Engl.*, **16**, 107 (1977).

44, 45. Dibenzyl, 44, and Diphenyl Phosphoramidate, 45: $R^1R^2NPO(OCH_2C_6H_5)_2$, 44, $R^1R^2NPO(OC_6H_5)_2$, 45

Compound **44**[a] has been prepared from an amino acid and the phosphoryl chloride, $(PhCH_2O)_2POCl$. Compound **45** has been prepared from a glucosamine; it is converted by transesterification into a dibenzyl derivative to facilitate cleavage.[b]

Cleavage

$$R^1R^2NPO(OCH_2Ph)_2 \xrightarrow{H_2/Pd\text{-}C^{\,a}} R^1R^2NPO(OH)_2 \xrightarrow[pH\ 4]{H_3O^+} R^1R^2NH$$

[a] A. Cosmatos, I. Photaki, and L. Zervas, *Chem. Ber.*, **94**, 2644 (1961).

[b] M. L. Wolfrom, P. J. Conigliaro, and E. J. Soltes, *J. Org. Chem.*, **32**, 653 (1967).

N–Si DERIVATIVES

46.   N-Trimethylsilylamine: $R^1R^2NSi(CH_3)_3$, 46

In spite of the high reactivity of N-silyl derivatives to moisture, they provide satisfactory protection during reactions run under anhydrous conditions (e.g., organometallic syntheses); they are readily cleaved by hydrolysis.[a]

[a]   J. R. Pratt, W. D. Massey, F. H. Pinkerton, and S. F. Thames, *J. Org. Chem.*, **40**, 1090 (1975).

N–S DERIVATIVES

N-SULFENYL DERIVATIVES

Sulfenamides, prepared from an amine and a sulfenyl halide,[a] are readily cleaved by acid hydrolysis and have been used in syntheses of peptides, penicillins, and nucleosides. They are also cleaved by nucleophiles,[b] and by Raney nickel desulfurization.[c]

[a]   For other methods of preparation, see F. A. Davis and U. K. Nadir, *Org. Prep. Proc. Int.*, **11**, 33 (1979).
[b]   W. Kessler and B. Iselin, *Helv. Chim. Acta*, **49**, 1330 (1966).
[c]   J. Meienhofer, *Nature*, **205**, 73 (1965).

47.   Benzenesulfenamide: $R^1R^2NSC_6H_5$, 47

48.   o-Nitrobenzenesulfenamide: $R^1R^2NSC_6H_4$-o-NO_2, 48

49.   2,4-Dinitrobenzenesulfenamide: $R^1R^2NSC_6H_3$-2,4-$(NO_2)_2$, 49

50.   Pentachlorobenzenesulfenamide: $R^1R^2NSC_6Cl_5$, 50

Benzenesulfenamide, **47**, and a number of substituted benzenesulfenamides (compounds **48, 49,** and **50**) have been prepared to protect the 7-amino group in cephalosporins. They are cleaved by sodium iodide ($CH_3OH/CH_2Cl_2/HOAc$, 0°, 20 min, 53% yield from compound **48**).[a]

o-Nitrobenzenesulfenamides, **48,** are also cleaved by acidic hydrolysis ($HOAc$/aq alcohol; HCl/Et_2O or EtOH, 0°, 1 h, 95% yield),[b] by nucleophiles (13 reagents, 5 min to 12 h, 90% cleaved[c]; PhSH or $HSCH_2CO_2H$, 22°, 1 h[d]; 2-mercaptopyridine/CH_2Cl_2, 1 min, 100% yield[e]), and by catalytic desulfurization [Raney Ni/DMF, column, few hours, satisfactory yields].[f]

[a]   T. Kobayashi, K. Iino, and T. Hiraoka, *J. Am. Chem. Soc.*, **99**, 5505 (1977).
[b]   L. Zervas, D. Borovas, and E. Gazis, *J. Am. Chem. Soc.*, **85**, 3660 (1963).
[c]   W. Kessler and B. Iselin, *Helv. Chim. Acta*, **49**, 1330 (1966).

[d] A. Fontana, F. Marchiori, L. Moroder, and E. Schoffone, *Tetrahedron Lett.*, 2985 (1966).

[e] M. Stern, A. Warshawsky, and M. Fridkin, *Int. J. Pept. Protein Res.*, **13**, 315 (1979).

[f] J. Meienhofer, *Nature*, **205**, 73 (1965).

51.   2-Nitro-4-methoxybenzenesulfenamide: $R^1R^2NSC_6H_3$-2-NO_2-4-OCH_3, 51

Compound **51,** prepared from an amino acid, the sulfenyl chloride and sodium bicarbonate, is cleaved by acid hydrolysis (HOAc/dioxane, 22°, 30 min, 95% yield).[a]

[a] Y. Wolman, *Isr. J. Chem.*, **5**, 231 (1967).

52.   Triphenylmethylsulfenamide: $R^1R^2NSC(C_6H_5)_3$, 52

Compound **52** can be prepared from an amino acid and the sulfenyl chloride; it is cleaved by hydrogen chloride in ether or ethanol (0°, 1 h, 90% yield).[a]

[a] L. Zervas, D. Borovas, and E. Gazis, *J. Am. Chem. Soc.*, **85**, 3660 (1963).

N-SULFONYL DERIVATIVES

Sulfonamides are prepared from an amine and sulfonyl chloride in the presence of pyridine or aqueous base.[a] Alkylsulfonamides are too stable to be used as protective groups. Arylsulfonamides are stable to alkaline hydrolysis and to catalytic reduction; they are cleaved by Na/NH_3,[b] Na/C_4H_9OH,[c] sodium naphthalenide,[d] or sodium anthracenide,[e] and by refluxing in acid (48% HBr/cat. phenol).[f] Some sulfonamides that have been used as protective groups are described here.

[a] E. Fischer and W. Lipschitz, *Ber.*, **48**, 360 (1915).

[b] V. du Vigneaud and O. K. Behrens, *J. Biol. Chem.*, **117**, 27 (1937).

[c] G. Wittig, W. Joos, and P. Rathfelder, *Liebigs Ann. Chem.*, **610**, 180 (1957).

[d] S. Ji, L. B. Gortler, A. Waring, A. Battisti, S. Bank, W. D. Closson, and P. Wriede, *J. Am. Chem. Soc.*, **89**, 5311 (1967).

[e] K. S. Quaal, S. Ji, Y. M. Kim, W. D. Closson, and J. A. Zubieta, *J. Org. Chem.*, **43**, 1311 (1978).

[f] H. R. Snyder and R. E. Heckert, *J. Am. Chem. Soc.*, **74**, 2006 (1952).

53.   Benzenesulfonamide: $R^1R^2NSO_2C_6H_5$, 53

Formation (→) / Cleavage (←)[a]

[a] W. A. Remers, R. H. Roth, G. J. Gibs, and M. J. Weiss, *J. Org. Chem.* **36**, 1232 (1971).

54. *p*-Methoxybenzenesulfonamide: $R^1R^2NSO_2C_6H_4$-*p*-OCH_3, **54**

Compound **54**, prepared to protect the guanidino group in arginine, and stable to catalytic reduction, is quantitatively cleaved by methanesulfonic acid and by boron(III) trifluoroacetate; a toluenesulfonamide is only partially cleaved by these reagents.[a]

[a] O. Nishimura and M. Fujino, *Chem. Pharm. Bull.,* **24**, 1568 (1976).

55. 2,4,6-Trimethylbenzenesulfonamide: $R^1R^2NSO_2C_6H_2$-2,4,6-$(CH_3)_3$, **55**

Compound **55**, prepared in 60% yield from the guanidino group in arginine, is quantitatively cleaved by strong acid hydrolysis (e.g., HF, 0°, 60 min; CH_3SO_3H, 20°, 30 min; CF_3SO_3H, 20°, 30 min). It is also cleaved by 25% HBr/HOAc (20°, 60 min, 71% yield) and $Na/NH_3(-15°, 1$ min, 60% yield). It is stable to CF_3CO_2H (20°, 60 min).[a]

[a] H. Yajima, M. Takeyama, J. Kanaki, O. Nishimura, and M. Fujino, *Chem. Pharm. Bull.,* **26**, 3752 (1978).

56. Toluenesulfonamide: $R^1R^2NSO_2C_6H_4$-*p*-CH_3, **56**

Compound **56** is prepared from amines, including α- and ϵ-amino acids, the guanidino group in arginine,[a] and the imidazole nitrogen in histidine,[b] and *p*-toluenesulfonyl chloride. *N*-Methyl-*N*-*p*-toluenesulfonylpyrrolidinium perchlorate selectively forms an *N*-tosyl derivative from an amino alcohol.[c] Compound **56** is cleaved by the reagents described on p. 284 that are used to cleave *N*-sulfonyl derivatives. It is also cleaved by the following conditions.

Cleavage

(1) **56** $\xrightarrow[\text{0°, 1 h}]{\text{HF/anisole}}$ R^1R^2NH [b]

(2) **56** $\xrightarrow[\text{70°, 8 h}]{\text{HBr/HOAc}}$ R^1R^2NH, 45–50% [d]

During the synthesis of L-2-amino-3-oxalylaminopropionic acid, a neurotoxin, cleavage with Na/NH_3 or $[C_{10}H_8]^{\cdot}Na^+$ gave a complex mixture of products.[d]

(3) **56** $\xrightarrow[\substack{C_6H_6 \text{ or } C_6H_5CH_3, \text{ reflux} \\ 20 \text{ h}}]{NaAlH_2(OCH_2CH_2OCH_3)_2}$ R^1R^2NH, 65–75% [e]

$$R^2 = \text{H, alkyl, aryl}$$

Note that $LiAlH_4$ does not cleave sulfonamides of primary amines; those from secondary amines must be heated at $120°.^e$

$$(4)\quad 56 \xrightarrow[\text{Me}_4\text{N}^+\text{Cl}^-,\ 5°]{\text{electrolysis}} R^1R^2NH,\ 65\text{–}98\%^f$$

$$(5)\quad 56 \xrightarrow[\text{6–20 h}]{hv,\ \text{Et}_2\text{O}} R^1R^2NH,\ 85\text{–}90\%^g$$

[a] J. R. Bell, J. H. Jones, D. M. Regester, and T. C. Webb, *J. Chem. Soc., Perkin Trans. 1*, 1961 (1974).

[b] T. Fujii and S. Sakakibara, *Bull. Chem. Soc. Jpn.*, **47**, 3146 (1974).

[c] T. Oishi, K. Kamata, S. Kosuda, and Y. Ban, *J. Chem. Soc., Chem. Commun.*, 1148 (1972).

[d] B. E. Haskell and S. B. Bowlus, *J. Org. Chem.*, **41**, 159 (1976).

[e] E. H. Gold and E. Babad, *J. Org. Chem.*, **37**, 2208 (1972).

[f] L. Horner and H. Neumann, *Chem. Ber.*, **98**, 3462 (1965).

[g] A. Abad, D. Mellier, J. P. Pète, and C. Portella, *Tetrahedron Lett.*, 4555 (1971).

57.   Benzylsulfonamide: $R^1R^2NSO_2CH_2C_6H_5$, 57

Compound **57**, prepared in 40–70% yield, is cleaved by reduction [Na/NH_3, 75% yield; H_2/Raney Ni, 65–85% yield, but not by H_2/PtO_2] and by acid hydrolysis (HBr or HI, slow).[a] Compound **57** can also be cleaved by photolysis (2–4 h, 40–90% yield).[b]

[a] H. B. Milne and C.-H. Peng, *J. Am. Chem. Soc.*, **79**, 639, 645 (1957).

[b] J. A. Pincock and A. Jurgens, *Tetrahedron Lett.*, 1029 (1979).

58.   *p*-Methylbenzylsulfonamide: $R^1R^2NSO_2CH_2C_6H_4$-*p*-CH_3, 58

Compound **58** has been prepared to protect the ϵ-amino group in lysine; it is quantitatively cleaved by anhydrous hydrogen fluoride/anisole ($-20°$, 60 min).[a]

[a] T. Fukuda, C. Kitada, and M. Fujino, *J. Chem. Soc., Chem. Commun.*, 220 (1978).

59.   Trifluoromethylsulfonamide: $R^1R^2NSO_2CF_3$, 59

A trifluoromethylsulfonamide can be prepared from a primary amine to allow monoalkylation of that amine.[a]

Formation

$$R^1R^2NH + (CF_3SO_2)_2O \xrightarrow[-78°]{CH_2Cl_2} 59,\ \sim\text{quant}^a$$

$R^2 = H$, alkyl, aryl

Cleavage

(1)   **59** $\xrightarrow[\text{C}_6\text{H}_6,\ \text{reflux, few min}]{\text{NaAlH}_2(\text{OCH}_2\text{CH}_2\text{OCH}_3)_2}$ $\text{R}^1\text{NH}_2,\ 95\%\,^a$

$\text{R}^2 = \text{H}$

(2)   **59** $\xrightarrow[\text{acetone, 12 h}]{p\text{-Br-C}_6\text{H}_4\text{COCH}_2\text{Br}/\text{K}_2\text{CO}_3}$ $\xrightarrow{\text{H}_3\text{O}^+}$ $\text{R}^1\text{NH}_2,\ 80\%\,^b$

$\text{R}^2 = \text{H}$

(3)   **59** $\xrightarrow[\text{reflux}]{\text{LiAlH}_4/\text{Et}_2\text{O}}$ $\text{R}^1\text{R}^2\text{NH},\ 90\text{–}95\%\,^a$

$\text{R}^2 = \text{alkyl, aryl}$

---

[a]   J. B. Hendrickson and R. Bergeron, *Tetrahedron Lett.*, 3839 (1973).

[b]   J. B. Hendrickson, R. Bergeron, A. Giga, and D. Sternbach, *J. Am. Chem. Soc.*, **95**, 3412 (1973).

60.   Phenacylsulfonamide: $\text{R}^1\text{R}^2\text{NSO}_2\text{CH}_2\text{COC}_6\text{H}_5$, 60

Like the trifluoromethylsulfonamides, phenacylsulfonamides are used to prevent dialkylation of primary amines. Compound **60** is prepared in 91–94% yield from the sulfonyl chloride, and cleaved in 66–72% yield by Zn/HOAc/trace HCl.[a]

---

[a]   J. B. Hendrickson and R. Bergeron, *Tetrahedron Lett.*, 345 (1970).

8

Reactivities, Reagents, and Reactivity Charts

REACTIVITIES

In the selection of a protective group it is of paramount importance to know the reactivity of the resulting protected functionality toward various reagents and reaction conditions. The number of reagents available to the organic chemist is large; approximately 3000 reagents are reviewed in the excellent series of books by the Fiesers.[a] In an effort to assess the effect of a wide variety of standard types of reagents and reaction conditions on the different possible protected function-alities, 108 prototype reagents have been selected and grouped into 16 categories[b]:

A. Aqueous
B. Nonaqueous Bases
C. Nonaqueous Nucleophiles
D. Organometallic
E. Catalytic Reduction
F. Acidic Reduction
G. Basic or Neutral Reduction
H. Hydride Reduction
I. Lewis Acids
J. Soft Acids
K. Radical Addition
L. Oxidizing Agents
M. Thermal Reactions
N. Carbenoids
O. Miscellaneous
P. Electrophiles

These 108 reagents are used in the Reactivity Charts that have been prepared for each class of protective groups. The reagents and some of their properties are described on the following pages.

REAGENTS

A. Aqueous

1. pH < 1, 100°	Refluxing HBr
2. pH < 1	1 N HCl
3. pH 1	0.1 N HCl
4. pH 2–4	0.01 N HCl; 1–0.01 N HOAc
5. pH 4–6	0.1 N H$_3$BO$_3$; phosphate buffer; HOAc–NaOAc
6. pH 6–8.5	H$_2$O
7. pH 8.5–10	0.1 N HCO$_3^-$; 0.1 N OAc$^-$; satd CaCO$_3$
8. pH 10–12	0.1 N CO$_3^{-2}$; 1–0.01 N NH$_4$OH; 0.01 N NaOH; satd Ca(OH)$_2$
9. pH > 12	1–0.1 N NaOH
10. pH > 12, 150°	

B. Nonaqueous Bases

11. NaH	
12. (C$_6$H$_5$)$_3$CNa	$pK_a = 32$
13. [C$_{10}$H$_8$]$^-$Na$^+$	$pK_a \cong 37$
14. CH$_3$SOCH$_2^-$Na$^+$	$pK_a = 35$
15. KO-t-C$_4$H$_9$	$pK_a = 19$
16. LiN(i-C$_3$H$_7$)$_2$	(LDA) $pK_a = 36$
17. Pyridine; Et$_3$N	$pK_a = 5$; 10
18. NaNH$_2$; NaNHR	$pK_a = 36$

C. Nonaqueous Nucleophiles

19. NaOCH$_3$/CH$_3$OH, 25°	$pK_a = 16$
20. Enolate anion	$pK_a = 20$
21. NH$_3$; RNH$_2$; RNHOH	$pK_a = 10$
22. RS$^-$; N$_3^-$; SCN$^-$	
23. OAc$^-$; X$^-$	$pK_a = 4.5$
24. NaCN, pH 12	
25. HCN, cat. CN$^-$, pH 6	$pK_a = 9$. For cyanohydrin formation

D. Organometallic

26. RLi

27. RMgX

28. Organozinc   Reformatsky reaction. Similar: R_2Cu; R_2Cd
29. Organocopper R_2CuLi
30. Wittig; ylide   Includes sulfur ylides

E. Catalytic Reduction

31. H_2/Raney Ni
32. H_2/Pt, pH 2–4
33. H_2/Pd–C
34. H_2/Lindlar
35. H_2/Rh–C or   Avoids hydrogenolysis of benzyl ethers
    H_2/Rh–Al_2O_3

F. Acidic Reduction

36. Zn/HCl
37. Zn/HOAc; $SnCl_2$/HCl
38. Cr(II), pH 5

G. Basic or Neutral Reduction

39. Na/l NH$_3$
40. Al(Hg)
41. $SnCl_2$/Py
42. H_2S or HSO_3^-

H. Hydride Reduction

43. $LiAlH_4$
44. Li-s-Bu$_3$BH, $-50°$ Li-Selectride
45. $[(CH_3)_2CHCH(CH_3)]_2BH$ Disiamylborane
46. B_2H_6, $0°$
47. $NaBH_4$
48. $Zn(BH_4)_2$ Neutral reduction
49. $NaBH_3CN$, pH 4–6
50. (i-C_4H_9)$_2$AlH, $-60°$ Dibal
51. Li(O-t-C_4H_9)$_3$AlH, $0°$

I. Lewis Acids (Anhydrous conditions)

52. $AlCl_3$, $80°$
53. $AlCl_3$, $25°$

54. $SnCl_4$, 25°; $BF_3 \cdot Et_2O$
55. $LiClO_4$; $MgBr_2$ — For epoxide rearrangement
56. TsOH, 80° — Catalytic amount
57. TsOH, 0° — Catalytic amount

J. Soft Acids

58. Hg(II)
59. Ag(I)
60. Cu(II)/Py — For example, for Glaser coupling

K. Radical Addition

61. HBr/initiator — "Acidic" HX addition; acidity $\cong$ TsOH, 0°
62. HX/initiator — Neutral HX addition; X = P, S, Se, Si
63. NBS/CCl_4, $h\nu$ or heat — Allylic bromination
64. $CHBr_3$; $BrCCl_3$; $CCl_4/In \cdot$ — Carbon-halogen addition

L. Oxidizing Agents

65. OsO_4
66. $KMnO_4$, 0°, pH 7
67. O_3, $-50°$
68. RCO_3H, 0° — Epoxidation of olefins; prototype for H_2O_2/H^+
69. RCO_3H, 50° — Baeyer-Villiger oxidation of hindered ketones
70. CrO_3/Py — Collins oxidation
71. CrO_3, pH 1 — Jones oxidation
72. H_2O_2/OH^-, pH 10–12
73. Quinone — Dehydrogenation
74. $^1O_2$ — Singlet oxygen
75. CH_3SOCH_3, 100° — (DMSO); HCO_3^- may be added to maintain neutrality
76. NaOCl, pH 10
77. Aq NBS — Nonradical conditions
78. I_2
79. C_6H_5SCl; C_6H_5SeX
80. Cl_2; Br_2
81. MnO_2/CH_2Cl_2
82. $NaIO_4$, pH 5–8
83. SeO_2, pH 2–4

84. SeO$_2$/Py   In EtOH/cat. Py
85. K$_3$Fe(CN)$_6$, pH 7–10   Phenol coupling
86. Pb(IV), 25°   Glycol and α-hydroxy acid cleavage
87. Pb(IV), 80°   Oxidative decarboxylation
88. Tl(NO$_3$)$_3$, pH 2   Oxidative rearrangement of olefins

M. Thermal Reactions

89. 150°   Some Cope rearrangements and Cope
                                        eliminations
90. 250°   Claisen or Cope rearrangement
91. 350°   Ester cracking; Conia "ene" reaction

N. Carbenoids

92. :CCl$_2$
93. N$_2$CHCO$_2$C$_2$H$_5$/Cu, 80°
94. CH$_2$I$_2$/Zn–Cu   Simmons-Smith addition

O. Miscellaneous

95. *n*-Bu$_3$SnH/initiator
96. Ni(CO)$_4$
97. CH$_2$N$_2$
98. SOCl$_2$
99. Ac$_2$O, 25°   Acetylation
100. Ac$_2$O, 80°   Dehydration
101. DCC   Dicyclohexylcarbodiimide,
                                        C$_6$H$_{11}$N=C=NC$_6$H$_{11}$
102. CH$_3$I
103. (CH$_3$)$_3$O$^+$BF$_4^-$ Or CH$_3$OSO$_2$F = Magic Methyl:
                                        SEVERE POISON
104. 1. LiN-*i*-Pr$_2$; 2. MeI   For C-alkylation
105. 1. K$_2$CO$_3$; 2. MeI   For *O*-alkylation

P. Electrophiles

106. RCHO
107. RCOCl
108. C$^+$ ion/olefin   For cation-olefin cyclization

REACTIVITY CHARTS

One requirement of a protective group is stability to a given reaction. The following charts were prepared as a guide to relative reactivities and thereby as an aid in the choice of a protective group. The reactivities in the charts were estimated by the individual and collective efforts of a group of synthetic chemists. *It is important to realize that not all the reactivities in the charts have been determined experimentally and considerable conjecture has been exercised.* For those cases in which a literature reference was available concerning the use of a protective group and one of the 108 prototype reagents, the reactivity is printed in italic type. However, an exhaustive search for such references has not been made; therefore the absence of italic type does not imply an experimentally unknown reactivity.

There are four levels of reactivity in the charts:

"H" (high) indicates that under the conditions of the prototype reagent the protective group is readily removed to regenerate the original functional group.

"M" (marginal) indicates that the stability of the protected functionality is marginal, and depends on the exact parameters of the reaction. The protective group may be stable, may be cleaved slowly, or may be unstable to the conditions. Relative rates are always important, as illustrated in the following example[e] (in which a hemithioacetal is cleaved in the presence of a dithiane), and may have to be determined experimentally.

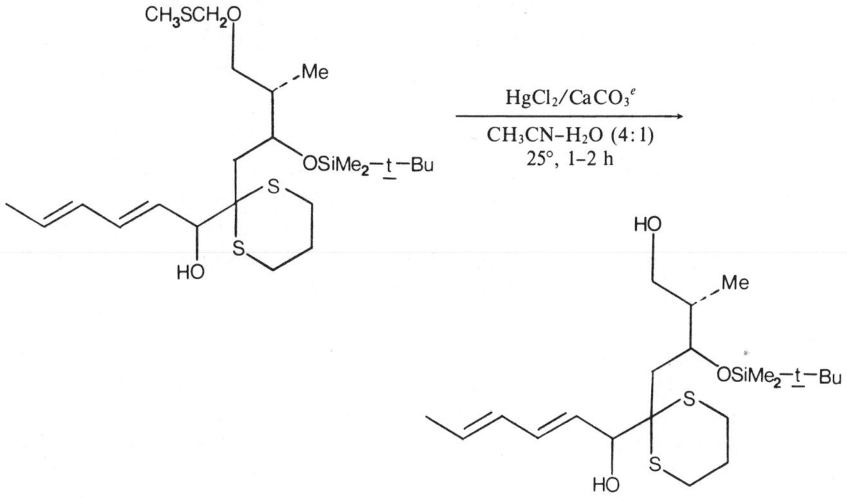

"L" (low) indicates that the protected functionality is stable under the reaction conditions.

"R" (reacts) indicates that the protected compound reacts readily, but that the original functional group is not restored. The protective group may be changed to a new protective group (eq. 1) or to a reactive intermediate (eq. 2), or the protective group may be unstable to the reaction conditions and react further (eq. 3).

(1) $ROCOC_6H_4\text{-}p\text{-}NO_2 \xrightarrow{\text{H}_2/\text{Pd–C}} ROCOC_6H_4\text{-}p\text{-}NH_2$

(2) $RCONR_2' \xrightarrow{\text{Me}_3\text{O}^+\text{BF}_4^-} [\underset{\displaystyle \overset{|}{OMe}}{RC}{=}\overset{+}{N}R_2'\ BF_4^-]$

(3) $RCH(OR')_2 \xrightarrow{\text{pH}<1,\ 100°} [RCHO] \rightarrow$ condensation products

The reactivities in the charts refer *only* to the protected functionality, not to atoms adjacent to the functional group; for example, $RCOOEt \xrightarrow{\text{LDA}}$: "L" (low) reactivity of PG (Et). However if the protected functionality is $R_2CHCOOEt$, this substrate obviously *will* react with LDA. Reactivity of the entire substrate must be evaluated by the chemist.

Five reagents [#25: HCN, pH 6; #88: $Tl(NO_3)_3$; #103: $Me_3O^+BF_4^-$; #104: LiN-i-Pr_2/MeI; and #105: K_2CO_3/MeI] were added after some of the charts had been completed; reactivities to these reagents are not included for all charts.

The number used to designate a protective group (PG) in a Reactivity Chart is the same as that used in the body of the text. Protective group numbers in the Reactivity Charts are not continuous, since not all of the protective groups described in the text are included in the charts. The protective groups that are included in the Reactivity Charts are in general those that have been used most widely; consequently, considerable experimental information is available for them.

The Reactivity Charts were prepared in collaboration with the following chemists, to whom I am most grateful: John O. Albright, Dale L. Boger, Dr. Daniel J. Brunelle, Dr. David A. Clark, Dr. Jagabandhu Das, Herbert Estreicher, Anthony L. Feliu, Dr. Frank W. Hobbs, Jr., Paul B. Hopkins, Dr. Spencer Knapp, Dr. Pierre Lavallée, John Munroe, Jay W. Ponder, Marcus A. Tius, Dr. David R. Williams, and Robert E. Wolf, Jr.

[a] L. F. Fieser and M. Fieser, *Reagents for Organic Synthesis*, Wiley-Interscience, New York, 1967, Vol. 1; M. Fieser and L. F. Fieser, Vols. 2–7, 1969–1979, M. Fieser, Vol. 8, 1980.

[b] The categories and prototype reagents used in this study are an expansion of an earlier set of 11 categories and 60 prototype reagents,[c] originally compiled for use in LHASA[d] (Logic and Heuristics Applied to Synthetic Analysis), a long-term research program at Harvard for Computer-Assisted Synthetic Analysis.

[c] E. J. Corey, H. W. Orf, and D. A. Pensak, *J. Am. Chem. Soc.*, **98**, 210 (1976).

[d] Selected references include: E. J. Corey, *Quart. Rev., Chem. Soc.*, **25**, 455 (1971); H. W. Orf, Ph.D. Thesis, Harvard University, 1976.

[e] E. J. Corey and M. G. Bock, *Tetrahedron Lett.*, 2643 (1975).

Reactivity Chart 1. Protection for the Hydroxyl Group: Ethers

1. Methyl Ether
2. Methoxymethyl Ether (MOM)
3. Methylthiomethyl Ether (MTM)
6. 2-Methoxyethoxymethyl Ether (MEM)
8. Bis(2-chloroethoxy)methyl Ether
9. Tetrahydropyranyl Ether (THP)
11. Tetrahydrothiopyranyl Ether
12. 4-Methoxytetrahydropyranyl Ether
13. 4-Methoxytetrahydrothiopyranyl Ether
15. Tetrahydrofuranyl Ether
16. Tetrahydrothiofuranyl Ether
17. 1-Ethoxyethyl Ether
18. 1-Methyl-1-methoxyethyl Ether
21. 2-(Phenylselenyl)ethyl Ether
22. *t*-Butyl Ether
23. Allyl Ether
26. Benzyl Ether
28. *o*-Nitrobenzyl Ether
35. Triphenylmethyl Ether
36. α-Naphthyldiphenylmethyl Ether
37. *p*-Methoxyphenyldiphenylmethyl Ether
41. 9-(9-Phenyl-10-oxo)anthryl Ether (Tritylone)
43. Trimethylsilyl Ether (TMS)
45. Isopropyldimethylsilyl Ether
46. *t*-Butyldimethylsilyl Ether (TBDMS)
48. *t*-Butyldiphenylsilyl Ether
51. Tribenzylsilyl Ether
53. Triisopropylsilyl Ether

(See chart, pp. 296–298)

Reactivity Chart 1. Protection for the Hydroxyl Group: Ethers

Column key (reagent conditions):

A. AQUEOUS — 1 pH<1, 100°; 2 pH<1; 3 pH 1; 4 pH 2-4; 5 pH 4-6; 6 pH 6-8.5; 7 pH 8.5-10; 8 pH 10-12; 9 pH>12; 10 pH>12, 150°
B. BASIC — 11 NaH; 12 Ph₃CNa; 13 (C₁₀H₈)⁻·Na⁺; 14 MeSOCH₂⁻Na⁺; 15 KO-t-Bu; 16 LiN-i-Pr₂; 17 Py; R₃N; 18 NaNH₂
C. NUCLEOPHILIC — 19 NaOMe; 20 Enolate; 21 NH₃; RNH₂; 22 RS⁻; N₃⁻; SCN⁻; 23 OAc⁻; X⁻; 24 NaCN, pH 12; 25 HCN, pH 6
D. ORGANOMET. — 26 RLi; 27 RMgX; 28 Organozinc; 29 Organocopper; 30 Wittig; ylide
E. CAT. REDN. — 31 H₂/Raney (Ni); 32 H₂/Pt pH 2-4; 33 H₂/Pd; 34 H₂/Lindlar; 35 H₂/Rh
F. — 36 Zn/HCl; 37 Zn/HOAc; 38 Cr(II), pH 5

PG	1	2	3	4	5	6	7	8	9	10	11	12	13	14	15	16	17	18	19	20	21	22	23	24	25	26	27	28	29	30	31	32	33	34	35	36	37	38
1	H	M	L	L	L	L	L	L	L	L	L	L	L	L	L	L	L	L	L	L	L	L	L	L	L	L	L	L	L	L	L	L	L	L	L	L	L	L
2	H	H	H	M	L	L	L	L	L	M	L	L	L	L	L	L	L	L	L	L	L	L	L	L	L	L	L	L	L	L	L	M	L	L	L	H	M	L
3	H	H	R	L	L	L	L	L	L	M	L	M	R	L	L	M	L	L	L	L	L	L	L	L	L	L	L	L	L	L	R	R	R	L	R	R	R	M
6	H	H	M	L	L	L	L	L	L	L	L	L	L	L	L	L	L	L	L	L	L	L	L	L	L	L	L	M	L	L	L	L	L	L	L	M	L	L
8	H	H	H	H	M	L	L	L	M	H	L	R	R	R	R	R	L	L	R	L	M	R	L	R		R	R	L	L	L	R	R	R	L	L	H	H	M
9	H	H	H	H	M	L	L	L	L	L	L	L	L	L	L	L	L	L	L	L	L	L	L	L	M	L	L	L	L	L	L	H	L	L	L	H	H	M
11	H	H	H	H	L	L	L	L	L	M	L	M	R	R	L	M	L	L	L	L	L	L	L	L		L	L	L	L	L	R	R	R	L	R	H	H	M
12	H	H	H	H	M	L	L	L	L	L	L	L	L	L	L	L	L	L	L	L	L	L	L	L		L	L	L	L	L	L	H	L	L	L	H	H	M
13	H	H	H	H	M	L	L	L	L	M	L	L	M	L	L	L	L	L	L	L	L	L	L	L		M	L	L	L	L	R	R	R	L	R	H	H	M
15	H	H	H	H	H	L	L	L	L	L	L	L	L	L	L	L	L	L	L	L	L	L	L	L		L	L	L	L	L	L	H	L	L	L	L	L	H
16	H	H	H	M	L	L	L	L	L	M	L	M	R	R	L	M	L	L	L	L	L	L	L	L		L	L	L	L	L	R	R	R	L	R	H	H	M
17	H	H	H	H	L	L	L	L	L	L	L	L	L	L	L	L	L	L	L	L	L	L	L	L		L	L	L	L	L	L	H	L	L	L	H	H	L
18	H	H	H	H	M	L	L	L	L	L	L	L	L	L	L	L	L	L	M	L	L	L	L	L		R	L	L	L	L	L	H	R	L	L	H	H	M
21	H	H	M	L	L	L	L	L	L	M	L	R	R	R	M	R	L	L	L	R	L	R	L	L	L	L	L	L	L	L	R	L	L	L	R	H	H	L
22	H	H	L	L	L	L	L	L	L	L	R	L	L	L	L	L	L	R	L	L	L	L	L	L		L	L	L	L	L	L	L	R	L	L	L	L	L
23	H	H	L	L	L	L	L	L	M	R	L	L	L	L	R	L	L	L	R	L	L	L	L	L	L	L	L	L	L	L	R	R	R	L	R	L	L	L
26	H	H	L	L	L	L	L	L	L	L	L	L	R	L	L	L	L	L	L	L	L	L	L	L	L	L	L	L	L	L	H	H	R	L	L	L	L	L
28	H	H	L	H	L	L	L	H	L	H	R	M	R	R	L	R	L	R	L	L	L	L	L	L	L	M	R	R	R	L	H	H	H	L	R	R	R	R
35	H	H	H	H	L	L	L	L	L	L	L	L	H	L	L	L	L	L	L	L	L	L	L	L		L	L	L	L	L	H	H	H	L	L	H	H	L
36	H	H	H	H	L	L	L	L	L	M	L	L	H	L	L	L	L	L	L	L	L	L	L	L		L	L	L	L	L	H	H	R	L	L	H	H	L
37	H	H	H	H	M	L	L	L	L	L	L	L	H	L	L	L	L	L	L	L	L	L	L	L		L	L	L	L	L	H	H	H	L	L	H	H	M
41	H	H	L	L	L	L	L	L	L	L	L	L	R	L	L	L	L	L	L	M	R	H	L	L	H	R	R	R	L	R	R	R	R	L	R	R	R	R
43	H	H	H	H	H	H	H	H	H	H	H	H	H	H	L	L	L	H	H	L	H	H	H	H	L	H	H	L	L	L	L	H	M	L	L	H	H	H
45	H	H	H	H	M	L	L	L	H	H	L	L	L	L	L	L	L	L	L	L	L	L	L	L		L	L	L	L	H	M	H	L	L	L	H	H	M
46	H	H	H	H	L	L	L	L	H	H	L	L	L	L	L	L	L	L	L	L	L	L	L	L		L	L	L	L	L	L	H	R	L	L	H	L	L
48	H	M	M	L	L	L	L	L	H	H	L	L	L	L	L	L	L	L	L	L	L	L	L	L		L	L	L	L	L	L	L	L	L	L	M	L	L
51	H	H	M	H	L	L	L	L	H	H	L	L	H	L	L	L	L	L	L	L	L	L	L	L		L	L	L	L	L	L	H	L	L	L	H	H	L
53	H	H	M	H	L	L	L	L	H	H	L	L	L	L	L	L	L	L	L	L	L	L	L	L		L	L	L	L	L	L	H	L	L	L	H	H	L

Reactivity Chart 1. Protection for the Hydroxyl Group: Ethers (Continued)

PG	39 Na/NH₃	40 Al(Hg)	41 SnCl₂/Py	42 HSO₃⁻;H₂S	43 LiAlH₄	44 Li-s-Bu₃BH	45 (C₅H₁₁)₂BH	46 B₂H₆,0°	47 NaBH₄	48 Zn(BH₄)₂	49 NaBH₃CN pH4-6	50 i-Bu₂AlH	51 Li(OsBu)₃AlH	52 AlCl₃,80°	53 AlCl₃,25°	54 SnCl₄;BF₃	55 LiClO₄;MgBr₂	56 TsOH,80°	57 TsOH,0°	58 Hg(II)	59 Ag(I)	60 Cu(II)/Py	61 HBr/In.	62 HX/In.	63 NBS/CCl₄	64 Br₃CCl/In.	65 OsO₄	66 KMnO₄ pH7,0°	67 O₃,-50°	68 RCO₃H,0°	69 RCO₃H,50°	70 CrO₃/Py	71 CrO₃,pH1	72 H₂O₂ pH10-12	73 Quinone	74 ¹O₂	75 DMSO,100°	76 NaOCl pH10	77 aq NBS
1	L	L	L	L	L	L	L	L	L	L	L	L	L	H	H	H	L	L	L	L	L	L	L	L	M	L	L	L	L	L	L	L	L	L	L	L	L	L	L
2	L	L	L	L	L	L	L	L	L	L	L	L	L	H	H	H	L	H	L	L	L	L	H	L	R	L	L	L	R	L	M	L	L	L	L	L	L	L	R
3	R	L	M	L	L	L	L	L	L	L	L	L	L	H	H	M	M	M	L	H	H	L	M	L	R	M	L	L	R	R	R	L	H	L	L	R	M	R	R
6	L	L	L	L	L	L	L	L	L	L	L	L	L	H	H	H	H	H	L	L	L	L	H	L	R	M	L	L	R	R	L	L	M	L	L	L	L	L	M
8	R	M	M	L	M	L	L	L	L	L	L	M	L	H	H	H	L	H	M	L	R	L	H	L	R	R	L	L	R	R	M	L	H	L	L	L	M	L	M
9	R	L	L	L	L	L	L	L	L	L	L	L	L	H	H	H	M	H	L	L	L	L	H	L	M	L	L	L	H	H	H	L	H	L	L	L	L	L	L
11	L	L	L	L	L	L	L	L	L	L	M	L	L	H	M	H	L	H	L	H	H	L	M	L	M	L	L	L	R	R	R	L	H	L	L	R	M	R	R
12	R	L	L	L	L	L	L	L	L	L	M	L	L	H	H	M	L	H	L	L	L	L	H	L	M	L	L	L	R	R	R	L	H	L	L	L	M	L	R
13	L	L	L	L	L	L	L	L	L	L	M	L	L	H	H	H	L	H	L	L	L	L	H	L	M	L	L	L	R	R	R	L	H	L	L	R	L	R	R
15	R	L	L	L	L	L	L	L	L	L	H	L	L	H	H	H	M	H	L	L	L	L	H	L	M	L	L	L	R	R	H	L	H	L	L	L	M	L	R
16	R	L	L	L	L	L	L	L	L	L	L	L	L	H	M	M	L	H	L	H	R	L	M	L	M	L	L	L	R	R	R	L	H	L	L	R	M	R	M
17	L	L	L	L	L	L	L	L	L	L	L	L	L	H	H	H	L	H	L	L	L	L	M	L	R	L	L	L	R	L	H	L	H	L	L	L	L	L	M
18	L	L	L	L	L	L	L	L	L	L	M	L	L	H	H	H	M	H	M	L	L	L	H	L	M	L	L	L	L	L	R	L	R	R	L	L	L	L	R
21	R	M	L	L	L	L	L	L	L	L	L	L	L	H	M	M	L	M	L	L	L	L	R	L	R	L	L	L	R	R	R	L	R	L	L	R	M	R	L
22	L	L	L	L	L	L	L	L	L	L	L	L	L	H	H	H	L	H	L	L	L	L	M	L	L	L	R	L	L	L	L	L	L	L	L	L	L	L	L
23	H	L	L	L	L	L	R	R	L	L	L	R	L	H	H	H	L	L	L	R	L	L	R	R	R	R	L	R	R	M	R	R	M	L	L	R	R	L	R
26	H	L	L	L	L	L	L	L	L	L	L	L	M	H	H	H	L	L	L	L	L	L	R	L	R	L	L	L	R	L	L	L	L	L	L	L	L	L	L
28	R	R	L	L	R	L	L	L	L	L	L	L	H	H	L	L	L	L	L	L	L	L	R	L	R	L	L	L	L	L	L	L	L	H	L	L	L	L	L
35	H	M	H	R	L	L	L	L	L	L	M	L	L	H	H	H	L	H	H	L	L	L	H	L	L	L	L	L	R	L	H	L	H	L	L	L	L	L	L
36	H	L	L	H	L	L	L	L	L	L	M	L	L	H	H	H	L	H	H	L	L	L	H	L	L	L	L	L	L	L	H	L	H	L	L	L	L	L	L
37	H	M	L	L	L	L	L	L	L	L	M	L	L	H	H	H	L	H	H	L	L	L	H	L	L	L	L	L	L	L	H	L	H	L	L	L	L	L	L
41	R	R	L	L	R	R	R	L	R	R	L	R	M	H	H	H	M	H	L	L	L	L	H	L	L	L	L	L	L	L	R	L	L	L	L	L	L	L	L
43	H	L	M	L	H	H	L	H	H	L	H	L	H	H	M	M	L	H	L	L	L	L	H	M	R	M	L	L	L	L	R	H	H	L	L	L	L	H	H
45	L	L	L	L	L	L	L	L	L	L	M	L	L	H	L	H	L	H	M	L	L	L	H	L	M	L	L	L	L	L	H	L	H	L	L	L	L	L	L
46	L	L	L	L	L	L	L	L	L	L	L	L	L	H	M	M	L	H	L	L	L	L	H	L	M	L	L	L	L	L	L	L	L	L	L	L	L	L	L
48	L	L	L	L	L	L	L	L	L	L	L	L	L	H	L	L	L	M	L	L	L	L	L	L	M	L	L	L	L	L	L	L	M	L	L	L	L	L	L
51	R	R	L	L	L	L	L	L	L	L	L	L	L	H	L	L	L	H	L	L	L	L	R	L	R	L	L	L	L	L	L	L	H	L	L	L	L	L	L
53	L	L	L	L	L	L	L	L	L	L	L	L	L	H	M	L	L	H	L	L	L	L	H	L	M	L	L	L	L	L	L	L	H	L	L	L	L	L	L

Column category groupings: **G.** (39–42) · **H. HYDRIDE REDN.** (43–51) · **I.** (52–57) · **J.** (58–60) · **K.** (61–64) · **L. OXIDANTS** (65–77)

Reactivity Chart 1. Protection for the Hydroxyl Group: Ethers (Continued)

PG	78 I₂	79 PhSeX;PhSCl	80 Br₂;Cl₂	81 MnO₂/CH₂Cl₂	82 NaIO₄ pH5-8	83 SeO₂ pH2-4	84 SeO₂/Py	85 K₃Fe(CN)₆,pH8	86 Pb(IV),25°	87 Pb(IV),80°	88 Tl(NO₃)₃	89 150°	90 250°	91 350°	92 :CCl₂	93 N₂CHCO₂R/Cu	94 CH₂I₂/Zn(Cu)	95 R₃SnH/In·	96 Ni(CO)₄	97 CH₂N₂	98 SOCl₂	99 Ac₂O,25°	100 Ac₂O,80°	101 DCC	102 MeI	103 Me₃O⁺BF₄⁻	104 1.LDA 2.MeI	105 1.K₂CO₃ 2.MeI	106 RCHO	107 RCOCl	108 C⁺/olefin
1	L	L	M	L	L	L	L	L	L	L	L	L	L	L	L	L	L	L	L	L	L	L	L	L	L	L	L	L	L	L	L
2	L	L	R	L	L	H	L	L	L	H	H	L	L	L	L	L	L	L	L	L	L	L	L	L	L	M	L	L	L	L	H
3	L	L	R	L	R	H	M	L	R	R	R	L	L	R	M	R	L	M	L	L	L	L	L	R	R	R	M	R	L	L	M
6	L	L	M	L	L	L	L	L	L	M	M	L	M	R	L	L	L	L	L	L	L	L	L	L	L	M	L	L	L	L	H
8	L	L	M	L	L	H	L	L	M	H		L	L	R	L	M	L	R	L	L	L	L	M	L	L	M					
9	L	L	M	L	L	H	L	L	M	H	H	L	M	H	L	L	L	L	L	L	L	L	L	L	L	M	L	L	L	L	R
11	L	L	R	L	R	H	M	M	H	H		M	M	H	M	M	L	M	L	L	L	L	L	R	R	R	R	R			
12	L	L	R	L	L	H	L	L	H	H		L	M	H	L	L	L	L	L	L	L	L	M	L	L	M					
13	L	L	R	L	R	H	M	M	R	H		L	M	H	M	M	L	M	L	L	L	L	H	L	R	R	R	R			
15	L	L	R	L	M	H	L	L	H	H		L	M	R	L	L	L	L	L	L	L	L	L	L	L	M					
16	L	L	R	L	R	H	M	M	H	H		R	R	R	M	M	L	M	L	L	L	L	L	R	R	R	R	R	L	L	H
17	L	L	M	L	L	H	L	L	M	H		L	L	R	L	M	L	L	L	L	L	L	L	L	L	M					
18	L	L	M	R	M	H	L	L	M	H		L	M	R	L	L	L	L	L	L	L	L	M	L	L	M					
21	R	R	R	R	R	R	R	L	R	R		R	R	R	M	L	M	M	L	L	L	L	L	R	R	R					
22	L	L	L	L	L	L	L	L	L	L	M	L	R	R	L	M	L	L	L	L	L	L	L	L	L	L					
23	M	R	R	L	L	R	R	L	L	R	R	L	R	R	R	R	R	M	M	L	L	L	L	L	L	M	L	L	L	L	H
26	L	L	R	L	L	L	L	L	L	L	L	L	M	R	L	R	L	L	L	L	L	L	L	L	L	M	L	L	L	L	H
28	L	L	M	L	L	L	L	L	L	L	L	L	M	R	L	L	L	L	L	L	L	L	L	L	L	M	M	L	L	L	L
35	L	L	L	L	L	H	L	L	L	H		L	M	R	L	L	L	L	L	L	L	L	M	L	L	L				L	L
36	L	L	L	L	L	H	L	L	L	H		L	M	R	L	L	L	L	L	L	L	L	L	L	L	L					
37	L	L	L	L	M	H	L	L	M	H		L	M	R	L	L	L	L	L	L	L	L	M	L	L	L	L	L	L	L	L
41	L	L	L	L	L	L	L	L	L	L	H	L	R	R	L	L	L	M	L	L	L	L	M	L	L	L	M		L	L	L
43	L	L	L	L	H	H	L	H	H	H		L	L	R	L	L	L	L	L	L	L	M	H	L	L	H					
45	L	L	L	L	M	H	L	L	H	H		L	L	R	L	L	L	L	L	L	L	L	H	L	L	M	M				
46	L	L	L	L	L	H	L	L	M	L	M	L	L	R	L	L	L	L	L	L	L	L	M	L	L	M					
48	L	L	L	L	L	L	L	L	L	M		L	L	R	L	L	L	L	L	L	L	L	L	L	L	L	L	L	L	L	L
51	L	L	M	L	L	H	L	L	M	H		L	L	R	L	L	L	L	L	L	L	L	M	L	L	L		L	L	L	L
53	L	L	L	L	L	H	L	L	M	H		L	L	R	L	L	L	L	L	L	L	L	M	L	L	M				L	

L. OXIDANTS (78–88) M. (89–91) N. (92–95) O. MISCELLANEOUS (96–105) P. (106–108)

Reactivity Chart 2.   Protection for the Hydroxyl Group: Esters

1. Formate Ester
3. Acetate Ester
6. Trichloroacetate Ester
10. Phenoxyacetate Ester
19. Isobutyrate Ester
22. Pivaloate Ester
23. Adamantoate Ester
27. Benzoate Ester
31. 2,4,6-Trimethylbenzoate (Mesitoate) Ester
34. Methyl Carbonate
36. 2,2,2-Trichloroethyl Carbonate
39. Allyl Carbonate
41. *p*-Nitrophenyl Carbonate
42. Benzyl Carbonate
46. *p*-Nitrobenzyl Carbonate
47. *S*-Benzyl Thiocarbonate
48. N-Phenylcarbamate
51. Nitrate Ester
53. 2,4-Dinitrophenylsulfenate Ester

(See chart, pp. 300–302)

Reactivity Chart 2. Protection for the Hydroxyl Group: Esters

Column legend (reagent/conditions):

A. AQUEOUS — 1: pH<1, 100°; 2: pH<1; 3: pH 1; 4: pH 2-4; 5: pH 4-6; 6: pH 6-8.5; 7: pH 8.5-10; 8: pH 10-12; 9: pH>12; 10: pH>12, 150°

B. BASIC — 11: NaH; 12: Ph_3CNa; 13: $(C_{10}H_8)^{\cdot-}Na^+$; 14: $MeSOCH_2^-Na^+$; 15: KO-t-Bu; 16: LiN-i-Pr2; 17: Py; R3N; 18: $NaNH_2$

C. NUCLEOPHILIC — 19: NaOMe; 20: Enolate; 21: NH_3; RNH2; 22: RS^-; N_3^-; SCN⁻; 23: OAc⁻; X⁻; 24: NaCN, pH 12; 25: HCN, pH 6

D. ORGANOMET. — 26: RLi; 27: RMgX; 28: organozinc; 29: organocopper; 30: Wittig; ylide

E. CAT. REDN. — 31: H_2/Raney (Ni); 32: H_2/Pt pH 2-4; 33: H_2/Pd; 34: H_2/Lindlar; 35: H_2/Rh

F. — 36: Zn/HCl; 37: Zn/HOAc; 38: Cr(II), pH 5

PG	1	2	3	4	5	6	7	8	9	10	11	12	13	14	15	16	17	18	19	20	21	22	23	24	25	26	27	28	29	30	31	32	33	34	35	36	37	38
1	H	H	H	M	L	L	H	H	H	H	H	L	L	H	H	H	L	H	H	H	H	H	L	M	L	H	H	M	M	H	M	M	M	L	L	M	M	L
3	H	M	L	L	L	L	M	H	H	H	R	H	H	H	H	H	L	R	H	R	H	L	H	H		H	H	L	L	L	L	L	L	L	L	L	L	L
6	H	M	L	L	L	L	H	H	H	H	L	H	H	H	H	L	L	L	H	R	M	H	L	R	L	H	H	H	H	L	R	R	R	L	L	R	R	H
10	H	M	L	L	L	L	M	H	H	H	M	M	H	H	H	H	L	R	H	R	H	H	L	L		H	H	L	L	L	L	L	L	L	R	L	L	L
19	H	M	L	L	L	L	L	M	H	H	H	H	H	H	L	M	L	H	M	M	M	L	L	L	M	H	H	L	L	L	L	L	L	L	L	L	L	L
22	H	M	L	L	L	L	L	L	M	H	L	L	H	M	H	L	L	M	M	M	H	L	L	L	L	H	M	L	L	L	L	L	L	L	L	L	L	L
23	H	H	L	L	L	L	L	L	M	H	L	L	H	H	L	L	L	L	L	L	L	L	L	L		H	H	L	L	L	L	L	L	L	L	L	L	L
27	H	M	L	L	L	L	L	M	H	H	L	L	H	H	L	L	L	H	M	M	M	L	L	L	M	L	L	L	L	L	L	L	L	L	L	L	L	L
31	H	M	L	L	L	L	L	L	L	H	L	M	H	L	M	L	L	L	L	L	L	L	L	L		L	H	L	L	L	L	L	L	L	L	L	L	L
34	H	M	L	L	L	L	L	L	H	H	L	H	H	H	H	L	L	H	M	L	M	L	L	L		R	R	L	H	M	R	L	L	L	L	R	H	H
36	H	M	L	L	L	L	M	H	H	H	L	H	H	H	H	H	L	H	R	R	M	H	L	R		R	R	L	H	M	R	L	L	L	R	H	H	H
39	H	M	L	L	L	L	L	H	H	H	R	H	H	H	M	H	L	R	M	M	M	L	L	L		R	R	L	H	L	R	R	R	L	R	L	L	L
41	H	M	L	L	L	L	H	H	H	H	L	L	H	H	R	L	L	H	R	R	M	L	L	M	M	R	R	M	M	M	R	R	R	L	R	R	R	R
42	H	M	L	L	L	L	L	H	H	H	L	H	H	H	R	H	L	H	R	M	M	L	L	L		R	R	L	L	L	H	H	H	L	M	L	L	L
46	H	M	L	L	L	L	L	H	H	H	L	H	H	H	R	H	L	H	R	R	M	L	L	L		R	R	L	L	L	H	H	H	L	R	R	R	R
47	H	L	L	L	L	L	L	H	H	H	L	H	H	H	R	H	L	H	R	R	M	L	L	L		R	R	M	M	H	R	R	R	L	H	L	L	L
48	H	H	L	L	L	L	L	L	M	H	R	H	H	H	R	R	L	H	M	L	M	L	L	L		R	R	L	L	L	L	L	L	L	L	L	L	L
51	H	H	M	L	L	L	L	L	H	H	L	L	H	H	H	M	L	H	H	H	H	M	L	L		H	H	M	M	M	H	H	H	L	H	H	H	H
53	H	M	M	L	L	L	L	H	H	H	L	L	H	H	H	H	L	H	H	H	H	H	H	H		M	M	H	H	H	H	H	H	L	H	H	H	H

300

Reactivity Chart 2. Protection for the Hydroxyl Group: Esters (Continued)

The following chart is oriented with the reagents listed as rows (number and description) and the protective groups (PG) as columns. Reactivity is indicated by L (low), M (marginal/medium), H (high), or R.

Reagent group headings (by reagent number): G. (39–42); H. HYDRIDE REDN. (43–51); I. (52–57); J. (58–60); K. (61–64); L. OXIDANTS (65–77).

Reagent	1	3	6	10	19	22	23	27	31	34	36	39	41	42	46	47	48	51	53
39 Na/NH₃	H	H	H	H	H	H	H	H	H	H	H	H	H	H	H	H	H	H	H
40 Al(Hg)	L	L	H	H	H	L	L	L	L	L	R	L	H	L	R	L	L	H	H
41 SnCl₂/Py	L	L	L	L	L	L	L	L	L	L	L	L	L	L	L	L	L	H	H
42 HSO₃; H₂S	L	L	L	L	L	L	L	L	L	L	L	L	L	L	L	L	L	M	H
43 LiAlH₄	H	H	H	H	H	H	H	H	H	H	H	H	H	H	H	H	H	H	H
44 Li-s-Bu₃BH	H	M	M	M	L	L	L	L	L	L	M	L	M	L	L	H	L	H	H
45 (C₅H₁₁)₂BH	M	L	L	L	L	L	L	L	L	L	L	H	L	L	L	L	M	H	H
46 B₂H₆, 0°	M	L	L	L	L	L	L	L	L	L	L	R	L	L	L	L	H	H	H
47 NaBH₄	M	L	L	L	L	L	L	L	L	L	L	M	L	L	L	M	L	R	H
48 Zn(BH₄)₂	M	L	M	L	L	L	L	L	L	L	L	L	M	L	L	M	L	H	H
49 NaBH₃CN pH 4-6	M	L	L	L	L	L	L	L	L	L	L	L	L	L	L	L	L	H	H
50 i-Bu₂AlH	H	H	H	H	L	L	L	H	L	H	H	H	H	H	H	H	H	H	H
51 Li(OtBu)₃AlH	M	L	L	L	L	L	L	L	L	L	L	L	R	L	L	M	L	H	H
52 AlCl₃, 80°	H	L	R	R	L	L	L	L	L	L	R	R	R	R	R	R	R	R	H
53 AlCl₃, 25°	H	L	R	R	L	L	L	L	L	L	R	R	M	R	M	M	L	R	H
54 SnCl₄; BF₃	L	L	L	L	L	L	L	L	L	L	L	H	L	L	L	L	L	R	R
55 LiClO₄; MgBr₂	L	L	L	L	L	L	L	L	L	L	L	L	L	L	L	L	L	L	L
56 TsOH, 80°	H	M	M	M	M	L	L	M	L	M	M	M	M	M	M	M	M	M	M
57 TsOH, 0°	L	L	L	L	L	L	L	L	L	L	L	L	L	L	L	L	L	L	L
58 Hg(II)	L	L	L	L	L	L	L	L	L	L	L	M	L	L	L	R	L	L	M
59 Ag(I)	L	L	R	L	L	L	L	L	L	L	R	L	L	L	L	M	L	L	M
60 Cu(II)/Py	L	L	L	L	L	L	L	L	L	L	L	L	L	L	L	L	L	L	L
61 HBr/In·	M	L	R	M	L	L	L	L	L	L	M	R	L	L	L	L	L	H	R
62 HX/In·	L	L	L	L	L	L	L	L	L	L	L	R	L	L	L	L	L	H	R
63 NBS/CCl₄	L	L	L	L	L	L	L	L	R	L	L	R	L	L	R	L	L	H	R
64 Br₃CCl/In·	L	L	L	L	L	L	L	L	L	L	L	R	L	L	L	L	L	H	R
65 OsO₄	L	L	L	L	L	L	L	L	L	L	L	H	L	L	L	L	L	L	L
66 KMnO₄, pH 7, 0°	L	L	L	L	L	L	L	L	L	L	L	H	L	L	L	R	L	L	R
67 O₃, -50°	L	L	M	L	L	L	L	L	L	L	L	H	L	L	L	R	L	L	R
68 RCO₃H, 0°	L	L	L	L	L	L	L	L	L	L	L	H	L	L	L	M	L	L	R
69 RCO₃H, 50°	L	L	L	L	L	L	L	L	L	L	L	H	L	L	L	R	L	L	R
70 CrO₃/Py	L	L	L	L	L	L	L	L	L	L	L	L	L	L	L	L	L	L	L
71 CrO₃, pH 1	L	L	L	L	L	L	L	L	L	L	L	L	L	L	L	R	M	R	
72 H₂O₂ pH 10-12	H	L	L	L	L	L	L	L	L	H	H	R	H	L	L	H	L	L	L
73 Quinone	L	L	L	L	L	L	L	L	L	L	L	L	L	L	L	L	L	L	L
74 1O₂	L	L	L	L	L	L	L	L	L	L	L	R	L	L	L	L	L	L	M
75 DMSO, 100°	M	L	H	L	L	L	L	L	L	L	M	L	M	L	L	L	L	L	H
76 NaOCl pH 10	H	L	H	M	L	L	L	L	L	L	M	H	H	L	L	H	L	L	H
77 aq NBS	L	L	L	L	L	L	L	L	L	L	L	R	L	L	L	R	L	L	H

301

Reactivity Chart 2. Protection for the Hydroxyl Group: Esters (Continued)

PG	L. OXIDANTS								M.						N.										O. MISCELLANEOUS				P.		
	78 I₂	79 PhSeX; PhSCl	80 Br₂; Cl₂	81 MnO₂/CH₂Cl₂	82 NaIO₄ pH 5-8	83 SeO₂ pH 2-4	84 SeO₂/Py	85 K₃Fe(CN)₆, pH 8	86 Pb(IV), 25°	87 Pb(IV)', 80°	88 Tl(NO₃)₃	89 150°	90 250°	91 350°	92 :CCl₂	93 N₂CHCO₂R/Cu	94 CH₂I₂/Zn(Cu)	95 R₃SnH/In·	96 Ni(CO)₄	97 CH₂N₂	98 SOCl₂	99 Ac₂O, 25°	100 Ac₂O, 80°	101 DCC	102 MeI	103 Me₃O⁺BF₄⁻	104 1.LDA 2.MeI	105 1.K₂CO₃ 2.MeI	106 RCHO	107 RCOCl	108 C⁺/olefin
1	L	L	L	L	L	M	M	M	L	L	L	L	M	H	L	L	L	L	L	L	M	L	L	L	L	L	R	M	L	L	L
3	L	L	L	L	L	L	L	L	L	L		L	M	H	L	L	L	L	L	L	L	L	L	L	L						
6	L	L	L	L	L	L	L	H	L	L		M	H	H	L	L	R	R	M	L	L	L	L	L	L						
10	L	L	R	L	L	L	M	L	L	L		L	L	H	L	L	L	L	L	M	L	L	L	L	L						
19	L	L	L	L	L	L	L	L	L	L		L	M	H	L	L	L	L	L	L	L	L	L	L	L						
22	L	L	L	L	L	L	L	L	L	L	L	L	M	H	L	L	L	L	L	L	L	L	L	L	L	L	L	L	L	L	L
23	L	R	L	L	L	L	L	L	L	L	L	L	L	H	R	L	L	L	L	L	L	L	L	L	L	L	L	L	L	L	L
27	L	L	L	L	L	L	L	L	L	L	L	L	M	H	L	L	L	L	L	L	L	L	L	L	L	R	L	M	L	L	L
31	L	L	L	L	L	L	L	L	L	L	L	L	M	H	L	L	L	L	L	L	L	L	L	L	L	R	R	H	L	L	L
34	L	L	L	L	L	L	L	L	L	L		M	H	H	L	L	L	L	L	L	L	L	L	L	L	R	M	M	L	L	L
36	L	L	L	L	L	L	L	L	L	L	L	M	H	H	L	L	H	R	M	M	L	L	L	L	L				L	L	L
39	M	R	R	L	L	H	H	L	L	R		M	H	H	R	R	R	H	H	L	L	R	L	L	L				L	L	L
41	L	L	L	L	L	L	M	L	L	L	L	M	H	H	L	L	L	R	L	L	L	L	L	L	L						
42	L	L	L	L	L	M	M	L	L	L		M	H	H	L	L	L	L	L	L	L	L	L	L	L						
46	L	M	L	L	L	H	H	L	L	L		M	H	H	L	L	L	H	L	L	L	R	L	L	L						
47	M	L	H	L	L	L	L	L	L	R		M	H	H	L	L	L	R	L	L	L	L	L	L	L		R				
48	L	L	L	L	L	L	L	L	L	L		M	M	H	L	L	L	L	L	L	L	R	L	L	L						
51	L	L	L	L	L	L	L	L	L	L		H	H	H	L	L	L	H	L	L	H	L	L	L	L		R	L			
53	M	L	R	L	M	M	M	L	R	R		H	H	H	L	L	M	H	M	L	M	L	L	L	H		L	L			

Reactivity Chart 3.   Protection for 1,2- and 1,3-Diols

1.   Methylenedioxy Derivative
2.   Ethylidene Acetal
6.   Acetonide Derivative
11.   Benzylidene Acetal
12.   *p*-Methoxybenzylidene Acetal
18.   Methoxymethylene Acetal
20.   Dimethoxymethylenedioxy Derivative
28.   Cyclic Carbonates
29.   Cyclic Boronates

(See chart, pp. 304–306)

Reactivity Chart 3. Protection for 1,2- and 1,3-Diols

Reagents:

A. AQUEOUS
1. pH<1, 100° · 2. pH<1 · 3. pH 1 · 4. pH 2-4 · 5. pH 4-6 · 6. pH 6-8.5 · 7. pH 8.5-10 · 8. pH 10-12 · 9. pH>12 · 10. pH>12, 150°

B. BASIC
11. NaH · 12. Ph₃CNa · 13. (C₁₀H₈)⁻·Na⁺ · 14. MeSOCH₂⁻Na⁺ · 15. KO-t-Bu · 16. LiN-i-Pr₂ · 17. Py; R₃N · 18. NaNH₂

C. NUCLEOPHILIC
19. NaOMe · 20. Enolate · 21. NH₃; RNH₂ · 22. RS⁻; N₃⁻; SCN⁻ · 23. OAc⁻; X⁻ · 24. NaCN, pH 12 · 25. HCN, pH 6

D. ORGANOMET.
26. RLi · 27. RMgX · 28. Organozinc · 29. Organocopper · 30. Wittig; ylide

E. CAT. REDN.
31. H₂/Raney (Ni) · 32. H₂/Pt pH 2-4 · 33. H₂/Pd · 34. H₂/Lindlar · 35. H₂/Rh

F.
36. Zn/HCl · 37. Zn/HOAc · 38. Cr(II)', pH 5

PG	1	2	3	4	5	6	7	8	9	10	11	12	13	14	15	16	17	18	19	20	21	22	23	24	25	26	27	28	29	30	31	32	33	34	35	36	37	38
1	H	H	L	L	L	L	L	L	L	L	L	L	L	L	L	L	L	L	L	L	L	L	L	L		L	L	L	L	L	L	L	L	L	L	L	L	L
2	H	H	H	M	L	L	L	L	L	L	L	L	L	L	L	L	L	L	L	L	L	L	L	L		L	L	L	L	L	L	M	L	L	L	H	M	L
6	H	H	H	M	L	L	L	L	L	L	L	L	L	L	L	L	L	L	L	L	L	L	L	L		L	L	L	L	L	L	M	L	L	L	H	H	L
11	H	H	H	H	H	L	L	L	L	M	L	L	R	L	L	L	L	L	L	L	L	L	L	L		L	L	L	L	L	L	M	H	L	L	H	H	L
12	H	H	H	H	M	L	L	L	L	M	L	L	R	L	L	L	L	L	L	L	L	L	L	L	L	L	L	L	L	L	H	H	H	L	L	H	H	M
18	H	H	H	H	H	L	L	L	L	M	L	L	L	M	L	L	L	L	L	L	L	L	L	L		M	H	H	L	L	L	H	L	L	L	H	H	H
20	H	H	H	H	H	L	L	L	L	M	L	L	L	M	L	L	L	L	M	L	L	L	L	L		M	H	H	L	L	L	H	L	L	L	H	H	H
28	H	L	L	L	L	L	L	L	H	H	L	L	H	H	H	L	L	H	M	M	M	M	M	L	L	H	H	L	L	L	L	L	L	L	L	L	L	L
29	H	H	H	H	H	H	H	H	H	H	L	L	H	H	H	L	L	H	H	H	L	L	L	H	M	H	H	H	L	H	L	H	L	L	L	H	H	M

304

Reactivity Chart 3. Protection for 1,2- and 1,3-Diols (Continued)

PG	39 Na/NH₃ (G.)	40 Al(Hg)	41 SnCl₂/Py	42 HSO₃⁻; H₂S	43 LiAlH₄ (H. HYDRIDE REDN.)	44 Li-s-Bu₃BH	45 (C₅H₁₁)₂BH	46 B₂H₆, 0°	47 NaBH₄	48 Zn(BH₄)₂	49 NaBH₃CN pH 4-6	50 i-Bu₂AlH	51 Li(OtBu)₃AlH	52 AlCl₃, 80° (I.)	53 AlCl₃, 25°	54 SnCl₄; BF₃	55 LiClO₄; MgBr₂	56 TsOH, 80°	57 TsOH, 0°	58 Hg(II) (J.)	59 Ag(I)	60 Cu(II)/Py	61 HBr/In. (K.)	62 HX/In.	63 NBS/CCl₄	64 Br₃CCl/In.	65 OsO₄ (L. OXIDANTS)	66 KMnO₄, pH 7, 0°	67 O₃, -50°	68 RCO₃H, 0°	69 RCO₃H, 50°	70 CrO₃/Py	71 CrO₃, pH 1	72 H₂O₂ pH 10-12	73 Quinone	74 ¹O₂	75 DMSO, 100°	76 NaOCl pH 10	77 aq NBS
1	L	L	L	L	L	L	L	L	L	L	L	L	L	H	H	H	L	M	L	L	L	L	R	L	L	L	L	L	R	L	L	L	L	L	L	L	L	L	L
2	L	L	L	L	L	L	L	L	L	L	L	L	L	H	H	H	L	M	L	L	L	L	M	L	L	L	L	L	R	L	M	L	H	L	L	L	L	L	L
6	L	L	L	L	L	L	L	L	L	L	L	L	L	H	H	H	L	M	L	L	L	L	M	L	L	L	L	L	L	L	H	L	H	L	L	L	L	L	L
11	H	L	L	L	L	L	L	L	L	L	L	L	L	H	H	H	L	H	L	L	L	L	H	L	R	L	L	L	R	L	H	L	H	L	L	L	L	L	M
12	H	L	L	L	L	L	L	L	L	L	L	L	L	H	H	H	L	M	L	L	L	L	H	L	R	L	L	L	R	M	H	L	H	L	L	L	L	L	M
18	L	L	L	L	R	L	L	L	L	L	M	L	L	H	H	H	M	R	H	L	L	L	H	L	H	L	L	M	L	H	H	L	H	L	L	L	L	L	M
20	L	L	L	L	R	L	L	L	L	L	M	L	L	H	H	H	M	R	H	L	L	L	H	L	L	L	L	M	L	H	H	L	H	L	L	L	L	L	M
28	H	L	L	L	H	L	L	L	L	L	L	H	L	H	H	L	L	M	L	L	L	L	L	L	L	L	L	L	L	L	L	L	L	H	L	L	L	M	L
29	H	M	L	L	H	H	H	H	H	H	H	H	H	H	H	L	L	L	L	H	H	L	L	L	L	L	L	H	L	H	H	L	H	H	L	L	L	H	H

305

Reactivity Chart 3. Protection for 1,2- and 1,3-Diols (Continued)

Reagent	PG 1	PG 2	PG 6	PG 11	PG 12	PG 18	PG 20	PG 28	PG 29
L. OXIDANTS									
78 I_2	L	L	L	L	L	L	L	L	L
79 PhSeX; PhSCl	L	L	L	L	L	L	L	L	L
80 Br_2; Cl_2	L	L	L	R	R	L	L	L	L
81 MnO_2/CH_2Cl_2	L	L	L	L	L	L	L	L	L
82 $NaIO_4$ pH 5–8	L	L	L	L	L	M	L	H	H
83 SeO_2 pH 2–4	L	M	M	H	H	H	R	H	H
84 SeO_2/Py	L	L	L	L	L	L	L	L	L
85 $K_3Fe(CN)_6$, pH 8	L	L	L	L	L	L	L	L	H
86 Pb(IV), 25°	L	L	L	L	L	L	L	L	H
87 Pb(IV), 80°	L	M	M	M	M	H	H	L	H
88 $Tl(NO_3)_3$	L	H						L	H
M.									
89 150°	L	L	L	L	L	L	L	L	L
90 250°	L	L	L	M	M	M	H	M	H
91 350°	M	M	L	H	H	H	H	H	H
N.									
92 :CCl_2	L	L	L	L	L	L	L	L	L
93 N_2CHCO_2R/Cu	L	L	L	L	M	L	L	L	L
94 CH_2I_2/Zn(Cu)	L	L	L	L	L	L	L	L	L
95 R_3SnH/In·	L	L	L	L	L	L	L	L	M
96 $Ni(CO)_4$	L	L	L	L	L	L	L	L	L
97 CH_2N_2	L	L	L	L	L	L	L	L	L
98 $SOCl_2$	L	L	L	L	L	L	L	L	M
99 Ac_2O, 25°	L	L	L	L	L	M	M	L	M
100 Ac_2O, 80°	M	M	M	M	M	R	R	L	H
O. MISCELLANEOUS									
101 DCC	L	L	L	L	L	L	L	L	L
102 MeI	L	L	L	L	L	L	L	L	L
103 $Me_3O^+BF_4^-$	M	M	M	M	R	R	R	L	R
104 1.LDA 2.MeI	L	L						L	L
105 1.K_2CO_3 2.MeI	L	L						H	H
P.									
106 RCHO	L	L						L	L
107 RCOCl	L	L	L					L	M
108 C^+/olefin	H	H						L	H

Reactivity Chart 4.   Protection for Phenols and Catechols

PHENOLS

1. Methyl Ether
2. Methoxymethyl Ether
3. 2-Methoxyethoxymethyl Ether
4. Methylthiomethyl Ether
6. Phenacyl Ether
7. Allyl Ether
8. Cyclohexyl Ether
9. *t*-Butyl Ether
10. Benzyl Ether
11. *o*-Nitrobenzyl Ether
12. 9-Anthrylmethyl Ether
13. 4-Picolyl Ether
15. *t*-Butyldimethylsilyl Ether
16. Aryl Acetate
17. Aryl Pivaloate
18. Aryl Benzoate
19. Aryl 9-Fluorenecarboxylate
20. Aryl Methyl Carbonate
21. Aryl 2,2,2-Trichloroethyl Carbonate
22. Aryl Vinyl Carbonate
23. Aryl Benzyl Carbonate
25. Aryl Methanesulfonate

CATECHOLS

27. Methylenedioxy Derivative
28. Acetonide Derivative
30. Diphenylmethylenedioxy Derivative
31. Cyclic Borates
32. Cyclic Carbonates

(See chart, pp. 308–310)

Reactivity Chart 4. Protection for Phenols and Catechols

Column key:

A. AQUEOUS — 1: pH<1, 100°; 2: pH<1; 3: pH 1; 4: pH 2-4; 5: pH 4-6; 6: pH 6-8.5; 7: pH 8.5-10; 8: pH 10-12; 9: pH>12; 10: pH>12, 150°

B. BASIC — 11: NaH; 12: Ph₃CNa; 13: (C₁₀H₈)⁻˙Na⁺; 14: MeSOCH₂⁻Na⁺; 15: KO-t-Bu; 16: LiN-i-Pr₂; 17: Py; R₃N; 18: NaNH₂

C. NUCLEOPHILIC — 19: NaOMe; 20: Enolate; 21: NH₃; RNH₂; 22: RS⁻; N₃⁻; SCN⁻; 23: OAc⁻; X⁻; 24: NaCN, pH 12; 25: HCN, pH 6

D. ORGANOMET. — 26: RLi; 27: RMgX; 28: Organozinc; 29: Organocopper; 30: Wittig; ylide

E. CAT. REDN. — 31: H₂/Raney (Ni); 32: H₂/Pt pH 2-4; 33: H₂/Pd; 34: H₂/Lindlar; 35: H₂/Rh

F. — 36: Zn/HCl; 37: Zn/HOAc; 38: Cr(II), pH 5

PG	1	2	3	4	5	6	7	8	9	10	11	12	13	14	15	16	17	18	19	20	21	22	23	24	25	26	27	28	29	30	31	32	33	34	35	36	37	38
1	H	M	L	L	L	L	L	L	L	L	L	L	L	L	L	L	L	L	L	L	L	H	L	H	L	L	L	L	L	L	L	L	L	L	L	L	L	L
2	H	H	H	M	L	L	L	L	L	L	L	L	L	L	L	L	L	L	L	L	L	M	L	L	L	L	L	L	L	L	L	M	L	L	L	H	M	L
3	H	H	L	L	L	L	L	L	L	L	L	L	L	L	L	L	L	L	L	L	L	M	L	L	L	L	L	L	L	L	L	M	L	L	L	L	L	L
4	H	M	M	M	L	L	L	L	L	L	L	L	R	R	L	L	L	L	L	L	L	M	L	L	L	M	L	L	L	L	R	R	R	L	R	H	H	M
6	H	L	L	L	L	L	L	L	L	H	R	R	R	R	M	R	L	R	R	R	L	M	L	M	L	R	R	M	M	R	M	R	R	L	R	H	H	H
7	H	L	L	L	L	L	L	M	M	H	L	L	R	L	R	L	L	L	R	L	L	M	L	M	L	R	M	L	M	L	R	R	R	L	R	M	L	L
8	H	H	L	L	L	L	L	L	L	L	L	L	L	L	L	L	L	L	L	L	L	L	L	L	L	L	L	L	L	L	L	L	L	L	L	M	L	L
9	H	H	H	L	L	L	L	L	L	L	L	L	L	L	L	L	L	L	L	L	L	L	L	L	L	L	L	L	L	L	L	L	L	L	L	H	M	L
10	H	H	L	L	L	L	L	L	L	L	L	L	R	L	L	L	L	L	L	L	L	M	L	L	L	M	L	L	L	L	H	H	H	L	H	L	L	L
11	H	L	L	L	L	L	L	L	L	R	L	R	R	R	L	L	L	R	L	L	L	M	L	L	L	R	R	M	R	L	H	H	H	L	H	R	R	R
12	H	M	L	L	L	L	L	L	L	L	L	L	R	L	L	L	L	L	L	L	L	H	L	M	L	R	M	L	L	L	H	H	L	L	L	L	L	L
13	H	L	L	L	L	L	L	L	L	L	L	R	R	L	L	R	L	L	L	L	L	M	L	L	L	R	M	L	L	L	H	H	L	L	L	M	L	L
15	H	H	H	H	M	L	L	L	M	H	R	L	L	L	L	L	L	L	L	L	L	L	L	L	M	L	L	L	L	L	L	H	L	L	L	H	H	M
16	H	H	L	M	L	L	H	H	H	H	L	R	R	H	R	R	L	H	H	L	H	M	L	H	L	R	R	L	L	L	L	L	L	L	L	L	L	L
17	H	M	L	L	L	L	L	L	M	H	L	L	R	M	L	L	L	M	M	L	L	M	L	M	L	L	L	L	L	L	L	M	L	L	L	L	L	L
18	H	H	H	L	L	L	L	M	H	H	R	R	R	H	L	L	H	H	M	M	L	M	L	H	L	R	R	L	L	L	L	L	L	L	L	L	L	L
19	H	H	L	L	L	L	L	M	H	H	L	L	R	R	R	H	L	M	M	M	M	M	L	H	L	R	R	L	L	L	L	L	L	L	L	L	L	L
20	H	M	H	H	L	L	L	M	H	H	R	R	R	R	M	H	R	M	R	R	M	H	L	H	L	H	H	L	L	L	L	H	L	L	L	H	H	M
21	H	M	L	L	L	L	M	H	H	H	L	L	R	R	R	H	L	R	R	R	M	M	L	H	L	H	H	L	L	L	R	L	L	L	R	H	H	H
22	H	M	L	L	L	L	L	H	H	H	L	L	R	H	L	L	L	H	M	R	M	M	L	H	L	H	H	M	L	M	R	R	R	L	R	H	L	L
23	L	H	M	L	L	L	L	M	H	H	R	R	R	H	M	L	H	H	M	M	M	M	L	H	L	H	H	L	L	L	H	H	H	L	L	M	L	L
25	H	L	L	L	L	L	L	L	H	H	L	L	R	H	L	R	R	M	M	L	M	L	L	M	L	H	M	L	R	M	R	L	L	L	L	L	L	L
27	H	L	L	L	L	L	L	L	L	L	L	L	L	L	L	L	L	L	L	L	L	L	L	L	L	L	M	L	L	L	L	L	L	L	L	L	L	L
28	H	M	H	M	L	L	L	L	L	L	L	L	L	L	L	L	L	L	L	L	L	L	L	L	L	L	L	L	L	L	L	M	L	L	L	H	M	L
30	H	H	H	H	L	L	L	L	L	L	L	L	R	L	L	L	L	L	L	L	L	L	L	L	L	L	L	L	L	L	H	H	H	L	L	H	M	M
31	H	H	H	L	L	L	L	L	H	H	R	R	H	H	R	R	H	H	H	H	H	H	L	H	M	H	H	H	L	H	L	H	L	L	L	H	H	M
32	H	H	H	L	L	L	M	H	H	H	L	L	R	H	L	L	H	M	M	M	M	H	L	H	L	R	R	L	L	L	L	L	L	L	L	M	L	L

308

Reactivity Chart 4. Protection for Phenols and Catechols (Continued)

Reagent groups: G. (39–42); H. HYDRIDE REDN. (43–51); I. (52–57); J. (58–60); K. (61–64); L. OXIDANTS (65–77)

PG	39 Na/NH₃	40 Al(Hg)	41 SnCl₂/Py	42 HSO₃;H₂S	43 LiAlH₄	44 Li-s-Bu₃BH	45 (C₅H₁₁)₂BH	46 B₂H₆,0°	47 NaBH₄	48 Zn(BH₄)₂	49 NaBH₃CN pH 4-6	50 i-Bu₂AlH	51 Li(OEt)₃AlH	52 AlCl₃,80°	53 AlCl₃,25°	54 SnCl₄;BF₃	55 LiClO₄;MgBr₂	56 TsOH,80°	57 TsOH,0°	58 Hg(II)	59 Ag(I)	60 Cu(II)/Py	61 HBr/In·	62 HX/In·	63 NBS/CCl₄	64 Br₃CCl/In·	65 OsO₄	66 KMnO₄ pH 7,0°	67 O₃,-50°	68 RCO₃H,0°	69 RCO₃H,50°	70 CrO₃/Py	71 CrO₃,pH 1	72 H₂O₂ pH 10-12	73 Quinone	74 ¹O₂	75 DMSO,100°	76 NaOCl,pH 10	77 aq NBS
1	R	L	L	L	L	L	L	L	L	L	L	L	L	H	M	L	L	L	L	L	L	L	L	L	R	L	L	L	L	L	L	L	L	L	L	L	L	L	L
2	R	L	L	L	L	L	L	L	L	L	L	L	L	H	H	H	M	H	L	L	L	L	H	L	R	L	L	L	L	L	M	L	R	L	L	L	L	L	L
3	R	L	L	L	L	L	L	L	L	L	L	L	L	H	M	L	L	H	L	L	L	L	M	L	R	L	L	L	L	L	L	L	M	L	L	L	L	L	L
4	R	L	L	L	L	L	L	L	L	L	L	M	L	H	H	M	L	M	L	H	M	L	L	L	R	L	L	R	R	R	R	L	L	L	L	R	L	R	R
6	R	R	R	L	R	R	R	R	R	R	R	R	R	H	H	L	L	M	L	L	L	L	R	L	L	L	L	L	L	L	R	L	L	L	L	L	L	L	L
7	R	L	L	L	L	L	R	R	L	L	L	L	L	H	H	L	L	L	L	R	L	L	R	R	R	R	R	R	R	R	R	L	L	L	L	R	R	L	R
8	R	L	L	L	L	L	L	L	L	L	L	L	L	M	L	L	L	L	L	L	L	L	M	L	L	L	L	L	L	R	L	L	L	L	L	L	L	L	L
9	R	L	L	L	L	L	L	L	L	L	L	L	L	H	H	H	L	H	L	L	L	L	M	L	L	L	L	L	L	L	L	L	R	L	L	L	L	L	L
10	R	L	L	L	L	L	L	L	L	L	L	L	L	H	M	L	L	L	L	L	L	L	R	R	R	R	L	L	L	L	M	L	L	L	L	L	L	L	L
11	R	R	R	L	R	R	L	L	L	L	L	R	L	H	M	L	L	L	L	L	L	L	M	M	M	M	L	L	L	L	M	L	L	L	L	L	L	L	L
12	R	L	L	L	L	L	L	L	L	L	L	L	L	H	M	L	L	M	L	L	L	L	R	R	R	R	L	L	L	L	M	L	L	L	L	L	L	L	L
13	R	L	L	L	L	L	L	L	L	L	L	L	L	H	H	L	L	M	L	L	L	L	R	R	R	R	L	R	R	R	M	L	M	L	L	L	L	L	L
15	R	L	L	L	L	L	L	L	L	L	M	M	L	H	M	L	L	H	L	L	L	L	H	L	L	L	L	L	L	L	L	L	R	L	L	L	L	L	L
16	R	L	L	L	H	H	L	L	M	L	L	H	M	R	R	M	L	H	L	L	L	L	L	L	L	L	L	L	L	L	L	L	L	H	L	L	L	H	L
17	R	L	L	L	H	H	L	L	L	L	L	M	L	R	R	L	L	M	L	L	L	L	L	L	L	L	L	L	L	L	L	L	L	L	L	L	L	L	L
18	R	L	L	L	H	M	L	L	L	L	L	H	M	R	R	M	L	H	L	L	L	L	L	L	L	L	L	L	L	L	L	L	L	H	L	L	L	M	L
19	R	L	L	L	H	H	L	L	L	L	M	H	M	H	H	L	L	H	L	L	L	L	R	L	R	L	L	L	L	L	L	L	L	M	L	L	L	M	L
20	R	M	L	L	H	L	L	L	L	L	L	H	M	H	H	L	L	H	M	L	L	L	H	L	L	L	L	L	L	L	L	L	H	M	L	L	L	M	L
21	R	L	M	L	H	M	L	L	M	L	L	H	M	H	H	L	L	H	L	L	R	L	M	L	L	L	L	L	L	L	L	L	L	R	L	L	L	R	L
22	R	L	L	L	H	L	R	R	L	L	L	H	M	H	H	L	L	H	L	H	L	L	R	R	R	R	R	R	R	R	R	L	R	R	L	R	L	R	R
23	R	L	L	L	H	L	L	L	L	L	L	H	M	H	H	L	L	H	L	L	L	L	M	L	R	L	L	L	L	L	M	L	M	M	L	L	L	L	L
25	R	L	L	L	L	L	L	L	L	L	L	L	L	H	L	L	L	L	L	L	L	L	L	L	L	L	L	L	L	L	L	L	L	L	L	L	L	L	L
27	R	L	L	L	L	L	L	L	L	L	L	L	L	H	H	L	L	M	L	L	L	L	M	L	R	L	L	L	L	L	L	L	M	L	L	L	L	L	L
28	R	L	L	L	L	L	L	L	L	L	M	L	L	H	H	M	M	M	L	L	L	L	H	L	L	L	L	L	L	L	L	L	R	L	L	L	L	L	L
30	R	M	M	L	L	L	L	L	L	L	L	L	L	H	H	M	M	M	L	L	L	L	H	L	L	L	L	L	L	L	L	L	R	L	L	L	L	L	L
31	R	L	M	L	H	H	H	H	H	H	H	H	H	H	H	H	L	H	L	L	L	L	H	L	L	L	L	L	L	L	R	L	R	L	L	L	L	L	L
32	R	L	L	L	H	L	L	L	L	L	L	H	M	H	H	L	L	H	L	L	L	L	H	L	L	L	L	L	L	L	L	L	R	R	L	L	L	M	L

309

Reactivity Chart 4. Protection for Phenols and Catechols (Continued)

Reagent legend (column numbers):

L. OXIDANTS: 78 I₂ · 79 PhSeX; PhSCl · 80 Br₂; Cl₂ · 81 MnO₂/CH₂Cl₂ · 82 NaIO₄ pH 5–8 · 83 SeO₂ pH 2–4 · 84 SeO₂/Py · 85 K₃Fe(CN)₆, pH 8 · 86 Pb(IV)', 25° · 87 Pb(IV)', 80° · 88 Tl(NO₃)₃

M.: 89 150° · 90 250° · 91 350°

N.: 92 :CCl₂ · 93 N₂CHCO₂R/Cu · 94 CH₂I₂/Zn(Cu) · 95 R₃SnH/In· · 96 Ni(CO)₄ · 97 CH₂N₂

O. MISCELLANEOUS: 98 SOCl₂ · 99 Ac₂O, 25° · 100 Ac₂O, 80° · 101 DCC · 102 MeI · 103 Me₃O⁺BF₄⁻ · 104 1.LDA 2.MeI · 105 1.K₂CO₃ 2.MeI

P.: 106 RCHO · 107 RCOCl · 108 C⁺/olefin

PG	78	79	80	81	82	83	84	85	86	87	88	89	90	91	92	93	94	95	96	97	98	99	100	101	102	103	104	105	106	107	108
1	L	L	L	L	L	L	L	L	L	L	L	L	L	L	L	L	L	L	L	L	L	L	L	L	L	L	L	L	L	L	L
2	L	L	L	L	L	M	L	L	M	R	R	L	M	H	L	L	L	L	L	L	L	L	L	L	L	L	L	L	L	L	H
3	L	L	L	L	L	L	L	L	L	L	L	L	M	H	L	L	L	L	L	L	L	L	L	L	L	L	L	L	L	L	L
4	L	L	R	L	R	M	L	L	R	R	R	L	M	H	M	L	L	L	L	L	L	L	L	L	R	R	L	L	L	L	M
6	L	M	M	L	L	R	L	L	M	R	R	L	L	M	R	R	L	M	L	L	L	L	L	L	L	M	R	M	L	L	L
7	L	R	R	L	L	H	M	L	L	R	R	M	R	R	R	R	R	M	M	L	L	L	L	L	L	M	L	L	L	L	R
8	L	L	L	L	L	L	L	L	L	L	L	L	L	L	L	L	L	L	L	L	L	L	L	L	L	L	L	L	L	L	L
9	L	L	L	L	L	L	L	L	L	M	L	L	M	H	L	L	L	L	L	L	L	L	L	L	L	L	L	L	L	L	H
10	L	L	L	L	L	L	L	L	L	L	L	L	L	L	L	L	L	L	L	L	L	L	L	L	L	L	L	L	L	L	L
11	L	L	L	L	L	L	L	L	L	L	L	L	L	L	L	L	L	L	L	L	L	L	L	L	L	L	L	L	L	L	L
12	L	L	M	L	L	L	L	L	L	L	L	L	L	L	L	L	L	L	L	L	L	L	L	L	L	L	L	L	L	L	L
13	L	L	L	L	L	L	L	L	L	L	L	L	L	L	L	L	L	L	L	L	L	L	L	L	R	R	L	L	L	L	L
15	L	L	L	L	L	M	L	L	M	R	R	L	L	L	L	L	L	L	L	L	L	L	M	L	L	L	L	L	L	L	L
16	L	L	L	L	L	M	L	R	L	L	M	L	L	L	L	L	L	L	L	L	L	L	L	L	L	L	R	H	L	L	L
17	L	L	L	L	L	L	L	L	L	L	L	L	L	H	L	L	L	L	L	L	L	L	L	L	L	L	L	L	L	L	L
18	L	L	L	L	L	L	L	L	L	L	L	L	L	M	L	L	L	L	L	L	L	L	L	L	L	L	L	M	L	L	L
19	L	L	L	L	L	L	L	L	L	L	L	L	L	M	L	L	L	L	L	L	L	L	L	L	L	L	R	M	L	L	L
20	L	L	L	L	L	H	L	L	L	M	R	L	M	H	L	L	L	L	L	L	L	L	L	L	L	R	L	M	L	L	M
21	L	L	L	L	L	L	L	M	L	L	L	L	M	H	L	L	R	R	M	L	L	L	L	L	L	M	R	H	L	L	L
22	M	R	R	L	L	L	L	L	M	R	R	L	M	H	R	R	R	R	R	L	L	L	L	L	L	R	L	H	L	L	R
23	L	L	L	L	L	L	L	L	L	L	R	L	M	H	L	L	L	L	L	L	L	L	L	L	L	R	L	M	L	L	L
25	L	L	L	L	L	L	L	L	L	L	L	L	L	M	L	L	L	L	L	L	L	L	L	L	L	L	R	L	L	L	L
27	L	L	L	L	L	L	L	L	L	L	L	L	L	L	L	L	L	L	L	L	M	L	L	L	L	L	L	L	L	L	L
28	L	L	L	L	L	M	L	L	L	M	M	L	M	H	L	L	L	L	L	L	L	L	L	L	L	L	L	L	L	L	L
30	L	L	L	L	L	R	L	L	L	R	R	L	M	H	L	L	L	L	L	L	L	L	L	L	L	L	L	L	L	L	L
31	L	L	L	L	L	L	L	L	L	R	R	L	M	H	L	L	L	L	L	L	R	R	R	L	R	R	R	R	R	R	H
32	L	L	L	L	L	L	L	M	L	L	M	L	L	M	L	L	L	L	L	L	L	L	L	L	L	R	L	L	L	L	L

310

Reactivity Chart 5. Protection for the Carbonyl Group

1. Dimethyl Acetals/Ketals
3. Bis(2,2,2-trichloroethyl) Acetals/Ketals
5. 1,3-Dioxanes
6. 5-Methylene-1,3-dioxanes
7. 5,5-Dibromo-1,3-dioxanes
8. 1,3-Dioxolanes
9. 4-Bromomethyl-1,3-dioxolanes
10. 4-*o*-Nitrophenyl-1,3-dioxolanes
11. *S,S'*-Dimethyl Acetals/Ketals
19. 1,3-Dithianes
20. 1,3-Dithiolanes
24. 1,3-Oxathiolanes
26. *O*-Trimethylsilyl Cyanohydrins
29. *N,N*-Dimethylhydrazones
30. 2,4-Dinitrophenylhydrazones
33. *O*-Phenylthiomethyl Oximes
34. Substituted Methylene Derivatives
43. Bismethylenedioxy Derivatives

(See chart, pp. 312–314)

Reactivity Chart 5. Protection for the Carbonyl Group

PG	1	3	5	6	7	8	9	10	11	19	20	24	26	29	30	33	34	43
A. AQUEOUS																		
1 pH<1, 100°	H	H	H	H	H	H	H	H	R	R	R	H	H	H	H	H	R	*H*
2 pH<1	H	H	H	H	H	H	H	H	*R*	M	*R*	H	H	L	L	H	M	H
3 pH 1	H	H	H	H	H	H	H	H	L	*L*	L	L	*H*	L	L	M	*L*	M
4 pH 2-4	L	L	L	L	L	M	M	M	L	*L*	L	*H*	*H*	L	L	L	L	M
5 pH 4-6	L	L	L	L	L	L	L	L	L	L	L	L	H	L	L	L	L	L
6 pH 6-8.5	L	L	L	L	L	L	L	L	L	L	L	L	H	L	L	L	L	L
7 pH 8.5-10	L	L	L	L	L	L	L	L	L	L	L	L	H	L	L	L	L	L
8 pH 10-12	L	L	L	L	M	L	L	L	L	L	L	L	H	L	L	L	L	L
9 pH>12	L	M	L	L	R	L	R	L	L	L	L	L	H	L	L	L	H	L
10 pH>12, 150°	L	R	L	R	R	L	R	*H*	L	L	H	R	H	H	*H*	R	H	L
B. BASIC																		
11 NaH	L	R	L	L	R	L	R	L	L	L	L	L	L	L	R	L	*H*	L
12 Ph₃CNa	*L*	R	L	L	R	L	R	M	L	L	L	L	L	L	R	R	*H*	L
13 (C₁₀H₈)⁻·Na⁺	L	H	L	H	H	L	H	H	L	L	L	L	R	M	R	H	R	L
14 MeSOCH₂⁻Na⁺	L	R	L	R	R	L	R	H	L	L	L	L	R	L	R	L	R	L
15 KO-t-Bu	L	R	L	R	R	L	R	L	L	L	L	L	M	L	R	L	R	L
16 LiN-i-Pr₂	L	L	*R*	L	*R*	L	R	R	L	L	L	L	L	*R*	*R*	M	R	L
17 Py; R₃N	L	L	L	L	L	L	L	L	L	L	L	L	L	L	L	L	L	L
18 NaNH₂	L	R	L	L	R	L	R	*M*	L	L	L	L	R	L	R	L	R	L
C. NUCLEOPHILIC																		
19 NaOMe	L	R	L	L	R	L	R	L	L	L	L	L	H	L	M	L	H	L
20 Enolate	L	R	L	L	R	L	R	M	L	L	L	L	L	L	M	L	R	L
21 NH₃; RNH₂	L	M	L	L	M	L	L	L	L	L	L	L	L	*R*	L	L	M	L
22 RS⁻; N₃⁻; SCN⁻	L	R	L	L	R	L	R	L	L	L	L	L	H	L	L	L	L	L
23 OAc⁻; X⁻	L	R	L	L	R	L	R	L	L	L	L	L	L	L	L	L	L	L
24 NaCN, pH 12	L	R	L	L	R	L	R	L	L	L	L	L	L	L	L	L	L	L
25 HCN, pH 6	L	R	L	L	R	L	R	L	L	L	L	L	L	L	L	L	L	L
D. ORGANOMET.																		
26 RLi	R	L	L	L	R	L	L	R	L	L	L	L	M	L	R	R	R	L
27 RMgX	L	L	R	L	R	L	R	R	L	L	L	L	M	L	R	L	R	L
28 Organozinc	L	L	R	L	R	L	R	M	L	L	L	L	H	L	R	L	R	L
29 Organocopper	L	L	R	L	R	L	M	R	L	L	L	L	L	L	L	L	R	L
30 Wittig; ylide	L	R	L	L	R	L	L	L	L	L	L	L	R	L	R	L	R	L
E. CAT. REDN.																		
31 H₂/Raney (Ni)	L	L	R	L	R	L	R	R	R	*R*	R	R	R	R	R	R	R	L
32 H₂/Pt pH 2-4	L	L	M	R	R	*R*	R	R	R	R	R	R	R	R	R	R	R	L
33 H₂/Pd	L	L	R	R	R	L	R	R	R	R	L	R	R	R	R	R	*L*	L
34 H₂/Lindlar	L	L	L	L	L	L	L	M	*H*	*H*	H	H	L	L	L	L	L	L
35 H₂/Rh	L	L	L	L	*R*	M	L	L	*R*	*R*	R	R	R	R	R	R	R	L
F.																		
36 Zn/HCl	H	H	H	H	H	H	H	H	L	L	L	H	H	H	R	R	M	H
37 Zn/HOAc	H	R	H	H	R	*L*	H	R	L	L	L	M	R	H	*R*	R	M	H
38 Cr(II), pH 5	L	M	M	M	M	L	M	R	L	L	L	L	R	M	*R*	M	M	L

Reactivity Chart 5. Protection for the Carbonyl Group (Continued)

Reagent	1	3	5	6	7	8	9	10	11	19	20	24	26	29	30	33	34	43
L. OXIDANTS																		
77 aq NBS	L	L	L	R	L	L	L	L	R	R	R	R	R	H	R	R	M	L
76 NaOCl pH 10	L	L	L	L	L	L	L	L	R	R	R	R	R	H	H	M	M	L
75 DMSO, 100°	L	H	L	L	H	L	R	L	M	M	M	M	R	L	L	M	M	L
74 ¹O₂	L	L	L	M	L	L	L	L	R	R	R	R	L	H	M	R	M	L
73 Quinone	L	L	L	L	L	L	L	L	L	L	L	L	L	L	L	L	L	L
72 H₂O₂ pH 10-12	L	L	L	L	L	L	L	L	L	L	L	L	R	H	L	L	L	L
71 CrO₃, pH 1	H	H	M	H	H	H	L	H	M	M	M	H	R	H	L	R	M	L
70 CrO₃/Py	L	L	L	L	L	L	L	L	L	L	L	L	L	L	L	L	L	L
69 RCO₃H, 50°	L	L	L	R	L	L	L	L	R	R	R	R	R	H	H	R	R	L
68 RCO₃H, 0°	L	L	L	R	L	L	L	L	R	R	R	R	R	H	H	R	R	L
67 O₃, -50°	L	L	L	R	L	L	L	L	R	R	R	R	H	H	H	H	H	L
66 KMnO₄, pH 7,0°	L	L	L	R	L	L	L	L	R	R	R	R	R	H	L	R	R	L
65 OsO₄	L	L	L	R	L	L	L	L	L	L	L	L	L	L	L	L	R	L
K.																		
64 Br₃CCl/In.	L	R	L	R	H	L	L	L	L	L	L	L	L	M	L	L	R	L
63 NBS/CCl₄	L	R	L	H	H	L	L	L	L	L	L	L	H	L	L	R	R	L
62 HX/In.	L	R	L	R	H	L	H	H	L	L	L	L	L	M	L	L	R	L
61 HBr/In.	H	R	H	H	H	H	H	H	H	H	H	H	R	R	R	R	R	M
J.																		
60 Cu(II)/Py	L	L	L	L	L	L	L	L	H	H	H	H	L	R	H	M	L	L
59 Ag(I)	L	R	L	L	R	L	R	L	H	H	H	H	L	L	L	R	L	L
58 Hg(II)	L	L	L	R	L	L	L	L	H	H	H	H	L	L	L	R	L	L
I.																		
57 TsOH, 0°	M	M	L	M	L	L	L	L	L	L	L	L	H	L	L	L	L	L
56 TsOH, 80°	H	H	L	H	L	L	L	L	L	L	L	H	H	L	L	L	L	L
55 LiClO₄; MgBr₂	L	L	L	L	L	L	L	L	L	L	L	L	M	L	L	L	L	M
54 SnCl₄; BF₃	H	H	H	H	H	H	H	H	L	L	L	M	L	L	L	L	L	M
53 AlCl₃, 25°	R	H	H	H	H	H	H	H	L	L	L	H	L	L	L	M	L	R
52 AlCl₃, 80°	H	H	H	H	H	H	H	H	H	H	H	H	H	H	L	H	M	R
H. HYDRIDE REDN.																		
51 Li(OtBu)₃AlH	L	L	L	L	M	L	L	L	L	L	L	L	R	L	M	M	M	L
50 i-Bu₂AlH	L	L	L	L	M	L	L	L	L	L	L	L	R	R	R	R	R	L
49 NaBH₃CN pH 4-6	L	L	L	L	L	L	L	L	L	L	L	L	L	R	R	R	R	L
48 Zn(BH₄)₂	L	L	L	L	L	L	L	L	L	L	L	L	M	L	M	L	L	L
47 NaBH₄	L	L	L	L	L	L	L	L	L	L	L	L	M	L	L	L	M	L
46 B₂H₆, 0°	M	M	M	R	L	M	M	M	L	L	L	L	R	L	L	R	R	L
45 (C₅H₁₁)₂BH	L	L	L	R	L	L	L	L	L	L	L	L	R	L	L	L	R	L
44 Li-s-Bu₃BH	L	L	L	L	L	L	L	R	L	L	L	L	R	L	R	L	R	L
43 LiAlH₄	L	R	L	L	R	L	M	R	L	L	L	L	R	R	R	L	R	L
G.																		
42 HSO₃⁻; H₂S	L	L	L	L	L	L	L	L	L	L	L	L	L	L	L	L	L	L
41 SnCl₂/Py	L	L	L	L	M	L	L	L	L	L	L	L	M	R	M	M	M	L
40 Al (Hg)	L	M	L	L	H	L	H	R	L	L	L	L	R	L	R	R	R	L
39 Na/NH₃	L	R	L	R	R	L	L	R	R	R	R	R	R	L	R	R	R	L

313

Reactivity Chart 5. Protection for the Carbonyl Group (Continued)

PG	78 I₂	79 PhSex; PhSCl	80 Br₂; Cl₂	81 MnO₂/CH₂Cl₂	82 NaIO₄ pH 5-8	83 SeO₂ pH 2-4	84 SeO₂/Py	85 K₃Fe(CN)₆, pH 8	86 Pb(IV), 25°	87 Pb(IV), 80°	88 Tl(NO₃)₃	89 150°	90 250°	91 350°	92 :CCl₂	93 N₂CHCO₂R/Cu	94 CH₂I₂/Zn(Cu)	95 R₃SnH/In·	96 Ni(CO)₄	97 CH₂N₂	98 SOCl₂	99 Ac₂O, 25°	100 Ac₂O, 80°	101 DCC	102 MeI	103 Me₃O⁺BF₄	104 1.LDA 2.MeI	105 1.K₂CO₃ 2.MeI	106 RCHO	107 RCOCl	108 C⁺/olefin
1	L	L	L	L	L	M	L	L	L	L	L	L	H	H	L	L	L	L	L	L	L	L	L	L	L	M			L	L	M
3	L	L	L	L	L	M	L	L	L	L		L	M	R	L	M	H	R	M	L	L	L	L	L	L	M			L	L	M
5	L	L	R	L	L	M	L	L	L	L		L	L	R	L	L	L	L	L	L	L	L	L	L	L	L			L	L	L
6	R	R	R	L	R	M	R	L	R	R		L	L	R	R	R	R	L	R	L	L	L	L	L	L	M			L	L	R
7	L	L	R	L	L	R	L	L	L	L		L	L	R	L	M	H	R	M	L	L	L	L	L	L	M			L	L	M
8	L	L	R	L	L	H	L	L	L	L		L	L	R	L	L	L	L	L	L	L	L	L	L	L	L			L	L	R
9	L	L	R	L	M	H	L	L	L	L		L	L	R	L	L	H	R	L	L	L	L	L	L	L	M			L	L	M
10	L	L	R	L	M	H	L	L	L	M		L	L	R	L	L	L	M	L	L	L	L	L	L	L	M			L	L	M
11	H	L	R	M	R	M	L	L	R	R		L	L	R	M	M	L	M	L	L	L	L	L	L	R	R			L	L	M
19	L	L	R	M	R	M	L	L	L	M		L	L	R	M	M	L	M	L	L	L	L	L	L	H	L			L	L	M
20	H	L	R	M	R	M	L	L	R	R		L	L	R	M	M	L	M	L	L	L	L	L	L	H	R			L	L	L
24	L	L	R	L	R	M	L	L	R	R		L	L	R	M	M	L	M	L	L	L	L	M	L	R	R			L	L	M
26	L	L	R	R	R	R	L	R	R	R		L	M	R	L	R	L	L	L	L	L	R	R	L	L	R			L	L	R
29	L	L	H	M	H	L	M	L	R	R		L	R	R	R	R	R	L	L	L	L	L	L	L	R	R			L	L	M
30	L	L	L	M	H	L	L	L	R	R		L	M	R	R	R	R	L	L	L	L	L	L	L	M	R			L	L	L
33	L	L	R	L	R	M	L	L	R	R		L	R	R	R	R	R	M	L	L	L	L	L	L	R	R			L	L	M
34	L	L	R	R	L	L	L	L	L	L		L	M	R	R	R	R	R	L	L	L	L	L	L	L	M			L	L	R
43	L	L	L	L	L	L	L	L	L	M		L	L	L	L	L	L	L	L	L	L	L	M	L	L	M			L	L	L

314

Reactivity Chart 6.   Protection for the Carboxyl Group

Esters

1. Methyl Ester
2. Methoxymethyl Ester
3. Methylthiomethyl Ester
4. Tetrahydropyranyl Ester
7. Benzyloxymethyl Ester
8. Phenacyl Ester
13. N-Phthalimidomethyl Ester
15. 2,2,2-Trichloroethyl Ester
16. 2-Haloethyl Ester
21. 2-(p-Toluenesulfonyl)ethyl Ester
23. t-Butyl Ester
27. Cinnamyl Ester
30. Benzyl Ester
31. Triphenylmethyl Ester
33. Bis(o-nitrophenyl)methyl Ester
34. 9-Anthrylmethyl Ester
35. 2-(9,10-Dioxo)anthrylmethyl Ester
42. Piperonyl Ester
45. Trimethylsilyl Ester
47. t-Butyldimethylsilyl Ester
50. S-t-Butyl Ester
59. 2-Alkyl-1,3-oxazolines

Amides and Hydrazides

64. N,N-Dimethylamide
68. N-7-Nitroindoylamide
71. Hydrazides
72. N-Phenylhydrazide
73. N,N'-Diisopropylhydrazide

(See chart, pp. 316–318)

Reactivity Chart 6. Protection for the Carboxyl Group

PG	A. AQUEOUS										B. BASIC								C. NUCLEOPHILIC							D. ORGANOMET.					E. CAT. REDN.					F. REDN.		
	1	2	3	4	5	6	7	8	9	10	11	12	13	14	15	16	17	18	19	20	21	22	23	24	25	26	27	28	29	30	31	32	33	34	35	36	37	38
	pH<1, 100°	pH<1	pH 1	pH 2-4	pH 4-6	pH 6-8.5	pH 8.5-10	pH 10-12	pH>12	pH>12, 150°	NaH	Ph₃CNa	(C₁₀H₈)⁻·Na⁺	MeSOCH₂⁻ Na⁺	KO-t-Bu	LiN-i-Pr₂	Py; R₃N	NaNH₂	NaOMe	Enolate	NH₃; RNH₂	RS⁻; N₃⁻; SCN⁻	OAc⁻; X⁻	NaCN, pH 12	HCN, pH 6	RLi	RMgX	Organozinc	Organocopper	Wittig; ylide	H₂/Raney (Ni)	H₂/Pt pH 2-4	H₂/Pd	H₂/Lindlar	H₂/Rh	Zn/HCl	Zn/HOAc	Cr(II)', pH 5
1	H	H	H	L	L	L	L	M	H	H	L	L	R	R	L	L	L	L	L	R	M	H	L	L	L	R	R	L	L	L	L	L	L	L	L	L	L	L
2	H	H	L	M	L	L	L	L	M	H	L	L	R	R	L	L	L	L	R	R	M	L	L	L	L	R	R	L	L	L	L	L	L	L	L	H	M	L
3	H	H	H	H	L	L	L	L	M	H	L	L	R	R	L	L	L	L	L	R	M	L	L	M	L	R	R	L	L	L	L	H	L	L	R	H	H	M
7	H	H	H	M	L	L	L	L	H	H	L	L	R	R	L	L	L	L	R	R	M	L	L	H	L	R	R	L	L	L	H	H	H	L	M	H	H	L
8	H	H	H	L	L	L	L	M	H	H	L	L	R	R	R	M	L	L	L	R	M	H	L	H	L	R	R	L	L	L	H	H	H	L	M	H	L	H
13	H	H	H	L	L	L	H	H	H	H	L	L	R	R	L	L	L	L	H	R	H	L	L	L	L	R	R	L	L	L	L	M	L	L	M	H	L	L
15	R	H	L	L	L	L	M	L	M	H	R	R	R	R	L	R	L	R	L	R	M	L	L	M	L	R	R	R	R	L	R	R	L	L	L	H	H	M
16	H	H	H	H	L	L	L	M	H	H	R	R	R	R	R	R	L	R	R	R	M	R	L	H	L	R	R	R	R	R	H	R	R	L	M	M	M	M
21	H	H	H	M	L	L	L	H	H	H	R	R	R	R	R	L	L	R	R	R	M	R	L	H	L	R	R	L	L	H	R	L	L	L	L	M	M	L
23	H	H	H	H	L	L	L	L	H	H	L	L	R	R	L	L	L	L	L	R	L	L	L	H	L	R	R	L	L	L	R	R	R	L	R	H	L	L
27	H	H	H	L	L	L	L	M	H	H	L	L	R	R	L	L	L	L	R	R	M	M	L	L	L	R	R	L	L	L	H	H	R	L	L	L	L	L
30	H	H	H	L	L	L	L	H	H	H	L	R	R	R	L	L	L	L	R	R	M	L	L	H	L	R	R	L	L	L	R	R	H	L	M	L	L	L
31	H	H	H	H	M	H	L	M	H	H	L	L	R	R	L	L	L	L	R	R	L	L	L	H	L	R	R	L	L	L	H	R	H	L	M	H	H	M
33	H	H	H	L	L	L	L	M	H	H	L	L	R	R	L	L	L	L	R	R	M	R	M	H	L	R	R	L	L	H	R	L	L	L	L	H	R	R
34	H	H	M	H	L	L	L	L	H	H	R	R	R	R	L	L	L	L	M	R	M	H	L	L	L	R	R	L	L	L	H	H	H	L	M	M	L	L
35	H	H	L	H	L	L	L	L	M	H	R	M	M	M	L	L	L	L	M	R	H	M	L	L	L	R	L	L	L	M	H	H	H	L	M	R	R	L
42	H	H	H	L	L	L	L	M	L	H	L	M	M	M	L	L	L	L	M	R	L	H	L	L	R	R	R	L	L	L	H	H	H	L	M	L	M	L
45	H	H	L	L	L	H	L	M	H	H	M	R	R	H	L	L	L	R	H	R	R	H	L	L	L	R	R	L	H	L	L	H	L	L	L	H	H	H
47	L	L	L	L	H	M	H	H	H	H	R	R	R	R	R	L	L	L	R	R	M	L	L	L	L	R	R	L	L	L	L	L	L	L	L	L	R	L
50	H	H	H	L	L	L	L	L	M	H	L	R	R	R	L	L	L	L	R	R	L	R	M	H	L	L	L	L	L	L	H	H	R	L	R	H	L	L
59	H	H	H	L	L	L	L	L	L	H	L	M	M	M	L	L	L	L	L	L	L	R	L	L	L	L	L	L	L	L	R	R	R	L	R	R	R	R
64	H	H	H	L	L	L	L	L	H	H	L	M	M	M	L	L	L	L	L	M	L	L	L	L	L	R	R	L	L	L	M	L	L	L	L	L	L	L
68	H	H	H	L	L	L	L	M	H	H	R	R	R	R	H	H	L	R	R	M	L	L	L	L	L	R	R	L	L	L	H	H	M	R	R	R	R	R
71	H	H	H	L	L	L	L	L	M	H	L	R	R	R	R	R	R	R	R	L	L	L	L	L	L	R	R	L	L	L	H	H	H	L	L	M	L	L
72	H	H	M	L	L	L	L	L	M	H	R	R	H	R	L	R	R	R	R	L	L	L	L	L	L	R	R	L	L	L	H	H	M	L	L	M	L	L
73	H	H	M	L	L	L	L	L	L	H	L	H	R	R	L	R	R	L	L	L	L	L	L	L	L	R	R	L	L	L	H	H	M	L	L	M	L	L

316

Reactivity Chart 6. Protection for the Carboxyl Group (Continued)

Reagent key (column numbers):

G. 39 Na/NH₃ · 40 Al(Hg) · 41 SnCl₂/Py · 42 HSO₃; H₂S
H. HYDRIDE REDN. 43 LiAlH₄ · 44 Li-s-Bu₃BH · 45 (C₅H₁₁)₂BH · 46 B₂H₆, 0° · 47 NaBH₄ · 48 Zn(BH₄)₂ · 49 NaBH₃CN pH 4-6 · 50 i-Bu₂AlH · 51 Li(OEtBu)₃AlH
I. 52 AlCl₃, 80° · 53 AlCl₃, 25° · 54 SnCl₄; BF₃ · 55 LiClO₄; 56 MgBr₂ · 56 TsOH, 80° · 57 TsOH, 0°
J. 58 Hg(II) · 59 Ag(I) · 60 Cu(II)/Py
K. 61 HBr/In· · 62 HX/In· · 63 NBS/CCl₄ · 64 Br₃CCl/In·
L. OXIDANTS 65 OsO₄ · 66 KMnO₄, pH 7,0° · 67 O₃, -50° · 68 RCO₃H, 0° · 69 RCO₃H, 50° · 70 CrO₃/Py · 71 CrO₃, pH 1 · 72 H₂O₂ pH 10-12 · 73 Quinone · 74 ¹O₂ · 75 DMSO, 100° · 76 NaOCl pH 10 · 77 aq NBS

PG	39	40	41	42	43	44	45	46	47	48	49	50	51	52	53	54	55	56	57	58	59	60	61	62	63	64	65	66	67	68	69	70	71	72	73	74	75	76	77
1	R	R	L	L	R	M	L	L	L	L	L	L	M	R	M	L	L	M	L	L	L	L	L	L	L	L	L	L	L	L	L	L	L	M	L	L	L	M	L
2	R	L	L	L	R	M	L	L	L	L	L	R	M	H	H	L	L	H	L	L	L	L	L	L	L	L	L	L	M	L	H	L	H	L	L	L	L	L	L
3	R	L	L	L	R	M	L	L	L	L	L	R	M	H	H	M	L	R	L	H	H	L	L	L	H	L	L	R	R	R	R	L	R	L	L	R	M	R	R
4	R	L	L	L	R	M	L	L	L	L	L	R	M	H	H	H	L	H	M	L	L	L	M	L	M	L	L	L	M	L	H	L	H	M	L	L	L	H	L
7	R	L	L	L	R	M	L	L	L	L	L	R	M	R	R	M	L	H	L	L	L	L	L	L	R	L	L	L	M	L	H	L	H	L	L	L	L	L	L
8	R	R	L	L	R	R	R	R	M	M	L	R	M	R	R	M	M	L	L	L	L	L	L	L	L	L	L	L	L	L	R	L	L	R	L	L	L	M	R
13	R	L	L	L	R	M	R	R	L	L	L	R	L	R	L	L	L	R	L	L	L	L	L	L	L	L	L	L	L	L	L	L	L	H	L	L	H	H	L
15	R	M	L	L	R	M	L	L	L	L	L	R	M	R	L	L	L	L	L	L	R	L	R	R	R	R	R	L	L	L	L	L	L	M	L	L	M	H	L
16	R	M	M	L	R	M	L	L	L	L	R	R	M	R	R	L	L	L	L	L	M	L	L	L	L	L	L	L	L	L	L	L	M	M	L	L	M	H	L
21	R	M	M	L	R	M	L	L	L	L	R	R	M	R	L	L	L	L	L	L	L	L	L	L	R	L	L	L	L	L	L	L	L	H	L	L	M	H	L
23	R	L	L	L	R	L	L	L	L	L	L	L	L	R	H	H	L	M	L	L	L	L	L	L	L	L	L	L	L	L	H	L	H	L	L	L	L	L	L
27	H	L	L	L	R	M	R	R	L	M	L	R	M	R	R	R	L	H	L	R	L	L	R	L	R	R	R	R	R	R	R	L	L	R	L	R	L	H	R
30	R	L	L	L	R	M	L	L	L	L	L	R	L	R	R	M	L	L	L	L	L	L	L	L	R	L	L	L	L	L	L	L	L	M	L	L	L	M	L
31	H	L	L	L	R	L	L	L	L	L	M	R	L	R	R	L	L	H	M	L	L	L	M	L	L	L	L	L	L	M	H	L	H	M	L	L	H	M	L
33	H	R	L	L	R	M	L	L	L	L	L	R	M	R	R	L	L	L	L	L	L	L	L	L	R	L	L	L	L	L	L	L	L	M	L	L	L	M	L
34	R	L	L	L	R	M	L	L	L	L	L	R	M	R	R	M	M	L	L	L	L	L	L	L	R	L	L	L	L	L	L	L	M	L	L	R	L	L	L
35	R	R	M	L	L	L	M	L	M	M	M	M	L	R	L	L	M	L	L	L	L	L	M	L	L	L	R	R	L	R	R	L	L	L	L	L	L	M	L
42	R	L	L	L	R	L	R	R	L	L	L	R	L	L	L	L	M	L	L	L	L	L	L	L	R	L	L	L	L	L	M	L	L	L	L	L	L	L	L
45	L	H	L	H	R	H	R	R	H	H	H	R	L	L	L	L	H	L	M	H	H	L	H	L	R	L	L	L	L	H	M	L	H	M	L	L	L	L	H
47	L	L	L	L	R	M	R	L	L	L	H	R	L	H	L	L	H	L	L	L	L	L	L	L	L	L	L	M	L	H	H	L	H	H	L	L	H	R	M
50	R	M	L	L	R	M	L	L	L	L	L	R	M	R	M	L	L	H	L	R	R	R	L	L	L	L	L	R	R	L	M	L	M	R	L	R	M	H	H
59	R	M	L	L	L	L	M	L	L	L	M	M	L	R	L	L	L	R	L	L	R	R	L	L	L	L	R	L	L	R	R	L	R	L	L	L	L	L	L
64	R	R	L	L	R	L	R	R	L	L	L	R	L	L	L	L	L	L	L	L	L	L	L	L	L	L	L	R	L	L	L	L	L	L	L	L	L	L	L
68	R	L	L	L	R	L	R	R	L	L	L	R	L	M	L	L	L	L	L	R	L	L	R	R	R	R	R	R	R	R	M	L	L	M	L	R	L	L	L
71	L	L	L	L	R	L	R	R	L	L	L	R	L	R	L	L	L	L	L	R	L	H	R	L	R	L	R	R	R	R	R	L	M	R	R	R	R	R	M
72	R	L	L	L	R	L	R	R	L	L	L	R	L	M	L	L	L	L	L	H	L	H	L	L	R	L	R	R	R	R	R	L	H	R	R	R	R	H	H
73	R	L	L	L	R	L	R	R	L	L	L	R	L	M	L	L	L	L	L	H	L	H	L	L	R	L	R	R	R	R	R	L	H	R	R	R	R	H	H

Reactivity Chart 6. Protection for the Carboxyl Group (Continued)

PG	78 I₂	79 PhSeX; PhSCl	80 Br₂; Cl₂	81 MnO₂/CH₂Cl₂	82 NaIO₄ pH 5-8	83 SeO₂ pH 2-4	84 SeO₂/Py	85 K₃Fe(CN)₆, pH 8	86 Pb(IV), 25°	87 Pb(IV), 80°	88 Tl(NO₃)₃	89 150°	90 250°	91 350°	92 :CCl₂	93 N₂CHCO₂R/Cu	94 CH₂I₂/Zn(Cu)	95 R₃SnH/In·	96 Ni(CO)₄	97 CH₂N₂	98 SOCl₂	99 Ac₂O, 25°	100 Ac₂O, 80°	101 DCC	102 MeI	103 Me₃O⁺BF₄⁻	104 1.LDA 2.MeI	105 1.K₂CO₃ 2.MeI	106 RCHO	107 RCOCl	108 C⁺/olefin
1	L	L	L	L	L	L	L	L	L	L	L	L	L	L	L	L	L	L	L	L	L	L	L	L	L	L			L	L	L
2	L	L	R	L	L	L	L	L	L	L	L	L	L	R	L	L	L	L	L	L	L	L	L	L	L	L			L	L	M
3	L	L	R	L	R	M	L	L	R	R	H	L	L	R	M	M	L	M	L	L	L	L	L	L	R	R			L	L	L
4	L	L	R	L	L	H	L	L	L	L	R	M	H	R	L	L	L	L	L	L	M	M	M	L	L	L			L	L	M
7	L	L	M	L	L	M	L	L	L	L	H	L	L	R	L	L	L	L	L	L	L	L	M	L	L	L			L	L	M
8	L	R	R	L	L	L	L	L	L	L	H	L	L	R	L	L	L	L	L	L	L	L	L	L	L	L			L	L	L
13	L	L	L	L	L	L	L	H	L	L	L	L	M	R	L	L	R	L	M	L	L	L	L	L	L	R			L	L	L
15	L	L	L	L	L	L	L	M	L	L	L	M	H	R	L	M	R	R	L	L	L	L	L	L	L	L			L	L	L
16	L	L	L	L	L	L	L	L	L	L	L	L	M	R	L	L	R	R	L	L	L	L	L	L	L	L			L	L	L
21	L	L	L	L	L	L	L	L	L	M	M	M	H	R	L	L	L	L	L	L	L	L	L	L	L	L			L	L	L
23	L	L	L	L	L	H	L	L	L	H	H	M	H	H	L	L	L	L	L	L	L	L	M	L	L	L			L	L	H
27	R	R	R	L	L	M	L	L	L	R	M	R	R	R	R	R	R	L	R	L	L	L	L	L	L	L			L	L	R
30	L	L	L	L	L	M	L	L	L	L	L	L	L	R	L	L	L	L	L	L	L	L	L	L	L	L			L	L	L
31	L	L	L	L	M	H	L	L	L	H	H	M	H	R	L	L	L	L	L	L	L	L	M	L	L	L			L	L	H
33	L	L	L	L	L	L	L	L	L	L	L	L	L	R	L	L	L	L	L	L	L	L	L	L	L	L			L	L	L
34	L	L	L	L	L	M	L	L	L	L	L	L	L	R	L	L	L	L	L	L	L	L	L	L	L	L			L	L	L
35	L	L	M	L	L	M	M	L	L	L	L	L	L	R	L	L	L	L	L	L	L	L	L	L	L	L			L	L	L
42	L	L	M	L	L	M	L	L	L	R	R	L	L	R	L	M	L	L	L	L	L	L	L	L	L	L			L	L	R
45	H	L	L	L	L	H	L	H	H	H	H	L	L	R	L	L	L	L	L	M	L	H	H	L	L	H			L	M	L
47	H	L	L	L	M	H	M	H	H	H	H	L	L	R	L	L	L	L	L	L	L	L	H	L	L	L			L	M	L
50	L	L	R	L	L	L	M	L	L	R	R	M	H	R	M	M	L	L	L	L	L	L	L	L	L	R			L	L	L
59	L	L	R	L	L	M	L	L	L	R	R	L	L	R	L	L	L	L	L	L	L	L	L	L	R	R			L	R	R
64	L	L	L	L	L	L	L	L	L	L	L	L	L	L	L	L	L	L	L	L	L	L	L	L	L	R			L	L	L
68	R	L	M	L	L	L	L	L	L	M	M	L	L	M	L	R	R	L	L	L	L	L	L	L	L	R			L	L	M
71	R	R	R	R	R	R	L	R	H	R	R	M	R	R	R	R	R	L	L	L	R	R	R	L	R	R			R	R	R
72	H	R	R	H	H	R	M	H	H	R	R	M	R	R	L	M	L	L	L	L	R	R	R	L	R	R			R	R	R
73	H	R	R	H	H	R	M	H	H	R	R	M	R	R	L	M	L	L	L	L	R	R	R	L	R	R			R	R	R

Reagent group headings: **L. OXIDANTS** (78–88) · **M.** (89–91) · **N.** (92–94) · **O. MISCELLANEOUS** (95–105) · **P.** (106–108)

Reactivity Chart 7.   Protection for the Thiol Group

1. *S*-Benzyl Thioether
3. *S*-*p*-Methoxybenzyl Thioether
5. *S*-*p*-Nitrobenzyl Thioether
6. *S*-4-Picolyl Thioether
7. *S*-2-Picolyl *N*-Oxide Thioether
8. *S*-9-Anthrylmethyl Thioether
9. *S*-Diphenylmethyl Thioether
10. *S*-Di(*p*-methoxyphenyl)methyl Thioether
12. *S*-Triphenylmethyl Thioether
15. *S*-2,4-Dinitrophenyl Thioether
16. *S*-*t*-Butyl Thioether
19. *S*-Isobutoxymethyl Hemithioacetal
20. *S*-2-Tetrahydropyranyl Hemithioacetal
23. *S*-Acetamidomethyl Aminothioacetal
25. *S*-Cyanomethyl Thioether
26. *S*-2-Nitro-1-phenylethyl Thioether
27. *S*-2,2-Bis(carboethoxy)ethyl Thioether
30. *S*-Benzoyl Derivative
36. *S*-(*N*-Ethylcarbamate)
38. *S*-Ethyl Disulfide

(See chart, pp. 320–322)

Reactivity Chart 7. Protection for the Thiol Group

Reaction	#	1	3	5	6	7	8	9	10	12	15	16	19	20	23	25	26	27	30	36	38
A. AQUEOUS																					
pH<1, 100°	1	H	H	H	H	H	H	H	H	H	H	H	H	H	H	H	H	H	H	H	H
pH<1	2	L	H	L	L	H	M	L	H	H	L	M	H	H	M	R	L	L	L	L	L
pH 1	3	L	M	L	L	M	L	L	L	M	L	L	L	L	L	L	L	L	L	L	L
pH 2-4	4	L	L	L	L	L	L	L	L	L	L	L	L	L	L	L	L	L	L	L	L
pH 4-6	5	L	L	L	L	L	L	L	L	L	L	L	L	L	L	L	L	L	L	L	L
pH 6-8.5	6	L	L	L	L	L	L	L	L	L	M	L	L	L	L	L	L	L	L	M	L
pH 8.5-10	7	L	L	L	L	L	L	L	L	L	H	L	L	L	L	L	H	L	M	H	L
pH 10-12	8	L	L	L	L	L	L	L	L	L	H	L	L	L	L	M	H	H	H	H	L
pH>12	9	L	L	M	L	L	L	L	L	L	H	L	M	M	L	M	H	H	H	H	H
pH>12, 150°	10	M	M	H	M	M	M	M	M	M	H	M	H	H	H	H	R	R	H	H	H
B. BASIC																					
NaH	11	L	L	M	L	L	L	L	L	L	R	L	L	L	R	R	H	H	L	R	R
Ph3CNa	12	R	R	R	R	R	R	R	R	L	R	L	L	L	R	R	H	H	L	R	R
(C10H8)·Na+	13	R	R	R	R	R	R	R	R	R	R	L	R	R	R	R	R	R	H	R	R
MeSOCH2·Na+	14	R	R	R	R	R	R	R	R	L	R	L	H	H	R	R	H	H	H	R	R
KO-t-Bu	15	L	L	L	L	L	L	L	L	L	R	L	L	L	L	M	H	H	H	H	L
LiN-i-Pr2	16	R	R	R	R	R	R	R	R	L	R	L	L	L	R	R	H	H	H	R	R
Py; R3N	17	L	L	L	L	L	L	L	L	L	L	L	L	L	L	L	H	L	H	H	L
NaNH2	18	L	L	R	L	L	L	M	M	L	R	L	L	L	R	R	H	H	H	H	H
C. NUCLEOPHILIC																					
NaOMe	19	L	L	M	L	L	L	L	L	L	M	L	L	L	R	R	H	H	H	H	R
Enolate	20	L	L	L	L	L	L	L	L	L	M	L	L	L	L	L	R	R	H	H	R
NH3; RNH2	21	L	L	L	L	L	L	L	L	L	L	L	L	L	L	L	H	H	H	H	L
RS-; N3-; SCN-	22	L	L	M	L	L	H	M	M	R	H	L	H	R	L	L	M	M	H	H	M
OAc-; X-	23	L	L	L	L	L	L	L	L	L	L	L	L	L	L	L	L	L	L	L	L
NaCN, pH 12	24	L	L	M	L	L	L	L	L	L	H	L	L	L	L	L	M	M	H	L	H
HCN, pH 6	25																				
D. ORGANOMET.																					
RLi	26	R	R	R	R	R	R	R	R	L	R	L	L	L	R	R	H	H	H	H	H
RMgX	27	R	R	R	R	R	R	R	R	L	R	L	L	L	R	R	R	R	H	H	H
Organozinc	28	L	L	L	L	L	L	L	L	L	L	L	L	L	L	L	L	R	H	L	H
Organocopper	29	L	L	R	L	L	L	L	L	L	R	L	L	L	L	L	R	R	L	L	M
Wittig; ylide	30	L	L	L	L	L	L	L	L	L	L	L	L	L	L	L	L	L	H	L	H
E. CAT. REDN.																					
H2/Raney (Ni)	31	R	R	R	R	R	R	R	R	R	R	R	R	R	R	R	R	R	R	R	R
H2/Pt pH 2-4	32	R	R	R	R	R	R	R	R	R	R	R	R	R	R	R	R	R	R	R	R
H2/Pd	33	R	R	R	R	R	R	R	R	R	R	M	R	R	R	R	R	R	R	R	R
H2/Lindlar	34	M	M	M	M	R	M	M	M	R	R	L	L	L	M	L	R	L	L	L	H
H2/Rh	35	R	R	R	R	R	R	R	R	R	R	M	R	R	R	R	R	R	R	R	R
F. REDN.																					
Zn/HCl	36	L	M	R	L	R	L	L	L	M	R	L	H	H	L	R	R	L	H	H	H
Zn/HOAc	37	L	L	R	L	R	L	L	L	L	R	L	L	L	L	R	R	L	L	L	H
Cr(II), pH 5	38	L	L	R	L	R	L	L	L	L	R	L	L	L	L	L	R	L	L	L	H

320

Reactivity Chart 7. Protection for the Thiol Group (Continued)

Reagent	PG →	1	3	5	6	7	8	9	10	12	15	16	19	20	23	25	26	27	30	36	38
G.																					
39 Na/NH₃		H	H	H	H	H	H	H	H	H	R	L	H	M	L	R	R	R	H	H	H
40 Al (Hg)		L	L	R	L	R	L	L	L	L	R	L	L	L	L	L	R	L	L	L	H
41 SnCl₂/Py		L	L	L	L	R	L	L	L	L	M	L	L	L	L	L	L	L	L	L	H
42 HSO₃⁻; H₂S		L	L	L	L	R	L	L	L	L	L	L	L	L	L	L	L	L	L	L	H
H. HYDRIDE REDN.																					
43 LiAlH₄		L	L	R	L	R	R	R	R	L	R	L	L	L	L	R	R	R	H	H	H
44 Li-s-Bu₃BH		L	L	M	L	R	L	L	L	L	M	L	L	L	L	M	M	R	H	M	H
45 (C₅H₁₁)₂BH		L	L	L	L	L	L	L	L	L	M	L	L	L	L	L	L	L	L	M	H
46 B₂H₆, 0°		L	L	L	L	R	L	L	L	H	M	L	L	L	R	R	L	L	M	H	H
47 NaBH₄		L	L	L	L	M	L	L	L	L	L	L	L	L	L	L	L	L	H	L	H
48 Zn(BH₄)₂		L	L	L	L	L	L	L	L	L	L	L	L	L	L	L	L	L	M	L	H
49 NaBH₃CN pH 4–6		L	L	L	L	L	L	L	L	L	L	L	L	L	L	L	L	L	L	L	M
50 i-Bu₂AlH		L	L	L	L	R	L	L	L	L	R	L	L	L	R	R	L	R	H	H	H
51 Li(OtBu)₃AlH		L	L	L	L	R	L	L	L	L	L	L	L	L	L	L	R	L	L	L	H
I.																					
52 AlCl₃, 80°		L	M	L	M	H	M	H	H	H	L	H	H	H	L	M	L	M	R	R	R
53 AlCl₃, 25°		L	L	L	L	M	L	M	M	H	L	M	H	H	L	L	L	L	M	H	H
54 SnCl₄; BF₃		L	L	L	L	L	L	L	M	L	L	L	M	M	L	L	L	L	L	L	L
55 LiClO₄; MgBr₂		L	L	L	L	L	L	L	L	L	L	L	L	L	L	L	L	L	L	L	L
56 TsOH, 80°		M	M	M	M	L	M	M	M	M	L	M	M	M	M	L	M	M	M	M	L
57 TsOH, 0°		L	L	L	L	L	L	L	L	L	L	L	L	L	L	L	L	L	L	L	L
J.																					
58 Hg(II)		L	R	L	L	L	M	M	R	R	L	R	R	R	R	L	L	L	R	R	R
59 Ag(I)		M	R	L	L	L	M	M	R	R	L	R	R	R	R	L	L	L	L	R	R
60 Cu(II)/Py		L	L	L	L	L	L	L	L	L	L	L	L	L	L	L	L	L	L	L	M
K.																					
61 HBr/In.		R	R	R	R	R	R	R	L	L	L	L	L	L	L	R	R	L	L	L	R
62 HX/In.		L	L	L	L	L	L	L	L	L	L	L	L	L	L	L	L	L	L	L	L
63 NBS/CCl₄		R	R	R	R	R	R	R	R	L	R	L	R	R	L	R	R	R	M	R	R
64 Br₃CCl/In.		R	R	R	R	R	R	R	R	L	L	L	L	L	L	L	R	L	L	L	R
L. OXIDANTS																					
65 OsO₄		L	L	L	L	L	L	L	L	L	L	L	L	L	L	L	L	L	L	L	M
66 KMnO₄, pH 7, 0°		R	R	R	R	R	R	R	R	R	R	R	R	R	R	R	R	R	R	R	R
67 O₃, −50°		R	R	M	R	R	R	R	R	R	L	R	R	R	R	R	R	R	M	R	R
68 RCO₃H, 0°		R	R	R	R	R	R	R	R	R	M	R	R	R	R	R	R	R	R	R	R
69 RCO₃H, 50°		R	R	R	R	R	R	R	R	R	R	R	R	R	R	R	R	R	R	R	R
70 CrO₃/Py		L	L	L	L	L	L	L	L	L	L	L	M	M	L	L	L	L	L	L	L
71 CrO₃, pH 1		R	R	R	R	R	R	R	R	R	R	R	R	R	R	R	R	R	R	R	R
72 H₂O₂ pH 10–12		L	L	L	L	L	L	L	L	L	R	L	L	L	L	R	R	R	H	H	L
73 Quinone		L	L	L	L	L	L	L	L	L	R	L	L	L	L	R	L	L	L	L	R
74 ¹O₂		R	R	R	M	M	R	R	M	M	M	M	M	R	R	R	R	M	R	R	R
75 DMSO, 100°		L	L	M	L	L	L	L	L	L	R	L	L	L	L	L	M	L	M	L	L
76 NaOCl pH 10		R	R	R	R	R	R	R	R	R	R	R	R	R	R	R	R	R	R	R	R
77 aq NBS		R	R	R	R	R	R	R	R	R	R	R	R	R	R	R	R	R	R	R	R

Reactivity Chart 7. Protection for the Thiol Group (Continued)

PG	78 I₂	79 PhSex; PhSCl	80 Br₂; Cl₂	81 MnO₂/CH₂Cl₂	82 NaIO₄ pH 5-8	83 SeO₂ pH 2-4	84 SeO₂/Py	85 K₃Fe(CN)₆, pH 8	86 Pb(IV), 25°	87 Pb(IV), 80°	88 Tl(NO₃)₃	89 150°	90 250°	91 350°	92 :CCl₂	93 N₂CHCO₂R/Cu	94 CH₂I₂/Zn(Cu)	95 R₃SnH/In·	96 Ni(CO)₄	97 CH₂N₂	98 SOCl₂	99 Ac₂O, 25°	100 Ac₂O, 80°	101 DCC	102 MeI	103 Me₃O⁺BF₄⁻	104 1.LDA 2.MeI	105 1.K₂CO₃ 2.MeI	106 RCHO	107 RCOCl	108 C⁺/olefin
					L. OXIDANTS							M.			N.			O.						MISCELLANEOUS					P.		
1	L	L	R	M	R	R	M	L	R	R		L	L	M	M	R	L	R	L	L	L	L	L	L	R	R	R	R	L	L	M
3	M	L	R	M	R	R	M	L	R	R		L	L	M	M	R	L	R	L	L	L	L	L	L	L	R	R	R	L	L	M
5	R	L	R	L	R	R	L	L	R	R		L	L	M	M	R	L	R	L	L	L	L	L	L	L	M	R	M	L	L	L
6	R	L	R	L	R	R	M	L	R	R		L	L	M	M	R	L	R	L	L	L	L	L	L	R	R	R	R	L	L	L
7	R	L	R	L	R	R	M	L	R	R		L	L	M	R	R	L	R	L	L	R	R	R	L	R	R	R	R	L	H	L
8	R	L	R	M	R	R	R	L	R	R		L	L	M	M	R	L	R	L	L	L	L	L	L	R	R	R	R	L	L	M
9	M	L	R	M	R	R	R	L	R	R		L	L	M	M	R	L	R	L	L	L	L	L	L	R	R	R	R	L	L	M
10	M	L	R	M	R	R	R	L	R	R		L	L	M	M	R	L	R	L	L	L	L	L	L	L	L	R	R	L	L	M
12	R	L	M	L	R	L	L	L	R	R		L	L	M	M	R	L	L	L	L	L	L	L	L	L	L	L	L	L	L	L
15	R	L	M	L	L	M	L	R	R	R		L	L	M	M	R	L	R	L	L	L	L	L	L	L	L	R	L	L	L	L
16	L	L	L	L	R	L	L	L	R	R		L	L	R	M	R	L	L	L	L	L	L	L	L	M	M	M	M	L	L	M
19	R	L	M	L	R	M	L	L	R	R		L	M	R	M	R	L	M	L	L	L	L	L	L	R	R	M	R	L	L	M
20	R	L	R	L	R	M	L	L	R	R		L	M	R	M	R	L	M	L	L	L	L	L	L	R	R	R	R	L	L	M
23	R	L	R	L	R	M	L	L	R	R		L	L	R	M	R	L	L	L	L	L	L	M	L	R	R	R	R	L	L	L
25	R	L	R	L	R	M	L	M	R	R		L	L	M	M	R	L	M	L	L	L	L	L	L	R	R	R	R	L	L	L
26	R	L	R	L	R	M	L	H	R	R		M	H	H	M	R	L	R	L	L	L	L	L	L	R	R	R	R	L	L	L
27	R	L	R	L	R	M	L	H	R	R		M	H	H	M	R	L	L	L	L	L	L	L	L	R	R	R	R	L	L	L
30	L	L	R	L	L	L	L	H	L	M		M	R	R	M	M	L	L	L	L	L	L	L	L	L	R	L	L	L	L	L
36	L	L	R	L	R	L	L	H	L	M		R	R	R	M	M	L	R	L	L	L	L	M	L	L	R	L	L	L	L	L
38	R	L	R	R	R	R	R	R	R	R		H	H	R	M	R	R	R	R	L	L	L	L	L	M	R	R	M	L	L	R

322

Reactivity Chart 8.   Protection for the Amino Group: Carbamates

1. Methyl Carbamate
5. 9-Fluorenylmethyl Carbamate
8. 2,2,2-Trichloroethyl Carbamate
11. 2-Trimethylsilylethyl Carbamate
16. 1,1-Dimethylpropynyl Carbamate
20. 1-Methyl-1-phenylethyl Carbamate
22. 1-Methyl-1-(4-biphenylyl)ethyl Carbamate
24. 1,1-Dimethyl-2-haloethyl Carbamate
26. 1,1-Dimethyl-2-cyanoethyl Carbamate
28. *t*-Butyl Carbamate
30. Cyclobutyl Carbamate
31. 1-Methylcyclobutyl Carbamate
35. 1-Adamantyl Carbamate
37. Vinyl Carbamate
38. Allyl Carbamate
39. Cinnamyl Carbamate
44. 8-Quinolyl Carbamate
45. *N*-Hydroxypiperidinyl Carbamate
47. 4,5-Diphenyl-3-oxazolin-2-one
48. Benzyl Carbamate
53. *p*-Nitrobenzyl Carbamate
55. 3,4-Dimethoxy-6-nitrobenzyl Carbamate
58. 2,4-Dichlorobenzyl Carbamate
65. 5-Benzisoxazolylmethyl Carbamate
66. 9-Anthrylmethyl Carbamate
67. Diphenylmethyl Carbamate
71. Isonicotinyl Carbamate
72. *S*-Benzyl Carbamate
75. *N*-(N′-Phenylaminothiocarbonyl) Derivative

(See chart, pp. 324–326)

Reactivity Chart 8. Protection for the Amino Group: Carbamates

Reagent categories: **A. AQUEOUS** (1–10) · **B. BASIC** (11–18) · **C. NUCLEOPHILIC** (19–25) · **D. ORGANOMET.** (26–30) · **E. CAT. REDN.** (31–35) · **F.** (36–38)

Reagent (no.)	1	5	8	11	16	20	22	24	26	28	30	31	35	37	38	39	44	45	47	48	53	55	58	65	66	67	71	72	75
1 pK<1, 100°	H	H	H	H	H	H	H	H	H	H	H	H	H	H	H	H	H	H	H	H	H	H	H	H	H	H	H	H	H
2 pK>1	H	M	H	H	H	H	H	H	L	H	H	H	H	H	H	H	H	L	L	H	H	M	H	H	H	H	L	L	M
3 pH 1	L	L	L	H	H	M	H	M	L	M	M	H	H	H	M	M	M	L	L	L	L	L	L	M	L	H	L	L	L
4 pH 2–4	L	L	L	M	L	L	M	L	L	M	L	M	M	H	M	M	L	L	L	L	L	L	L	L	L	M	L	L	L
5 pH 4–6	L	L	L	L	L	L	L	L	L	L	L	L	L	M	L	L	L	L	L	L	L	L	L	L	L	L	L	L	L
6 pH 6–8.5	L	L	L	L	L	L	L	L	L	L	L	L	L	M	L	L	L	L	L	L	L	L	L	L	L	L	L	L	L
7 pH 8.5–10	L	L	L	L	L	L	L	L	L	M	L	L	L	L	L	L	L	L	L	L	L	L	L	L	L	L	L	L	L
8 pH 10–12	L	L	L	L	L	L	L	L	R	H	L	L	L	L	L	L	L	L	L	L	L	L	L	L	L	L	L	L	L
9 pH>12	M	M	M	L	L	L	L	L	H	L	L	L	L	L	L	L	H	L	L	L	L	L	L	M	L	L	L	H	H
10 pH>12, 150°	H	H	M	M	M	M	M	H	H	M	H	H	M	M	M	M	H	M	M	H	M	M	M	H	M	M	M	H	H
11 NaH	H	L	R	L	L	L	L	L	M	L	L	L	L	L	L	L	L	L	L	L	L	L	L	L	L	L	L	L	L
12 Ph3CNa	L	H	R	L	R	L	L	L	R	L	L	L	L	L	L	L	L	L	L	R	R	L	L	R	R	R	R	L	L
13 (C10H8)⁻·Na⁺	L	L	R	L	R	L	L	H	M	L	L	L	L	L	L	L	L	M	H	L	R	L	L	H	L	L	L	H	L
14 MeSOCH2⁻Na⁺	L	L	R	L	R	L	L	L	H	L	L	L	L	L	L	L	L	L	L	L	L	L	L	L	L	L	L	L	L
15 KO-t-Bu	L	L	R	L	L	L	L	L	H	L	L	L	L	L	L	L	L	L	L	L	L	L	L	L	L	L	L	L	L
16 LiN-i-Pr2	L	L	R	L	R	L	L	L	H	L	L	L	L	L	L	L	L	L	L	L	L	L	L	L	L	L	L	L	L
17 Py; R3N	L	M	L	L	L	L	L	L	H	L	L	L	L	L	L	L	L	L	L	L	H	L	H	R	L	L	L	L	L
18 NaNH2	L	H	H	R	L	L	L	M	H	L	L	L	L	L	L	L	L	L	L	H	H	L	M	H	L	L	L	L	L
19 NaOMe	L	M	M	L	L	L	L	R	H	L	L	L	L	L	L	L	L	L	M	L	M	M	M	H	L	L	H	H	L
20 Enolate	L	L	R	L	L	L	L	R	H	L	L	L	L	L	L	L	L	L	M	L	M	M	M	H	L	L	L	L	M
21 NH3; RNH2	L	H	M	L	L	L	L	M	H	L	L	L	L	L	H	H	L	L	L	H	H	M	H	H	L	H	H	M	H
22 RS⁻; N3⁻; SCN⁻	H	L	M	L	L	L	L	L	L	L	L	L	L	L	L	L	L	L	L	H	L	L	L	L	L	L	L	L	L
23 OAc⁻; X⁻	L	L	L	L	L	L	L	L	M	L	L	L	L	L	L	L	L	L	L	L	L	L	L	L	L	L	L	L	L
24 NaCN, pH 12	L	M	L	L	L	L	L	L	L	L	L	L	L	L	L	L	L	L	L	L	L	L	L	L	L	L	L	L	L
25 HCN, pH 6	L	L	L	L	L	L	L	L	L	L	L	L	L	L	L	L	L	L	L	L	L	L	L	L	L	L	L	L	L
26 RLi	H	H	H	H	H	H	H	H	H	H	H	H	H	H	H	H	H	H	H	H	H	H	H	H	H	H	H	H	H
27 RMgX	L	H	H	H	H	H	H	H	H	H	H	H	H	H	H	H	H	H	H	H	H	H	H	H	H	H	H	H	H
28 Organozinc	L	M	R	L	R	L	L	M	H	L	L	L	L	L	L	L	L	L	L	L	L	L	M	L	L	M	L	L	L
29 Organocopper	L	M	R	L	R	L	L	M	H	L	L	L	L	L	L	L	L	L	L	L	R	L	L	L	L	L	L	L	L
30 Wittig; ylide	L	L	R	L	R	L	L	M	M	L	L	L	L	L	L	L	L	L	L	L	L	L	L	L	L	L	L	L	L
31 H2/Raney (Ni)	L	L	L	L	H	M	M	M	R	L	L	L	L	H	H	H	L	L	R	H	H	H	H	H	H	H	H	H	H
32 H2/Pt pH 2–4	L	L	L	L	H	H	H	H	R	H	M	M	M	H	H	H	L	H	H	H	H	H	H	H	H	H	H	R	R
33 H2/Pd	L	M	L	H	L	L	L	R	R	L	L	L	L	H	H	H	L	H	H	H	H	H	H	H	H	H	H	H	R
34 H2/Lindlar	L	L	L	L	L	L	L	L	L	L	L	L	L	L	L	L	L	L	L	L	L	L	L	L	L	L	L	L	L
35 H2/Rh	L	R	L	L	R	R	R	L	R	L	L	L	L	R	R	R	L	H	R	R	R	R	R	R	R	R	R	R	R
36 Zn/HCl	L	L	H	M	H	H	H	H	R	H	H	H	H	H	H	H	H	H	L	M	R	R	L	H	M	H	H	L	H
37 Zn/HOAc	L	L	H	L	H	H	H	H	R	H	M	M	H	H	H	H	H	H	L	L	R	R	L	M	M	M	H	L	H
38 Cr(II), pH 5	L	L	H	L	L	L	H	H	L	M	L	L	L	M	L	L	L	L	L	L	R	R	L	L	L	L	L	L	R

Reagents (column numbers):

- G.
  - 39 Na/NH₃
  - 40 Al(Hg)
  - 41 SnCl₂/Py
  - 42 HSO₃; H₂S
- H. HYDRIDE REDN.
  - 43 LiAlH₄
  - 44 Li-s-Bu₃BH
  - 45 (C₅H₁₁)₂BH
  - 46 B₂H₆, 0°
  - 47 NaBH₄
  - 48 Zn(BH₄)₂
  - 49 NaBH₃CN pH 4-6
  - 50 i-Bu₂AlH
  - 51 Li(OtBu)₃AlH
- I.
  - 52 AlCl₃, 80°
  - 53 AlCl₃, 25°
  - 54 SnCl₄; BF₃
  - 55 LiClO₄; MgBr₂
  - 56 TsOH, 80°
  - 57 TsOH, 0°
- J.
  - 58 Hg(II)
  - 59 Ag(I)
  - 60 Cu(II)/Py
- K.
  - 61 HBr/In·
  - 62 HX/In·
  - 63 NBS/CCl₄
  - 64 Br₃CCl/In·
- L. OXIDANTS
  - 65 OsO₄
  - 66 KMnO₄, pH 7, 0°
  - 67 O₃, -50°
  - 68 RCO₃H, 0°
  - 69 RCO₃H, 50°
  - 70 CrO₃/Py
  - 71 CrO₃, pH 1
  - 72 H₂O₂ pH 10-12
  - 73 Quinone
  - 74 ¹O₂
  - 75 DMSO, 100°
  - 76 NaOCl pH 10
  - 77 aq NBS

PG	39	40	41	42	43	44	45	46	47	48	49	50	51	52	53	54	55	56	57	58	59	60	61	62	63	64	65	66	67	68	69	70	71	72	73	74	75	76	77
1	L	L	L	L	R	L	L	L	L	L	L	R	L	R	L	L	L	L	L	L	L	L	L	L	L	L	L	L	L	L	L	L	H	L	L	L	L	L	L
5	H	L	L	L	M	L	L	L	L	L	L	M	L	R	L	L	L	M	L	L	L	H	R	R	R	L	L	L	L	L	L	H	H	L	L	L	L	L	L
8	R	M	H	L	R	L	L	L	L	L	L	M	L	R	R	R	L	H	L	L	L	H	M	M	M	M	L	L	L	L	L	L	L	L	L	L	R	L	L
11	L	L	L	L	R	L	L	L	L	L	L	M	L	R	L	L	L	R	H	L	L	L	L	L	L	L	L	L	L	L	L	L	H	L	L	L	L	L	R
16	H	L	L	L	R	L	R	R	L	L	L	M	L	R	H	L	L	H	H	R	L	R	R	R	L	R	R	R	R	M	R	R	H	L	L	R	L	L	R
20	R	L	L	L	R	L	L	L	L	L	L	M	L	R	H	M	L	R	H	L	L	L	L	L	L	L	L	L	L	L	L	L	H	L	L	L	L	L	L
22	R	L	L	L	R	L	L	L	L	M	M	M	L	R	H	M	L	R	H	L	L	L	H	L	L	L	L	L	L	L	M	L	H	L	L	L	L	L	L
24	H	L	H	L	R	L	L	L	L	L	L	M	L	R	H	M	L	M	H	L	L	L	M	L	L	L	L	L	L	L	L	L	M	L	L	L	R	L	L
26	H	L	L	L	R	M	R	R	L	L	L	M	L	R	H	M	L	R	L	L	M	H	H	L	L	L	L	L	L	L	L	H	H	R	L	R	M	M	L
28	L	L	L	L	M	L	R	R	L	L	M	L	L	R	H	M	L	R	H	L	L	L	L	L	L	R	L	L	R	L	M	L	H	L	L	M	L	L	L
30	L	L	L	L	M	L	R	R	L	L	L	M	L	R	M	L	L	M	L	R	L	L	M	L	L	R	L	L	R	L	R	L	M	L	L	R	L	L	L
31	L	L	L	L	M	L	L	L	L	L	L	M	L	R	M	M	L	R	M	L	M	L	M	L	L	L	L	L	R	L	R	L	H	L	L	L	L	L	L
35	L	L	M	L	M	L	L	L	L	L	L	M	L	R	L	L	L	L	L	L	L	L	M	L	L	L	L	L	R	L	R	L	L	L	L	L	L	L	R
37	L	L	L	L	R	L	M	R	L	L	M	M	L	R	L	L	L	R	L	R	L	L	R	R	R	L	R	R	M	R	H	L	M	L	L	R	L	L	R
38	L	L	L	L	R	L	L	L	L	L	L	L	L	R	H	H	L	R	L	R	L	L	R	R	R	R	R	R	L	R	L	L	H	L	L	L	L	L	R
39	H	R	L	L	R	L	L	L	L	L	L	M	L	R	H	H	L	R	L	R	L	L	R	R	R	L	R	R	R	R	L	H	M	L	L	L	L	R	R
44	H	R	L	L	M	L	L	L	L	L	L	M	L	R	L	L	L	L	L	L	M	H	R	L	R	L	L	L	R	R	L	L	H	L	L	L	L	M	L
45	H	L	L	L	M	L	L	L	L	L	L	M	L	R	L	L	L	L	L	L	L	L	L	L	R	L	L	L	R	H	L	H	H	L	L	L	L	L	R
47	H	L	M	L	M	L	L	L	L	L	L	M	L	R	L	L	L	M	L	L	L	L	R	R	R	L	M	M	H	H	R	R	H	L	L	M	L	L	L
48	H	L	L	L	R	L	L	L	L	L	L	M	L	R	L	L	L	R	L	L	L	H	H	R	R	R	L	L	L	L	L	L	M	L	L	R	L	L	L
53	H	L	L	L	R	L	L	L	L	L	L	M	L	R	L	L	L	L	L	L	L	H	M	L	R	L	L	L	L	L	L	L	L	L	L	L	L	L	L
55	H	L	L	L	R	L	L	L	L	L	L	M	L	R	R	L	L	R	L	L	L	L	M	L	R	L	L	L	L	R	L	L	L	L	L	L	L	R	L
58	H	L	M	L	R	L	L	L	L	L	L	M	L	R	H	L	L	M	L	L	L	H	M	L	R	L	L	L	L	H	L	L	L	L	L	L	L	R	R
65	H	L	L	L	M	L	L	L	L	L	L	M	L	R	H	L	L	R	L	L	L	H	H	R	R	L	R	R	R	R	R	R	M	L	L	L	L	L	L
66	H	L	L	L	M	L	L	L	L	L	L	M	L	R	H	L	L	R	L	L	L	L	R	R	R	L	L	L	L	L	L	L	L	L	L	L	L	L	L
67	H	L	L	L	M	L	L	L	L	L	L	M	L	R	H	L	L	R	L	L	L	L	H	R	R	L	L	L	L	L	M	L	H	L	L	L	L	L	L
71	H	L	L	L	M	L	L	L	L	L	L	M	L	R	H	L	L	R	L	L	L	L	H	M	R	L	L	L	R	R	R	L	H	L	L	L	L	R	L
72	H	L	M	L	M	L	L	L	L	L	L	M	L	R	M	L	L	R	L	H	L	L	M	M	R	L	L	R	H	H	H	L	L	M	L	L	L	R	R
75	H	R	L	L	M	L	L	L	L	L	L	M	L	R	H	M	L	L	L	R	L	M	R	R	R	L	L	L	H	H	H	L	H	M	L	L	R	R	R

325

Reactivity Chart 8. Protection for the Amino Group: Carbamates (Continued)

PG	78 I₂	79 PhSeₓ; PhSCl	80 Br₂; Cl₂	81 MnO₂/CH₂Cl₂	82 NaIO₄ pH 5-8	83 SeO₂ pH 2-4	84 SeO₂/Py	85 K₃Fe(CN)₆, pH 8	86 Pb(IV)', 25°	87 Pb(IV)', 80°	88 Tl(NO₃)₃	89 150°	90 250°	91 350°	92 :CCl₂	93 N₂CHCO₂R/Cu	94 CH₂I₂/Zn(Cu)	95 R₃SnH/In·	96 Ni(CO)₄	97 CH₂N₂	98 SOCl₂	99 Ac₂O, 25°	100 Ac₂O, 80°	101 DCC	102 MeI	103 Me₃O⁺BF₄	104 1.LDA 2.MeI	105 1.K₂CO₃ 2.MeI	106 RCHO	107 RCOCl	108 C⁺/olefin
																															L. OXIDANTS / M. / N. / O. MISCELLANEOUS / P.
1	L	L	L	L	L	L	L	L	L	L	L	L	L	M	L	L	L	L	L	L	L	L	L	L	L	R	L	L	L	L	L
5	L	L	L	L	L	M	L	L	L	L	L	L	M	H	L	L	L	L	L	L	L	L	L	L	L	R	L	L	L	L	L
8	L	L	L	L	L	L	L	L	L	L	L	L	M	H	L	L	R	R	M	L	L	L	L	L	L	R	L	L	L	L	R
11	L	L	L	L	L	M	L	L	L	L	L	L	M	H	L	L	L	L	L	L	L	L	L	L	L	R	L	L	L	L	L
16	R	R	R	L	L	R	L	L	R	R	R	L	M	H	R	R	R	R	R	L	L	L	L	L	L	R	R	L	L	L	R
20	L	L	L	L	L	H	L	L	L	L	L	L	M	H	L	L	L	L	L	L	L	L	L	L	L	R	L	L	L	L	M
22	L	L	L	L	L	H	L	L	L	L	L	L	M	H	L	L	L	L	L	L	L	L	L	L	L	R	L	L	L	L	M
24	L	L	L	L	L	L	H	M	L	L	L	M	H	H	L	L	M	R	L	L	L	L	L	L	L	R	L	H	L	L	H
26	L	L	L	L	L	L	L	L	L	R	L	H	H	H	L	L	L	L	L	L	L	L	L	L	L	R	R	H	L	L	M
28	L	L	L	L	L	H	L	L	L	R	L	H	H	H	L	L	L	L	L	L	L	L	L	L	L	R	L	L	L	L	H
30	L	R	R	L	L	M	L	L	L	L	L	L	M	H	L	L	L	L	L	L	L	L	L	L	L	R	L	L	L	L	M
31	L	L	L	L	L	M	L	L	L	L	L	L	M	H	L	L	L	L	L	L	L	L	L	L	L	R	L	L	L	L	M
35	L	L	L	L	L	M	L	L	L	L	L	L	L	M	L	L	L	L	L	L	L	L	L	L	L	R	L	L	L	L	H
37	L	R	R	L	L	H	L	L	M	R	R	L	M	H	R	R	L	R	H	L	L	L	L	L	L	R	L	L	L	L	R
38	L	R	R	L	L	M	L	L	M	R	R	L	M	R	R	R	R	R	H	L	L	L	L	L	L	R	L	L	L	L	R
39	L	R	R	L	L	M	M	L	M	R	R	L	M	R	R	R	R	R	H	L	L	L	L	L	L	R	L	L	L	L	R
44	L	L	L	L	L	L	L	L	L	L	L	L	L	M	L	L	R	L	H	L	L	L	L	L	R	R	L	R	L	L	M
45	L	L	L	L	L	L	L	L	L	L	L	L	M	H	L	L	L	L	L	L	L	L	L	L	R	R	L	R	L	L	L
47	L	L	R	L	L	L	L	L	M	R	M	L	L	M	R	R	R	R	R	L	L	L	L	L	L	R	L	L	L	L	L
48	L	L	L	L	L	L	L	L	L	L	L	L	L	M	L	L	L	L	L	L	L	L	L	L	L	R	L	L	L	L	M
53	L	L	L	L	L	L	R	L	L	L	L	L	L	M	L	L	L	R	L	L	L	L	L	L	L	R	L	L	L	L	M
55	L	L	L	L	L	L	M	L	L	L	L	L	L	M	M	M	M	R	L	L	L	L	L	L	L	R	L	L	L	L	M
58	L	L	L	L	L	L	R	L	L	L	L	L	L	M	L	L	L	R	L	L	L	L	L	L	L	R	L	L	L	L	L
65	L	L	R	L	L	L	R	L	M	R	M	L	L	M	M	M	M	M	L	L	L	L	L	L	R	R	L	L	L	L	H
66	L	L	L	L	L	L	L	L	L	L	L	L	L	M	L	L	L	L	L	L	L	L	L	L	L	R	R	L	L	L	M
67	L	L	L	L	L	L	L	L	L	L	L	L	L	M	L	L	L	L	L	L	L	L	L	L	L	R	L	L	L	L	H
71	L	L	L	L	L	L	L	L	L	L	L	L	L	M	L	L	M	L	R	L	L	L	L	L	R	R	L	R	L	L	L
72	L	L	R	L	R	L	L	L	M	M	L	L	L	M	L	L	R	R	L	L	L	L	L	L	L	R	L	L	L	L	L
75	R	R	R	L	R	M	L	L	L	R	L	L	M	H	R	R	M	R	L	R	L	L	M	L	R	R	R	R	L	L	M

326

Reactivity Chart 9.   Protection for the Amino Group: Amides

1. *N*-Formyl
2. *N*-Acetyl
3. *N*-Chloroacetyl
5. *N*-Trichloroacetyl
6. *N*-Trifluoroacetyl
7. *N-o*-Nitrophenylacetyl
8. *N-o*-Nitrophenoxyacetyl
9. *N*-Acetoacetyl
12. *N*-3-Phenylpropionyl
13. *N*-3-(*p*-Hydroxyphenyl)propionyl
15. *N*-2-Methyl-2-(*o*-nitrophenoxy)propionyl
16. *N*-2-Methyl-2-(*o*-phenylazophenoxy)propionyl
17. *N*-4-Chlorobutyryl
19. *N-o*-Nitrocinnamoyl
20. *N*-Picolinoyl
21. *N*-(N'-Acetylmethionyl)
23. *N*-Benzoyl
29. *N*-Phthaloyl
31. *N*-Dithiasuccinoyl

(See chart, pp. 328–330)

Reactivity Chart 9. Protection for the Amino Group: Amides

Reagent key (column numbers):

A. AQUEOUS — 1. pH>1, 100°; 2. pH<1; 3. pH 1; 4. pH 2–4; 5. pH 4–6; 6. pH 6–8.5; 7. pH 8.5–10; 8. pH 10–12; 9. pH>12; 10. pH>12, 150°

B. BASIC — 11. NaH; 12. Ph_3CNa; 13. $(C_{10}H_8)^{-}\cdot Na^{+}$; 14. $MeSOCH_2^{-}Na^{+}$; 15. $KO\text{-}t\text{-}Bu$; 16. $LiN\text{-}i\text{-}Pr_2$; 17. Py; R_3N; 18. $NaNH_2$

C. NUCLEOPHILIC — 19. NaOMe; 20. Enolate; 21. NH_3; RNH_2; 22. RS^-; N_3^-; SCN^-; 23. OAc^-; X^-; 24. NaCN, pH 12; 25. HCN, pH 6

D. ORGANOMET. — 26. RLi; 27. RMgX; 28. Organozinc; 29. Organocopper; 30. Wittig; ylide

E. CAT. REDN. — 31. H_2/Raney (Ni); 32. H_2/Pt pH 2–4; 33. H_2/Pd; 34. H_2/Lindlar; 35. H_2/Rh

F. — 36. Zn/HCl; 37. Zn/HOAc; 38. Cr(II), pH 5

PG	1	2	3	4	5	6	7	8	9	10	11	12	13	14	15	16	17	18	19	20	21	22	23	24	25	26	27	28	29	30	31	32	33	34	35	36	37	38
1	H	H	H	L	L	L	L	M	H	H	H	H	R	L	L	L	L	L	L	L	H	L	L	M	L	H	H	L	L	L	L	H	L	L	L	H	L	L
2	H	M	L	L	L	L	L	L	M	H	L	L	R	R	L	R	L	R	L	L	H	L	L	L	L	H	M	L	L	L	L	L	L	L	L	M	L	L
3	H	L	L	L	L	L	L	L	H	H	M	R	R	R	L	R	L	L	R	L	H	R	L	R	L	H	M	M	R	L	M	R	R	L	L	H	H	H
5	H	L	L	L	L	L	M	H	H	H	L	L	R	L	L	R	L	L	R	R	H	M	L	R	L	H	M	M	R	M	R	R	R	L	R	H	H	R
6	H	L	L	L	L	L	M	H	H	H	L	L	R	L	L	L	L	L	R	R	H	M	L	R	L	H	M	L	M	M	R	R	L	L	L	H	H	M
7	H	L	L	L	L	L	L	L	M	H	M	R	R	R	L	R	L	R	L	L	H	L	L	L	L	H	H	M	L	L	R	H	R	L	R	H	H	R
8	H	L	L	L	L	L	L	L	M	H	M	R	R	R	L	L	L	R	M	M	M	M	L	L	L	H	R	L	L	L	R	R	R	L	R	H	H	R
9	H	L	L	L	L	L	L	L	H	H	R	R	R	R	R	R	L	R	R	R	H	L	L	M	R	H	R	R	R	R	M	R	L	L	M	L	L	L
12	H	L	L	L	L	L	L	L	M	H	M	R	R	R	L	R	L	R	L	L	M	L	L	L	L	H	M	L	L	L	L	L	L	L	R	L	L	L
13	H	L	L	L	L	L	L	L	M	H	R	R	R	R	R	R	L	R	R	L	M	L	L	L	L	H	R	L	L	L	L	L	L	L	R	L	L	L
15	H	L	L	L	L	L	L	L	M	H	L	L	R	L	L	L	L	L	L	L	M	L	L	L	L	H	R	L	L	L	R	H	R	L	R	H	H	R
16	H	L	L	L	L	L	L	L	H	H	L	L	R	M	L	H	L	L	L	L	M	L	L	L	L	H	R	R	L	L	R	H	R	R	R	H	H	H
17	H	L	L	L	L	L	L	L	L	H	M	R	R	R	R	R	L	R	R	L	M	R	L	R	L	H	M	L	R	L	M	R	R	L	L	L	L	L
19	H	L	L	L	L	L	L	L	M	H	L	L	R	L	L	L	L	L	M	R	M	M	L	M	L	H	M	R	L	L	R	H	R	L	R	H	H	R
20	H	L	L	L	L	L	L	L	M	H	L	L	R	L	L	L	L	L	L	L	H	L	L	L	L	H	R	L	L	L	L	L	L	L	R	L	L	L
21	H	L	L	L	L	L	L	L	M	H	R	R	R	R	L	R	L	R	L	L	H	L	L	L	L	H	R	L	L	L	R	R	R	R	R	L	L	L
23	H	H	L	L	L	L	L	L	H	H	L	L	R	L	L	L	L	L	L	L	H	L	L	M	L	H	M	L	L	L	L	R	L	L	R	L	L	L
29	H	L	L	L	L	L	L	L	R	H	L	L	R	L	L	L	L	L	L	L	H	L	L	L	L	H	M	L	L	L	L	L	L	L	R	L	L	L
31	M	L	L	L	L	L	L	M	H	H	L	L	R	R	L	L	L	L	H	L	H	H	L	H	L	H	R	L	M	L	R	L	L	R	R	L	R	R

328

Reactivity Chart 9. Protection for the Amino Group: Amides (Continued)

PG	39 Na/NH$_3$	40 Al(Hg)	41 SnCl$_2$/Py	42 HSO$_3^-$; H$_2$S	43 LiAlH$_4$	44 Li-s-Bu$_3$BH	45 (C$_5$H$_{11}$)$_2$BH	46 B$_2$H$_6$, 0°	47 NaBH$_4$	48 Zn(BH$_4$)$_2$	49 NaBH$_3$CN pH 4-6	50 i-Bu$_2$AlH	51 Li(OtBu)$_3$AlH	52 AlCl$_3$, 80°	53 AlCl$_3$, 25°	54 SnCl$_4$; BF$_3$	55 LiClO$_4$; MgBr$_2$	56 TsOH, 80°	57 TsOH, 0°	58 Hg(II)	59 Ag(I)	60 Cu(II)/Py	61 HBr/In·	62 HX/In·	63 NBS/CCl$_4$	64 Br$_3$CCl/In·	65 OsO$_4$	66 KMnO$_4$, pH 7, 0°	67 O$_3$, -50°	68 RCO$_3$H, 0°	69 RCO$_3$H, 50°	70 CrO$_3$/Py	71 CrO$_3$, pH 1	72 H$_2$O$_2$ pH 10-12	73 Quinone	74 ^{1}O$_2$	75 DMSO, 100°	76 NaOCl pH 10	77 aq NBS
	G.				H. HYDRIDE REDN.									I.						J.			K.				L. OXIDANTS												
1	R	L	L	L	R	L	H	R	L	L	L	H	L	L	L	L	L	L	L	L	L	L	L	L	L	L	L	L	L	L	M	L	H	M	L	L	L	L	L
2	R	L	L	L	R	L	H	R	L	L	L	H	L	L	L	L	L	L	L	L	L	L	L	L	L	L	L	L	L	L	L	L	L	L	L	L	L	L	L
3	R	M	L	L	R	M	H	R	L	L	L	H	L	M	M	L	L	L	L	L	H	L	L	L	L	L	L	L	L	L	L	L	L	L	L	L	R	L	L
5	R	L	L	L	H	M	H	R	H	M	M	H	M	M	M	L	L	L	L	L	H	L	R	R	R	R	L	L	L	L	L	L	L	L	L	L	L	R	L
6	R	L	L	L	H	M	H	R	H	M	M	H	M	L	L	L	L	L	L	L	M	L	L	L	L	L	L	L	L	L	R	L	L	L	L	L	L	R	L
7	R	R	L	L	R	M	H	R	L	L	L	H	L	L	L	L	L	L	L	L	L	L	L	L	M	L	L	L	L	L	L	L	L	L	L	L	L	L	L
8	R	R	L	L	R	M	H	R	L	L	L	H	L	L	L	L	L	L	L	L	L	L	L	L	L	L	L	L	L	L	L	L	L	L	L	L	L	L	L
9	R	L	L	L	R	R	H	R	R	R	R	H	R	R	R	M	L	L	L	L	L	L	L	L	L	L	L	H	L	L	M	L	H	L	L	L	L	R	R
12	R	L	L	L	R	L	H	R	L	L	L	H	L	L	L	L	L	L	L	L	L	L	L	L	R	L	L	L	L	L	L	L	L	L	L	L	L	L	L
13	R	L	L	L	R	L	H	R	L	L	L	H	L	L	L	L	L	L	L	L	L	L	R	R	R	R	L	R	L	L	L	M	R	M	M	L	L	R	H
15	R	R	L	L	R	M	H	R	L	L	L	H	L	L	L	L	L	L	L	L	L	L	L	L	L	L	L	L	L	L	L	L	L	L	L	L	L	L	L
16	R	H	R	L	R	L	H	R	R	R	L	H	M	R	L	L	L	L	L	L	R	L	R	R	R	R	L	L	R	M	R	L	R	R	L	R	L	L	L
17	R	L	L	L	R	M	H	R	L	L	L	H	L	M	M	L	L	L	L	L	H	L	L	L	L	L	L	L	L	L	L	L	L	L	L	L	R	L	L
19	R	R	L	L	R	M	H	R	L	L	L	H	L	M	M	L	L	L	L	L	L	L	R	R	L	R	R	R	R	L	M	L	L	R	L	L	L	L	R
20	R	L	L	L	R	L	H	R	L	L	L	H	L	L	L	L	L	L	L	H	L	H	L	L	R	L	L	L	L	M	R	M	M	L	L	L	L	L	L
21	R	L	L	L	R	L	H	R	L	L	L	H	L	L	L	L	L	L	L	L	L	L	L	L	L	L	L	R	R	R	R	L	R	L	L	R	L	R	R
23	R	L	L	L	R	L	H	R	L	L	L	H	L	L	L	L	L	L	L	L	L	L	L	L	L	L	L	L	L	L	L	L	L	L	L	L	L	L	L
29	R	L	L	L	R	L	H	R	L	L	L	H	L	L	L	L	L	L	L	L	L	H	L	L	L	L	L	L	L	L	L	L	L	L	L	L	L	L	L
31	R	R	R	L	R	R	H	R	R	L	L	H	R	R	R	L	L	R	L	R	M	L	R	R	R	R	M	R	R	R	R	L	R	R	R	R	R	R	R

329

Reactivity Chart 9. Protection for the Amino Group: Amides (Continued)

PG	78 I₂	79 PhSeₓ; PhSCl	80 Br₂; Cl₂	81 MnO₂/CH₂Cl₂	82 NaIO₄ pH 5-8	83 SeO₂ pH 2-4	84 SeO₂/Py	85 K₃Fe(CN)₆, pH 8	86 Pb(IV), 25°	87 Pb(IV), 80°	88 Tl(NO₃)₃	89 150°	90 250°	91 350°	92 :CCl₂	93 N₂CHCO₂R/Cu	94 CH₂I₂/Zn(Cu)	95 R₃SnH/In·	96 Ni(CO)₄	97 CH₂N₂	98 SOCl₂	99 Ac₂O, 25°	100 Ac₂O, 80°	101 DCC	102 MeI	103 Me₃O⁺BF₄⁻	104 1.LDA 2.MeI	105 1.K₂CO₃ 2.MeI	106 RCHO	107 RCOCl	108 C⁺/olefin
1	L	L	M	L	L	L	L	L	L	L	L	L	L	L	L	L	L	L	L	L	L	L	L	L	L	R	R	L	L	H	L
2	L	L	L	L	L	L	L	L	L	L	L	L	L	L	L	L	L	L	L	L	L	L	L	L	L	R	R	L	L	L	L
3	L	L	L	L	L	L	L	L	L	L	L	L	L	L	L	L	M	R	L	L	L	L	L	L	L	R	R	L	L	L	L
5	L	L	L	L	L	L	L	M	L	L	L	L	M	R	L	L	H	R	M	L	L	L	L	L	L	R	L	M	L	L	L
6	L	L	L	L	M	L	L	M	L	L	L	L	M	R	L	L	L	R	L	L	L	L	L	L	L	R	L	L	L	L	L
7	L	L	L	L	L	L	L	L	L	L	L	L	L	L	L	L	L	L	L	L	L	L	L	L	L	R	R	L	L	L	L
8	L	L	L	L	L	L	L	L	L	L	L	L	L	L	L	L	L	L	L	L	L	L	L	L	L	R	R	L	L	L	L
9	L	L	R	L	M	H	M	L	M	R	M	L	L	M	L	M	L	R	L	M	L	M	M	L	L	R	R	R	L	M	M
12	M	L	L	L	L	L	H	L	L	L	L	L	L	L	L	L	L	L	L	L	L	L	L	L	L	R	R	L	L	L	L
13	L	L	H	M	M	M	M	R	M	R	L	L	L	L	L	L	L	L	L	R	L	R	R	R	L	R	R	R	L	R	L
15	L	L	L	L	L	L	L	L	L	L	L	L	L	L	L	L	L	L	L	L	L	L	L	L	L	R	L	L	L	L	L
16	L	L	L	L	L	L	L	R	L	L	R	R	R	R	M	R	R	R	L	L	L	L	L	L	L	R	L	M	L	L	R
17	L	L	L	L	L	L	L	L	L	L	L	R	R	R	L	L	R	R	L	L	L	L	L	L	L	R	R	L	L	L	L
19	L	L	M	L	L	L	L	L	L	R	M	L	L	L	R	R	R	L	L	L	L	L	L	L	L	R	L	L	L	L	M
20	L	L	L	L	L	L	L	L	L	L	L	L	L	L	L	M	M	L	L	L	L	L	L	L	M	R	L	M	L	L	L
21	L	L	R	L	R	M	L	L	R	R	L	M	H	H	M	M	L	L	L	L	L	L	L	L	R	R	R	R	L	L	L
23	L	L	L	L	L	L	L	L	L	L	L	L	L	L	L	L	L	L	L	L	L	L	L	L	L	R	L	L	L	L	L
29	L	L	L	L	L	L	L	L	L	L	L	L	L	L	L	L	L	L	L	L	L	L	L	L	L	R	L	L	L	L	L
31	M	M	R	R	M	R	M	R	R	R	L	L	M	H	M	M	H	R	R	L	L	L	L	L	M	R	M	L	L	L	L

Reagent groupings: L. OXIDANTS (78–88); M. (89–91); N. (92–94); O. MISCELLANEOUS (95–105); P. (106–108)

330

Reactivity Chart 10.   Protection for the Amino Group: Special —NH Protective Groups

1. *N*-Allyl
2. *N*-Phenacyl
3. *N*-3-Acetoxypropyl
5. Quaternary Ammonium Salts
6. *N*-Methoxymethyl
8. *N*-Benzyloxymethyl
9. *N*-Pivaloyloxymethyl
12. *N*-Tetrahydropyranyl
13. *N*-2,4-Dinitrophenyl
14. *N*-Benzyl
16. *N*-*o*-Nitrobenzyl
17. *N*-Di(*p*-methoxyphenyl)methyl
18. *N*-Triphenylmethyl
19. *N*-(*p*-Methoxyphenyl)diphenylmethyl
20. *N*-Diphenyl-4-pyridylmethyl
21. *N*-2-Picolyl *N'*-Oxide
24. *N*,*N'*-Isopropylidene
25. *N*-Benzylidene
27. *N*-*p*-Nitrobenzylidene
28. *N*-Salicylidene
33. *N*-(5,5-Dimethyl-3-oxo-1-cyclohexenyl)
37. *N*-Nitro
39. *N*-Oxide
40. *N*-Diphenylphosphinyl
41. *N*-Dimethylthiophosphinyl
47. *N*-Benzenesulfenyl
48. *N*-*o*-Nitrobenzenesulfenyl
55. *N*-2,4,6-Trimethylbenzenesulfonyl
56. *N*-Toluenesulfonyl
57. *N*-Benzylsulfonyl
59. *N*-Trifluoromethylsulfonyl
60. *N*-Phenacylsulfonyl

(See chart, pp. 332–334)

Reactivity Chart 10. Protection for the Amino Group: Special -NH Protective Groups

PG	pH<1, 100°	pH<1	pH 2-4	pH 4-6	pH 6-8.5	pH 8.5-10	pH 10-12	pH>12	pH>12, 150°	NaH	Ph₃CNa	(C₁₀H₈)⁻·Na⁺	MeSOCH₂⁻Na⁺	KO-t-Bu	LiN-i-Pr₂	Py; R₃N	NaNH₂	NaOMe	Enolate	NH₃; RNH₂	RS⁻; N₃⁻; SCN⁻	OAc⁻; X⁻	NaCN, pH 12	HCN, pH 6	RLi	RMgX	Organozinc	Organocopper	Wittig; ylide	H₂/Raney(Ni)	H₂/Pt pH 2-4	H₂/Pd	H₂/Lindlar	H₂/Rh	Zn/HCl	Zn/HOAc	Cr(II), pH 5
	1	2	3	4	5	6	7	8	9	10	11	12	13	14	15	16	17	18	19	20	21	22	23	24	25	26	27	28	29	30	31	32	33	34	35	36	37	38
	A. AQUEOUS										B. BASIC								C. NUCLEOPHILIC							D. ORGANOMET.					E. CAT. REDN.					F.		
1	H	L	L	L	L	L	L	L	R	L	R	R	R	R	R	R	L	R	L	L	L	L	L	L	L	L	L	L	L	L	R	R	H	L	R	L	L	L
2	H	L	L	L	L	L	L	L	R	R	R	R	R	R	R	L	R	M	M	M	L	L	L	R	R	R	R	M	L	R	L	R	R	L	R	H	H	L
3	H	R	L	L	L	L	M	R	R	R	L	R	R	R	L	R	L	R	R	R	R	L	L	L	L	R	R	L	L	L	L	L	L	L	L	L	L	L
5	H	L	L	L	L	L	L	L	L	R	L	R	L	M	M	R	L	R	L	L	L	H	L	M	L	R	R	L	L	R	L	L	L	L	L	H	L	L
6	H	H	L	L	L	L	L	L	L	L	L	L	L	L	L	L	L	L	L	L	L	L	L	L	L	L	L	L	L	L	L	L	L	L	L	L	L	L
8	H	H	L	L	L	L	L	L	L	L	L	L	L	L	L	L	L	L	L	L	L	L	L	L	L	L	L	L	L	L	H	H	R	L	L	L	L	L
9	H	H	M	L	L	L	M	H	H	H	L	L	R	R	L	L	L	H	H	M	H	L	L	L	L	R	R	L	L	L	L	L	L	L	L	M	L	L
12	H	H	H	H	H	L	L	L	L	L	L	L	L	L	L	L	L	L	L	L	L	L	L	L	L	L	L	L	L	L	L	L	L	L	L	H	H	H
13	H	L	L	L	L	L	L	H	H	L	M	R	L	L	L	L	L	H	M	M	L	H	L	R	L	R	R	M	M	L	R	R	R	R	L	R	R	R
14	H	L	L	L	L	L	L	L	L	L	L	L	L	L	L	L	L	L	L	L	L	L	L	L	L	L	L	L	L	L	H	M	M	M	H	L	L	L
16	H	L	L	L	L	L	L	L	R	L	M	R	L	L	L	L	L	L	L	L	L	L	R	L	H	H	M	M	L	H	H	H	L	R	R	R	R	R
17	H	H	H	H	L	L	L	L	L	L	L	L	L	L	L	L	L	L	L	L	L	L	L	L	L	L	L	L	L	L	H	H	H	L	L	H	H	L
18	H	H	H	M	L	L	L	L	L	L	L	L	L	L	L	L	L	L	L	L	L	L	L	L	L	L	L	L	L	L	H	H	H	L	H	H	H	L
19	H	H	H	M	M	L	L	L	L	L	L	L	L	L	L	L	L	L	L	L	L	L	L	L	L	L	L	L	L	L	H	H	H	L	H	H	H	L
20	M	L	L	L	L	M	L	L	L	L	L	L	L	L	L	L	L	L	L	L	L	L	L	L	L	R	R	L	L	L	H	H	H	L	L	H	H	L
21	H	L	L	L	L	L	L	L	R	L	L	R	L	L	H	L	H	L	L	L	L	L	L	L	L	L	L	L	L	L	R	R	R	R	R	R	R	R
24	H	H	H	M	L	L	L	L	L	H	L	L	R	L	L	L	L	L	L	L	H	L	L	L	R	R	R	R	M	R	R	R	R	L	R	R	R	R
25	H	H	H	M	L	L	L	L	M	H	L	L	L	L	L	L	L	L	L	L	H	L	L	L	R	R	R	R	M	R	R	R	H	L	R	R	R	R
27	H	H	H	M	L	L	L	L	L	H	L	M	R	L	L	L	L	L	L	L	H	L	L	L	R	R	R	R	M	R	R	R	R	L	R	R	R	R
28	H	H	H	M	L	L	L	L	M	H	R	R	R	R	R	R	L	R	R	R	H	L	L	R	R	R	R	R	R	R	R	R	R	L	R	R	R	R
33	H	H	L	L	L	L	L	L	L	H	L	R	R	R	R	R	L	R	H	R	H	R	L	R	L	R	R	R	M	M	R	R	R	L	R	R	R	R
37	H	L	L	L	L	L	L	L	L	H	L	L	L	L	L	L	L	H	H	H	R	L	L	L	L	H	H	H	M	L	R	H	H	L	H	H	H	H
39	H	M	L	L	L	L	L	L	L	H	L	L	R	L	L	M	L	L	L	L	L	L	L	L	L	R	R	L	L	H	R	H	H	L	H	H	H	H
40	H	H	H	M	L	L	L	L	L	H	L	L	L	L	L	L	L	L	L	L	L	L	L	L	L	L	L	L	L	L	L	M	L	L	L	H	M	L
41	H	H	H	L	L	L	L	L	L	H	L	L	R	L	L	L	L	L	L	L	L	L	L	L	L	M	M	M	L	L	R	R	R	L	R	H	L	L
47	H	M	L	L	L	L	L	L	M	H	L	L	L	L	L	L	L	L	L	R	H	H	H	H	H	L	L	L	L	L	R	R	R	L	R	L	L	L
48	H	H	H	H	L	L	L	L	M	H	L	M	R	L	L	L	L	L	L	R	H	H	H	H	H	R	R	M	M	L	R	R	R	L	R	H	H	R
55	H	L	L	L	L	L	L	L	L	L	L	L	H	L	L	L	L	L	L	L	L	L	L	L	L	H	H	H	L	L	L	L	L	L	L	H	L	L
56	H	L	L	L	L	L	L	L	L	L	L	L	H	L	L	L	L	L	L	L	L	L	L	L	L	H	H	H	L	L	L	L	L	L	L	H	L	L
57	H	M	L	L	L	L	L	L	H	R	R	R	R	R	R	L	R	R	R	L	R	L	L	L	H	H	H	L	L	H	H	H	L	L	H	H	M	M
59	H	L	L	L	L	L	L	L	H	L	L	H	L	L	L	L	L	L	L	L	L	L	L	L	L	H	H	H	H	L	L	L	L	L	L	L	L	M
60	H	L	L	L	L	L	L	L	R	H	R	R	R	R	R	R	L	R	R	R	R	L	R	L	R	R	H	H	R	R	H	H	H	L	R	H	H	M

Reactivity Chart 10. Protection for the Amino Group: Special -NH Protective Groups (Continued)

Reagent key (column number / reagent):

G. 39 Na/NH₃ · 40 Al(Hg) · 41 SnCl₂/Py · 42 HSO₃⁻; H₂S
H. HYDRIDE REDN. 43 LiAlH₄ · 44 Li-s-Bu₃BH · 45 (C₅H₁₁)₃BH · 46 B₂H₆, 0° · 47 NaBH₄ · 48 Zn(BH₄)₂ · 49 NaBH₃CN pH 4-6 · 50 i-Bu₂AlH · 51 Li(OEt)₃AlH
I. 52 AlCl₃, 80° · 53 AlCl₃, 25° · 54 SnCl₄; BF₃ · 55 LiClO₄; MgBr₂ · 56 TsOH, 80° · 57 TsOH, 0°
J. 58 Hg(II) · 59 Ag(I) · 60 Cu(II)/Py
K. 61 HBr/In. · 62 HX/In. · 63 NBS/CCl₄ · 64 Br₃CCl/In.
L. OXIDANTS 65 OsO₄ · 66 KMnO₄, pH 7, 0° · 67 O₃, -50° · 68 RCO₃H, 0° · 69 RCO₃H, 50° · 70 CrO₃/Py · 71 CrO₃, pH 1 · 72 H₂O₂ pH 10-12 · 73 Quinone · 74 ¹O₂ · 75 DMSO, 100° · 76 NaOCl pH 10 · 77 aq NBS

PG	39	40	41	42	43	44	45	46	47	48	49	50	51	52	53	54	55	56	57	58	59	60	61	62	63	64	65	66	67	68	69	70	71	72	73	74	75	76	77
1	R	L	L	L	L	L	R	R	R	L	L	L	L	L	L	L	L	L	L	R	L	L	R	R	R	R	R	R	R	R	R	R	L	R	R	R	M	M	R
2	R	L	L	L	R	L	R	R	L	L	R	M	R	L	L	L	L	L	L	L	L	L	L	L	L	L	R	R	R	R	R	R	L	R	R	R	M	M	R
3	R	L	L	L	R	R	L	L	L	R	L	L	M	R	L	L	L	R	L	L	L	L	L	L	L	L	R	R	R	R	R	R	L	R	R	R	M	R	R
5	L	L	L	L	L	L	L	L	L	L	L	L	R	R	L	L	L	L	L	L	L	L	L	L	L	L	L	L	L	L	L	L	L	L	L	L	H	L	L
6	L	M	L	L	L	L	L	L	L	L	L	L	L	H	H	H	L	M	L	L	L	L	L	L	L	L	R	R	R	R	R	R	L	R	R	R	M	M	R
8	H	L	L	L	L	L	L	L	L	L	L	L	L	H	H	H	L	M	L	L	L	L	L	L	R	L	R	R	R	R	R	R	L	R	R	R	M	M	R
9	R	L	L	L	R	L	L	L	L	L	L	R	M	H	H	M	L	M	L	L	M	L	L	L	L	L	R	R	R	R	R	R	M	R	R	R	M	R	R
12	L	L	L	L	L	L	L	L	L	L	L	L	L	H	L	H	L	L	L	L	L	L	L	L	L	L	R	R	R	R	R	R	H	R	R	R	M	M	R
13	R	R	L	L	R	L	L	L	L	L	L	L	M	L	L	L	L	L	L	L	M	L	L	L	L	L	L	M	M	M	H	H	L	L	L	L	L	M	L
14	H	M	L	L	L	L	L	L	L	L	L	L	L	L	L	L	L	L	L	L	L	L	L	L	L	L	R	R	R	R	R	R	L	R	R	R	M	M	R
16	H	L	L	L	R	M	L	L	L	L	L	M	M	L	L	L	L	L	L	L	L	L	L	L	L	L	R	R	R	R	R	R	L	R	R	R	R	R	R
17	H	L	L	L	L	L	L	L	L	L	L	L	L	R	R	L	M	M	L	L	L	L	L	L	L	L	R	R	R	R	R	R	H	R	R	R	R	R	R
18	H	L	L	L	L	L	L	L	L	L	L	L	L	L	L	L	L	M	L	L	L	L	L	L	L	L	R	R	R	R	R	R	H	R	R	L	R	R	R
19	H	L	R	L	L	L	L	L	L	L	L	L	L	R	R	L	M	M	L	L	L	L	L	L	L	L	R	R	R	R	R	R	H	R	R	L	R	R	R
20	H	L	M	L	L	L	L	L	L	L	L	L	L	L	L	L	L	R	L	L	L	L	L	L	L	L	R	R	R	R	R	R	L	R	R	M	R	R	R
21	R	R	R	L	R	R	R	R	M	L	L	M	M	R	L	L	L	H	L	L	L	R	L	L	R	L	R	R	R	R	R	R	L	R	R	R	R	R	R
24	M	L	L	M	R	L	R	R	L	L	L	R	R	L	L	L	L	L	L	M	M	H	R	R	R	R	R	R	R	R	R	R	H	M	R	M	R	R	R
25	R	L	L	R	R	M	R	R	M	M	M	R	R	L	L	L	L	L	L	M	M	H	R	R	R	R	R	R	R	R	R	R	H	M	R	L	R	R	R
27	R	R	L	L	R	L	R	R	M	M	M	R	R	L	L	L	L	M	L	M	M	H	R	R	R	R	R	R	R	R	R	R	H	M	R	L	R	R	R
28	R	L	L	M	R	L	R	R	M	M	M	R	R	L	L	L	L	M	L	M	M	H	R	R	R	R	R	R	R	R	R	R	H	M	R	L	R	R	R
33	R	L	L	L	R	R	R	R	L	L	L	M	M	R	L	L	L	H	L	L	L	H	R	R	R	R	R	R	R	R	R	R	L	L	R	R	L	L	R
37	R	R	R	R	R	L	R	R	L	L	L	M	L	R	L	L	L	L	L	L	H	H	L	L	L	L	L	R	R	L	L	R	R	R	L	M	R	R	L
39	H	R	R	L	H	M	L	L	L	L	R	L	H	H	L	L	L	L	L	L	L	L	L	L	L	L	L	L	L	L	L	L	L	L	L	L	H	L	L
40	H	H	H	H	M	L	L	L	L	L	R	L	R	M	L	L	L	M	L	H	H	H	L	L	L	L	L	L	L	L	L	L	H	L	L	L	L	L	L
41	H	H	H	L	M	L	L	L	L	L	L	L	L	M	L	L	L	M	L	H	L	H	L	L	L	L	L	L	L	L	L	L	H	L	L	L	L	L	L
47	H	H	M	M	H	M	R	R	R	R	R	M	H	H	H	L	L	M	L	H	H	H	L	L	R	L	R	R	R	R	R	R	L	L	L	R	L	R	R
48	H	H	M	H	H	H	R	R	R	R	R	M	H	H	L	L	L	M	L	H	H	H	L	L	L	L	L	R	R	R	R	R	L	L	L	M	L	R	R
55	H	H	H	L	L	L	L	L	L	L	L	L	L	L	L	L	L	L	L	L	L	H	L	L	R	L	R	L	R	L	L	L	L	L	L	L	L	M	L
56	H	H	H	L	L	L	L	L	L	L	L	L	L	L	L	L	L	L	L	L	L	H	L	L	R	L	R	L	L	L	L	L	L	L	L	L	L	M	L
57	H	H	H	L	L	L	L	L	L	L	L	L	L	L	L	L	L	L	L	L	L	H	L	L	M	L	R	L	L	L	L	L	L	L	L	L	L	M	L
59	M	L	L	L	R	L	L	L	L	L	L	L	L	L	L	L	L	L	L	L	L	L	L	L	L	L	L	L	L	L	L	L	L	L	L	L	L	M	L
60	H	L	H	L	R	R	R	L	M	M	M	M	R	L	L	L	L	L	L	L	L	L	L	L	L	L	L	L	L	L	L	L	L	L	L	R	L	M	R

Reactivity Chart 10. Protection for the Amino Group: Special -NH Protective Groups (Continued)

PG	78 I₂	79 PhSeX; PhSCl	80 Br₂; Cl₂	81 MnO₂/CH₂Cl₂	82 NaIO₄ pH 5-8	83 SeO₂ pH 2-4	84 SeO₂/Py	85 K₃Fe(CN)₆, pH 8	86 Pb(IV), 25°	87 Pb(IV), 80°	88 Tl(NO₃)₃	89 150°	90 250°	91 350°	92 :CCl₂	93 N₂CHCO₂R/Cu	94 CH₂I₂/Zn(Cu)	95 R₃SnH/In.	96 Ni(CO)₄	97 CH₂N₂	98 SOCl₂	99 Ac₂O, 25°	100 Ac₂O, 80°	101 DCC	102 MeI	103 Me₃O⁺BF₄	104 1. LDA 2. MeI	105 1. K₂CO₃ 2. MeI	106 RCHO	107 RCOCl	108 C⁺/olefin
				L. OXIDANTS								M.			N.			O. MISCELLANEOUS											P.		
1	L	R	R	R	R	R	R	R	R	R	R	L	L	M	R	R	R	L	R	L	L	L	L	L	R	R	L	R	L	L	R
2	M	L	R	R	R	M	R	R	R	R	R	L	L	M	L	L	L	R	L	R	R	L	L	L	R	R	R	R	L	L	R
3	L	L	R	R	R	R	M	R	L	L	R	L	R	R	L	L	L	L	L	R	R	L	L	L	R	R	M	R	L	R	L
5	L	L	R	L	R	L	L	R	R	R	L	R	L	R	L	L	L	L	L	L	L	L	L	L	L	R	L	R	L	L	R
6	L	L	R	R	R	M	R	R	R	R	R	L	L	M	L	L	L	L	L	L	L	L	L	L	R	R	L	R	L	L	R
8	L	L	R	R	R	R	R	R	R	R	R	L	L	H	L	L	L	L	L	L	L	L	L	L	M	R	L	R	L	L	R
9	L	L	R	R	R	R	R	R	R	R	R	L	L	H	L	L	L	L	L	L	L	L	L	L	R	R	L	M	L	L	R
12	L	L	R	R	R	R	R	R	R	R	R	L	L	H	L	L	L	L	L	L	L	L	L	L	R	R	L	L	L	L	R
13	L	L	R	M	R	L	L	L	R	R	L	L	M	H	L	L	L	M	L	L	L	L	L	L	M	R	L	R	L	L	M
14	L	L	R	R	R	R	R	R	R	R	R	L	M	M	L	L	L	L	M	L	L	L	L	R	R	R	R	R	L	L	M
16	L	L	R	R	R	R	M	R	R	R	R	L	R	R	L	L	L	M	L	L	M	L	L	L	M	R	L	L	L	L	R
17	R	L	R	R	R	R	M	R	R	R	R	L	M	H	R	R	R	L	R	L	L	L	R	L	R	R	L	M	L	R	R
18	R	R	R	L	R	L	L	R	R	R	L	L	R	H	R	R	R	R	R	L	L	R	R	L	M	R	L	L	L	L	R
19	R	R	R	L	R	M	L	R	R	R	R	L	L	H	R	R	R	R	L	L	L	R	L	L	L	R	L	R	L	L	R
20	R	R	R	R	R	M	L	R	R	R	R	L	L	H	R	R	R	R	M	M	M	R	L	R	R	R	R	R	L	L	M
21	L	R	R	L	R	R	R	M	R	R	R	R	R	H	L	R	R	M	L	L	M	H	H	L	M	R	R	R	L	R	R
24	R	R	R	R	R	R	L	R	R	R	R	R	M	R	R	R	R	L	R	L	L	R	R	L	R	R	R	M	R	L	R
25	R	R	R	R	R	L	M	R	R	R	R	L	R	H	R	R	R	L	R	L	L	R	R	R	M	R	R	R	L	L	R
27	R	R	R	L	R	M	M	R	R	R	R	L	R	H	R	R	R	R	R	R	R	R	L	L	M	R	R	L	L	L	R
28	R	R	R	R	R	M	M	R	R	R	R	L	L	R	R	R	R	R	R	M	M	R	R	L	R	R	R	R	L	L	M
33	R	L	R	L	R	R	R	R	R	R	R	R	M	R	R	R	R	R	R	L	R	H	H	L	L	R	R	R	L	R	R
37	R	R	R	R	R	R	R	M	R	R	R	L	R	H	R	R	R	L	R	L	L	L	L	L	M	R	R	R	R	R	R
39	L	L	L	R	R	L	L	R	R	R	L	L	R	R	R	R	R	R	L	L	R	L	L	L	L	L	R	L	L	L	M
40	L	L	L	L	R	M	M	M	L	L	L	L	R	H	L	L	L	L	L	L	M	L	L	R	L	L	R	R	L	L	M
41	L	L	L	L	R	R	R	M	L	L	L	L	R	M	L	L	L	R	R	L	M	L	L	R	R	L	R	R	L	L	M
47	R	L	R	R	R	R	L	M	R	R	R	M	L	H	L	L	L	L	L	L	M	M	M	L	R	R	R	R	L	L	R
48	R	R	R	R	R	M	M	R	L	L	L	M	R	R	L	L	L	L	L	L	R	R	R	L	R	R	R	R	L	R	M
55	L	L	L	L	R	M	M	L	L	L	L	L	M	H	L	L	L	L	L	L	L	L	L	L	L	R	L	L	L	L	M
56	L	L	L	L	L	R	R	L	L	L	L	L	M	H	L	L	L	L	L	L	L	L	L	L	L	R	L	L	L	L	M
57	L	R	M	L	L	M	R	M	L	L	L	M	M	H	L	L	L	L	R	L	R	R	R	L	L	R	R	R	L	L	M
59	L	L	L	L	L	L	L	L	L	L	L	M	M	R	L	L	L	L	L	L	L	L	L	L	L	R	R	R	L	L	M
60	L	R	R	L	L	R	R	L	L	L	L	M	M	H	L	L	L	R	R	R	R	R	R	L	M	R	R	R	L	L	M

Index